The Economics of Standards

The Economics of Standards

Theory, Evidence, Policy

Knut Blind

Senior Researcher and Deputy Head, Department of Technology Analysis and Innovation Strategy, Fraunhofer Institute for Systems and Innovation Research, Karlsruhe and Reader, University of Kassel, Germany

Edward Elgar
Cheltenham, UK • Northampton, MA, USA

Published by
Edward Elgar Publishing Limited
Glensanda House
Montpellier Parade
Cheltenham
Glos GL50 1UA
UK

Edward Elgar Publishing, Inc.
136 West Street
Suite 202
Northampton
Massachusetts 01060
USA

A catalogue record for this book
is available from the British Library

Library of Congress Cataloguing in Publication Data

Blind, Knut, 1965 -
    The economics of standards : theory, evidence, policy / Knut Blind.
        p. cm.
    Includes bibliographical references and index.
        1. Standardization–Economic aspects. I. Title.

    HD62.B55 2004
    330–dc22

                                                                2004041177

ISBN 1-84376-793-7

Printed and bound in Great Britain by MPG Books Ltd, Bodmin, Cornwall

Contents

D   THE ECONOMIC IMPACTS OF STANDARDS:
     EMPIRICAL EVIDENCE

Figures

Tables

Acknowledgments

The book on 'The Economics of Standards: Theory, Evidence, Policy' collects and synthesises the research I have done in the area of standardization within the last years. It is nearly identical to the habilitation thesis 'Driving Forces and Economic Impacts of Standardisation' which I submitted to the University of Kassel. The publication of this book was only possible due to a sequence of lucky circumstances and to the constructive support of several people, some of whom I would like to mention explicitly.

The starting point for this book was a project on the economic benefits of standards which the Fraunhofer Institute for Systems and Innovation Research performed on behalf of the German Institute for Standardization (DIN), the German Federal Ministry of Economic Affairs and several companies, in co-operation with the Technical University of Dresden. In this context, I like to thank Professor Hariolf Grupp who supported me in the acquisition and performance of the project. Furthermore, the constructive support of our client in person of Dr Torsten Bahke and Dr Bernd Hartlieb encouraged me to continue my research in this field. Further projects on behalf of the German Federal Ministry of Education and Research, the Swiss State Secretariat for Economic Affairs, and the Research and Enterprise Directorates–General of the European Commission followed, which allowed a continuation of my research activities, an exception within a contract–research organization. Besides the thanks to the clients, most of the work was undertaken in teams and in collaboration with other institutes. Among the numerous colleagues and partners, I would like to mention explicitly the contribution of Dr Andre Jungmittag in the application of sophisticated econometrics.

Based on this continuity of research in the field of standardization, Professor Jochen Michaelis encouraged me to write a habilitation thesis when he became professor at the University of Kassel. I have to thank him for supervising my research, giving me constructive comments and being the first referee of the work. Professor Hans G. Nutzinger provided the second evaluation of the work.

All the work was conducted within the department 'Technology Analysis and Innovation Strategy' of the Fraunhofer Institute for Systems and Innovation Research, which gave me the freedom and the atmosphere to follow my research interest and to produce this book. Since I was involved in several other research projects, sometimes colleagues had to bear a higher burden. I would like to mention especially the support of my colleague and friend Dr Jakob Edler. For the proofreading of the whole text and for the translation of parts, I like to thank Chris Mahler-Johnston. Renate Klein and Sabine Wurst were responsible for the very time-consuming and puzzling preparation of the manuscript. However, all remaining errors are my sole responsibility.

Last but not least, I have to thank my wife Ulrike for being patient and understanding, but also strong especially in the last phase of finishing the book, when I was seldom available and she had to cope with three children under three years and a son in early puberty. Finally, our four children Jan Ole, Antonia, Pauline and Julius always gave me the feeling that there are other things more important than writing a book on standardization.

Knut Blind, Karlsruhe, Germany
February 2004

1. General Introduction

1.1 BACKGROUND

Standardization has been crucial for the development of the industrial society. The rapid progression of the globalized and increasingly complex knowledge economy not only demands new standards, but is also challenging the form of standardization. In particular governmental regulations or formal standards released by standards development organizations are confronted with an increasing tendency towards informal and de facto industry standardization.

The rapid technical progress in information and communication technology since the 1970s changed the economic and social situation worldwide. Some milestones of this development are closely related to standards. One early milestone was the establishment of the so-called IBM standard for personal computers at the beginning of the 1980s. The expansion of the use of computers depends not only on the progress of hardware technologies, but also on the parallel development of software programs. Some software packages, especially the Office programs provided by Microsoft, became more and more popular and are used by millions of private and commercial users, who share all the same features of these programs and are able to exchange data files without performing any file converting. Finally, the radical innovations in telecommunication technology, accompanied by the liberalization of the respective markets, allow nowadays the worldwide connectivity of Internet users and telecommunication network participants. The progress of all these technologies and respective industries was facilitated by the existence of common specifications in form of compatibility and interface standards.

On the one hand, technical progress in general has widened the spectrum of available technologies and therefore also extended the possible product assortment. On the other hand, further improvements of process technologies make mass production more attractive and allow the realization of economies of scale. Consequently the tension in the trade-off between variety and cost reduction became stronger and the need for variety-reducing standards has increased.

Technical progress has also caused an increased complexity in products and services, which has made their characteristics less transparent to the

1

consumers and users. Furthermore, new technologies are accompanied by new kinds of risks which may affect not only the users of these technologies, but also other groups in society or the environment as a whole. The negative effects occur not only during the usage period, but also can arise much later. It is again progress in science and technology which allows the more precise detection of a broader set of these often hidden harmful results. Finally, the progress of economic welfare also increases the demand for health, safety and environmental protection.

Taking all of these aspects together, it is obvious that effective and efficient regulation is more and more crucial and that the demand for the regulation of these issues rises. Besides direct public intervention by prohibitions or financial incentives, minimum quality and safety standards are also an instrument for improving the efficiency of markets or even to make their emergence and functioning possible at all.

All these various developments increase the need for standards and their relevance as a policy instrument. From an institutional and supply side perspective we differentiate three kinds of standards. First, standards can be regulated by governmental bodies and receive an obligatory status in the form of a regulation. Second, standards development organizations offer platforms to interested companies, users and other interested groups to develop voluntary standards, often called formal standards, according to a number of principles such as transparency, openness and consensus. Finally, individual companies or small groups of companies in the form of informal consortia can set their own, often proprietary, industry standards and try to reach their acceptance or even dominance within the market process. Although we observe an increasing tendency to the latter form, the following study concentrates mostly on the formal standards released by official standard development organizations (SDOs) at the national, European or international level.[1]

Following the development in technology, the literature on the economics of standards experienced its peak roughly in the decade between 1985 and 1995, starting with the pioneering contributions of the author couples Katz and Shapiro and Farrell and Saloner. In the context of the new paradigm of network industries, the need for compatibility within classical networks, like telephone networks, or between components of complementary technologies like hard- and software was acknowledged and perceived by economists. Furthermore, beside price and quality, the decision about compatibility became a further instrument of companies to secure their market shares within a competitive framework. And if companies decide to reach compatible solutions, then the issue of standardization comes up.

In the economic literature several perspectives have been analyzed. Primarily, the strategic interactions between companies were in the focus of economic theory. In the first stage, the question is, under what circumstances do companies seek compatibility and therefore a common standard? Then, if compatibility is efficient for them, it has to be asked whether companies are able to realize compatible solutions in the form of a standard. Since there are game-theoretic constellations like the prisoner dilemma or the battle of sexes which are likely to generate inferior outcomes in the form of no standards at all, because of a lack of incentives or incompatible solutions due to failed communication or co-ordination, the question of the adequate institutional setting comes up again. Is co-ordination by market mechanisms superior to committee solutions, or is intervention by public authorities even necessary? Regarding the committee solutions, we differentiate between informal industry consortia not following specific rules, like transparency and openness, and the work of standards development organizations, like the British Standards Institute (BSI), the German Standards Institute (DIN), the European Standards Institute (CEN), or the International Organization for Standardization (ISO). After a common solution in the form of a formal standard is found, its impact on competition and its parameters of price, market volumes and quality must be analyzed. Finally, the interrelationship between standardization and market structure is an important aspect to investigate, since the market structure influences the chosen institutional setting, the process and the outcome of standardization. In the long run, the results of standardization processes may have impacts on the market structure, especially if proprietary technology is leveraged by standard setting resulting in a closed, privately owned de facto standard. In summary, economic theory was successful in working out the central incentives and mechanisms relevant in standardization, although relatively little emphasis was placed on acknowledging the institutional setting of formal standard development in formal modeling and on the interplay between formal and informal standardization processes.

Parallel to the analysis of company strategies and tactics regarding standardization, economic theory also had a focus on the user side. One central issue has been the tension between the preference for product variety and enjoying low prices due to economies of scale or positive network externalities. The latter aspect in particular provides a reason to argue for less variety in order to exploit these externalities. However, this reduction in variety may provide the critical mass and therefore the ground for a new generation of a base technology, which allows a broad set of new products or services to build on, such as in case of mobile telecommunications, where the development of the European standard for a Global System for Mobile Communications (GSM) was the starting point for the mobile communication industry to flourish. Therefore, from an intertemporal perspective, the reduction of variety at crucial points in time leading to the breakthrough of a new base technology or platform may guarantee a broader product variety in the long run. Furthermore, the reduction of variety regarding interfaces between complementary components of a system technology increases the number of combinations and therefore the variety. It is also helpful to solve another problem users may face if they make a commitment towards a specific company standard: the phenomenon of being locked-in in a specific standard. If there are interface standards between different components of a whole system, the user may switch more easily between various suppliers, compared to a constellation where a whole system must be replaced by an alternative system from another supplier. Finally, the involvement of users in standardization processes has been focused by representatives of political economists, since they are not well represented in standardization and often not able to follow the process, due to limited knowledge or resources. Here the question of public intervention arises in order to correct this under-representation.

Leaving aside the perspective of the individual actors and taking a comprehensive macroeconomic view, it is evident that standards help to reduce costs of transactions between the different actors and therefore the emergence and organization of markets. This perspective is taken from the so-called institutional economists. Finally, standards can also be used as instruments to solve market failures, as in the case of obligatory environmental, safety or health standards. They are able to correct both negative externalities and information asymmetries.

This brief outline should provide a rough background to the variety of standards aspects tackled by economic theory. It has to be noted that since the end of the 1990s, a significant stagnation in theoretical analyses has been observed. Empirical investigations are mainly based on case studies and anecdotal evidence. Hawkins (1996) investigated for the first time the general feasibility of using standards as indicators for technological change or strength. Then, Swann et al. (1996) tried simultaneously to assess quantitatively the impact of formal standards on trade, although classical trade literature has already dealt with non-tariff trade barriers for a long time. The obvious gap between the vast amount of theoretical literature on the economics of standards and relatively little empirical work is caused both by the rigidity of the theoretical models based on strong and very restrictive assumptions and the lack of adequate data, which does not allow adequate empirical tests of the hypotheses derived from the theoretical models. The stagnation of the theoretical work in standardization may therefore also be caused by the lack of empirical work, which is not able to challenge and to give adequate feedback to the theoretical considerations. Consequently, empirical investigations have been of little relevance in inspiring further theoretical work in standardization.

1.2 OBJECTIVES AND BASIC APPROACH

The objectives of the following work have to be seen in the context of the slow-down of theoretical approaches analyzing standardization and of the lack of adequate data which prohibits progress in empirical investigations. Therefore, the general challenge to be tackled by this study is to inspire further research both in theory and empirics in the field of standardization.

In order to achieve general progress in empirical standardization research, we follow a comprehensive approach regarding several dimensions. We consider the micro-perspective of the single firm, which is the driving unit both for the development and the implementation of new standards. We also analyze the relationships on a sectoral level, because besides companies other 'interested groups', such as researchers, consumer organizations, unions and environmental groups, are involved, especially in the standardization process, and therefore also indirectly in the implementation of formal standards. Finally, we investigate for the first time the role of formal standards, as one form of codified technical information, for macroeconomic growth. Especially in the knowledge economy, standards play a crucial role as a source of codified and publicly available and usable information. Besides the internal perspective, we analyze the external dimension of standards and their role in international trade flows also on the micro- and the sectoral level. The three levels of analysis are connected by comparisons of their respective results, taking into account the sometimes limited comparability.

Besides the comprehensive three level perspective, the main objectives of the study can be divided into two main separate, but interdependent categories. In the first stage of analysis, we look for the driving forces for standardization in standards development organizations. Besides impulses from new developments in science and technology, we consider also aspects of the market structure, such as company size and concentration, and the external aspects measured by export and import ratios.

Very closely linked to the explanatory factors for the participation in and the output of standardization processes are the various impact dimensions of formal standards. Regarding the impact dimensions, we try to clarify firstly whether the interrelationship between technical change and standardization is a virtuous or a vicious circle, because the traditional view presupposes that standards in general hinder technical change. The results of this analysis also allow us to evaluate the performance of the standards development organizations, in the sense of whether they are able to absorb recent trends in science and technology and to produce standards reflecting the state of the art in science and technology. Second, we investigate which of the diverging economic theories about the role of formal standards for trade is reflected in the results of the quantitative empirical analysis. The objective is to answer the question whether their trade-fostering effects outweigh their trade-hindering effects, especially by differentiating the analysis into national and international standards. Finally, we try for the first time to assess the impact of formal standards and other sources of technological know-how in quantitative terms for growth in Germany.

Traditionally, complex economic phenomena are analyzed firstly at the micro level of the individual agent, for example worker, household or firm, because often the relationships cannot be investigated at an aggregated level, due to data restrictions. In this study, we analyze the driving forces for standardization and the impact dimensions of formal standards, at first based on aggregated sectoral data using the single standard document as a unit of observation to identify both the output of standardization and to quantify the stock of standards as an indicator for the quantity of codified technical knowledge. For bivariate and multivariate analyses, this new kind of technology indicator is embedded into patent statistics and traditional economic statistics. For the service sector, this kind of analysis is not feasible, because significant numbers of formal standards specifically applicable in the service sector are not available and only currently under preparation.

The relevance of a second micro-based company analysis is justified by the following reasons. Companies are the most important, but not the only group influencing standardization processes, because governmental institutions, unions, environmental and consumer groups have the right to intervene and to change specifications according to their preferences until a consensus among all parties is reached. Consequently, it is essential to analyze the output of standardization processes at the aggregate level. Nevertheless, we focus particularly on the company level and analyze separately both the company-specific driving factors for standardization and the companies' perceptions of impact dimensions of standards and standardization processes, which cannot be derived from data in official national statistics. The additional, but narrow focus on the companies is justified, because they are numerous actors – in contrast to single representatives of other interest groups – in the standardization processes and they are mainly responsible for the economic impacts of standards. Therefore, their specific assessments and attitudes are crucial for the development of new and better standardization related policies. For the micro-based analysis, we rely on survey data collected among German and European manufacturing companies during the performance of the studies mentioned in Section 1.3. As far as possible, we compare the outcomes of the sectoral and micro-based analyses in order to check for consistency and find explanations for diverging results. For the service sector, we just perform a micro-analysis focusing on the implementation of the famous series of ISO 9000 quality standards within German service companies, since analyses based on sectoral level are not yet feasible due to the lack of adequate data.

The difficulties of performing empirical research on standards and standardization are aggravated by a set of recent trends in science, technology, economy and society. First, both scientific and technological development are characterized by an increasing complexity and interdisciplinarity. Thus, as already mentioned, the complexity of technological systems has increased. This trend also influences standardization, because on the one hand the need for standards grows, but on the other hand the identification of the object of research becomes more difficult and the influencing factors and the impact dimensions broader and more blurred, which complicate the empirical analysis. Second, we observe both a trend towards standardization processes within informal industry consortia or other user groups, and stronger intermingled relationships between informal and formal standards and between formal standards and governmental regulations. In order to tackle this difficult challenge, we restrict our quantitative empirical analysis to formal standards released by standards development organizations being aware of the substitutive and complementary relations to informal industry standards. Finally, the globalization of the economy, in terms of the increasing number of multinational companies and growing transnational trade volumes, increases the need for uniform international standards and questions the existence of idiosyncratic standards. Within the European context, the target of the completion of the Single Market and the Cassis-de-Dijon decision – confirming the origin principle and also the mutual recognition of national requirements – reflect this trend by institutional transformations and adjustments. In the long run, national standards will lose importance as barriers for imports and international standards will crowd out national standards almost totally, leaving just a small fraction of standardization activity on the national level. However, this new stable equilibrium has not yet been reached and we are still in a phase of transition. Therefore, the time-series analyses conducted in this study cover a period until the middle of the 1990s, until which the completion of the Single Market and the substitution of national standards by European and international standards were just at their beginnings and their impacts on technical change, trade and growth not yet effective. An extension of the time series in order to cover the final impacts of the completion of the Single Market would be superior, but this is not possible based on the available data. However, the actual restriction of the times series enables us also to differentiate between the effects of national and European or international standards, a distinction which will be not feasible any more after a complete harmonization of standards specifications at least within Europe (Blind 2002c), but very useful with respect to the trade relations with the United States of America.

Although we claim to follow a comprehensive approach in analyzing the empirics of standards and standardization, the following aspects besides the other informal types of standards are not covered by the study. First, we have no empirical data about the standardization process itself and concentrate either on the decision to participate in the process or on the successful output of the process: the standard document. Consequently, the aspect of exchanging tacit knowledge within standardization processes is not integrated in our analysis. Secondly, the perspective of the user is taken into account only to a limited degree by analyzing the decision to implement formal standards and describing some selected impact dimensions within the firm.

The ground for the study is provided by several studies which the Fraunhofer Institute for Systems and Innovation Research has been commissioned to perform since 1997. The starting point for the whole approach was a study on the economic benefits of standardization on behalf of the German Institute for Standardization (DIN), the German Federal Ministry for Economic Affairs and several large companies (Blind et al. 1999a; Blind and Grupp 2000; DIN 2000b). Some working steps of this project, which lasted three years, have also been applied to the situation in Switzerland funded by the Swiss State Secretariat for Economic Affairs (Blind et al. 2000). In the year 2000, the Research Directorate General of the European Commission appointed a consortium of research institutes co-ordinated by the Fraunhofer Institute for Systems and Innovation Research to perform a study on the relationship between intellectual property rights and standardization (Blind et al. 2002). All these studies were focused on the manufacturing sector. However, a specific analysis of the first German Innovation Survey (Blind and Hipp 2003) and an ongoing study funded by a research grant of the Enterprise Directorate General provided also first insights into the emerging role of standards in the service sector (Blind 2002b).

1.3   STRUCTURE AND MAIN RESULTS OF THE STUDY

The study is divided into four parts. Part A attempts to give an overview of the theoretical literature about the economic impacts of standards by providing a general classification of standards and their economic effects, and some in-depth surveys regarding some specific impact dimensions. Due to their various often contradicting impacts, incontestable statements about the direction of all impacts together cannot be made. This vagueness has crucial consequences for the empirical analyses.

In Part B, the institutional perspective or the supply side of regulations in a broader sense is presented by giving some basic definitions as a basis for narrowing down the focus of the study to technical rules in the form of formal standards. Then the legal implications of formal standards and the institutional framework for their elaboration are described. Finally, the feasibility to use formal standards as a new indicator for technical change is discussed and approved.

The major part, Part C, concentrates on the demand for standards released by standardization development organizations. After an introduction explaining the selected approach, theoretical hypotheses are presented about the driving forces both at sectoral and firm level. First, the test of the sectoral model follows including a special focus on the role of intellectual property rights, then the firm-based model is tested and compared with the results of the sectoral model. We find that the sectoral output of standards is mainly influenced by the sectoral R&D intensities and the propensities to patent. Market concentration has a positive impact on de jure standardization up to a certain degree. Finally, export-intensive sectors are likely to produce more standards. The micro-econometric investigation makes clear that large companies are more likely to be active in standardization than small ones. The export intensity of a company is a further crucial factor for the participation in standardization processes. Part C concludes with an analysis of quality standards in the service sector and finds that besides sector- and size-specific differences, the use of 'risky' technologies positively influences the probability of introducing the quality standard ISO 9000. This quality standard has another twofold impact: first, as expected, it is a quality seal for the customers of the service company, especially in markets with homogeneous products and average qualities. Second, its introduction has impacts on the internal processes of the service companies.

The second major part, Part D, concentrates on the impact dimension of formal standards. Three aspects are covered. First, we try to elucidate the intertwined relationship between technical change and standardization empirically in a quantitative manner. Second, the impact of formal standards on foreign trade is assessed quantitatively by several analyses on different levels of aggregation. Finally, the influence of standardization on macroeconomic growth is identified in quantitative terms.

Based on time-series analyses using patents and standards documents, a positive impact of technical change and innovation on standardization activities can be found. The other way round, increasing stocks of standards have only a weak, but still positive impact on innovation, respective to the growth of patent applications. Analyzing answers from a company survey, a more critical attitude is expressed on the adequacy of the stock of standards especially in relation to innovation. However, compared to other obstacles for innovation, formal standards have only a small importance.

Reflecting the contradicting theories about the role of standards in foreign trade, the impact of standards on trade flows are mostly ambiguous and differ between the bilateral trade flows analyzed, although intra-industry trade benefits in general from the stocks of standards. However, it is more promising to analyze the impact of standards on trade focusing on specific technologies and sectors, and not on total trade or on specific bilateral relationships between two countries. The analysis focused on measurement and testing confirmed this and made it obvious that trade flow analyses have to consider the characteristics of a technology in order to detect significant impacts of standards on trade. Since the assumption that standards have the same impact across sectors or technologies is rather restrictive, it should be given up and be substituted by sector- or technology-related analyses taking into account the rather different impact dimensions of standards.

In extension and refinement to most other empirical studies, the role of standardization with its important functions, for example for the diffusion of technological innovations, is integrated in the production function of the German business sector covering a period from 1960 until 1996. This estimated production function is then used to assess the effects of the stock of standards as well as the impact of the usual production factors on growth of real gross value added. Altogether, it became evident that standards were at least as important for economic growth as the innovation indicator patents. It is clear that the innovation potential is not the only deciding factor in economic development, but that it must also be broadly disseminated by means of formal standards. However, the decreasing contribution of official standards to economic growth since the 1980s hints at the increasing importance of de facto and industry standards in times of shorter product cycles.

Based on the various insights, Part D concludes the study with an overall summary, future research challenges and respective policy considerations, which are based on sound scientific insights instead and not on the personal interests of stakeholders.

NOTES

[1]   In the study, we demarcate this kind of formal or de jure standards against regulations and informal industry or de facto standards. In German, the respective term is 'norm' as a subset of the general term standard.

PART A

The Economic Impacts of Standards: Theory

2. Introduction to Part A

Although the main focus of this work is devoted to the empirical analysis of standardization, standards and their economic impact, we start with a survey of the economic impacts of standards based on theoretical considerations. Since the empirics of standards are so complex, we do not aim in this chapter to develop theoretical models, which are the basis for the derivation of theoretical hypotheses. However, we discuss the most important theoretical aspects of standards and their economic impact in order to present a theoretical framework, which allows a better understanding of the hypotheses we test in the empirical parts which are based more on plausible reflections than on derivations from formal mathematical models.

In Chapter 3, we present a simple taxonomy of standards based on their economic impacts, although such a clear-cut distinction does not reflect the actual impacts of standards implemented. However, this distinction builds the basis for the discussion of the role of standards in three different areas, on which we focus in our empirical analysis of the economic impacts of standards. In Chapter 4, we focus on four standard-specific problem areas, which will also be objects of the empirical analyses. First, we analyze the role of standards within technological change. This relationship is most challenging both from the theoretical point of view and presents the core of our empirical analysis, on which we base two other empirical investigations. Very closely linked to the first complex relationship is the question whether standards have a positive or negative impact on competition. The third issue we focus on is the ambivalent role of technical standards in foreign trade. Finally, we briefly discuss the role of standards for economic growth in general.

3. A Classification of Standards by their Economic Effects

3.1 INTRODUCTION

A classification of standards can be performed in several ways, not only depending on the respective scientific discipline, but also within the same discipline.[1] David (1987) proposed a categorization based on the economic effect of a standard. This is a useful approach in order to analyze both the economic driving forces for standardization and the economic impact dimensions.[2]

However, it should also be taken into account that there are other ways to structure the great variety of standard types. Consequently, some economists favor a categorization based on the process of standard building (that is formal or *de facto*) (David and Greenstein 1990; see also Chapter 5.2 in Part B), or a typology based on whether the standard relates to products, services, or processes (or indeed if it is a meta standard), although the issue of service standards is only an emerging field.[3]

The David (1987) approach classifies standards according to the economic problems they solve. It has been widely accepted and used (for example Nicolas and Repussard 1988), though some later writers have extended the number of categories (for example Tassey 2000). In our classification we use the dimensions of compatibility, quality, variety-reducing and information standards. However, even if standards are developed just to serve one purpose they often fulfill multiple functions. Therefore, it is not likely that each standard will fall exactly and exclusively into a single category, but very often several economic aspects are touched by a single standard. This feature presents a challenge for the empirical analysis. However, a distinction is important for the theoretical discussion, because standards have different economic effects, and the analytical models used to analyze and understand these effects are different.

3.2 COMPATIBILITY AND INTERFACE STANDARDS

Starting with the development of the first network industries, that is the railways, the importance of standards for compatibility and interfaces grew significantly. The rapid progress in information and communications technologies in the last few decades has demonstrated the huge economic importance of this kind of standards, which also attracted the majority of economists interested in standardization.

Economic theory focuses on two particular economic phenomena, which influence suppliers and customers in so-called network industries (Shy 2001). First, producer and customer decisions are influenced by what are called *network effects*, or sometimes, *network externalities* (Farrell and Saloner 1985, 1986; Katz and Shapiro 1985, 1986, 1994). The basic idea behind this concept is that it is preferable to choose a system that is widely used by others. Second, producers and customers face *switching costs* (Farrell and Shapiro 1988; von Weizsäcker 1982) after the decision for one and against another system. Before they are committed to a particular system with its respective interfaces or standards, they are relatively free to choose between different specifications. After this decision both producers and users invest in the particular system or standard. And the longer they stay with the system or standard, the more expensive they find it to switch to another comparable system or standard.

When these phenomena, network externalities and switching costs, exist simultaneously, there is a risk that markets can get locked into inferior designs because both sides are reluctant to switch to something better. Also switching is costly, and it becomes even more expensive if one cannot be sure how many participants of the existing network will switch too. Only with complete information and identical preferences among firms are industries not trapped in obsolete or inferior standards (Farrell and Saloner 1985). Katz and Shapiro (1992) are able to provide conditions, like exponential market growth, under which there is a tendency to rush too fast into new, incompatible technologies. In addition, companies supplying these technologies earn greater profits than their entries contribute to social welfare.

This phenomenon has been described as a problem of technological lock-in (David 1985; Cowan 1990). The empirical importance of lock-in is ambivalent, and some authors question sustainable lock-ins (Liebowitz and Margolis 1990, 1994, 1999), because various incentives exist to change to superior technologies even taking switching costs into account.

Compatibility or interface standards help to expand market opportunities because they foster network effects (or externalities), which are benefits that follow from being part of a large network of users. There are two main categories of network externalities (Katz and Shapiro 1985): *direct* and *indirect*. The value of being a subscriber to a telephone network depends in an obvious and *direct* way on the number of other subscribers. If there are few other users, then the utility of the network is limited.[4] The utility function of an individual consists of a component independent from the network and of a component depending on the size of the network. The independent parameter can take positive or negative values and differs by consumer. It represents the utility of the network, if nobody else will join the network. The part depending on the network increases continuously with the expected size of the network, represented by the number of the participants. Each additional new participant creates positive network externalities for all other participants. Rohlfs (1974) proved the existence of multiple equilibria, because individual demand depends both on the price and the decision of all other individuals regarding their participation. The planned individual demand also depends on the price for access to the network and the decisions of all other individuals in the total population. The interdependence of the demand function is directly derived from the interdependence of the individual utility functions. The demand decreases with increasing price and increases with additional access to the network. An equilibrium user number is defined as Nash-equilibrium, under which for all individuals the planned demand is equal to the realized demand. Given the price and the decisions of all other individuals, no individual has an incentive to change his decision regarding participation. Rohlfs (1974) shows the existence of multiple equilibria assuming a fixed price. Even at a price equal to zero, the equilibrium number of users can either be zero or the total population. If the former equilibrium is realized, we observe complete market failure and every consumer expects that nobody else will join the network – a self-fulfilling prophecy.

In contrast to direct network externalities, indirect network externalities are generated in a paradigm in which each user must possess two or more components to derive benefits from this system. The most popular example of indirect network benefits and compatibility between complementary products is the so-called hardware–software paradigm (Katz and Shapiro 1994). Under a static framework, the decision of a user to buy such a system or single components has no impact on the utility of other users given prices and varieties of the different components. Indirect network effects arise only under a dynamic framework when users make their purchase decisions, including repairs over time, or consumers enter the market continuously. These indirect externalities are caused by the incremental impact of one user's adoption decision on the future assortment and prices of products and services belonging to the whole system. Under the hardware–software paradigm, the consumer chooses the hardware to buy in the first period forming expectations about the second period, especially about the supply of compatible software packages. A similar constellation applies to the decision to buy a car, since the owner of a very popular car model can enjoy *indirect* network benefits. Although he may not care about the large size of the network *per se*, he expects to benefit from the consequence of a good and broad repair and maintenance network, including a sufficient supply of spare parts.

When direct or indirect network externalities are appreciated by the customers, the suppliers generally have incentives to produce a product or to provide a service that conforms to the accepted standard of the respective networks. If the product or the service is company-specific and does not conform to the industry standard, the demand side will generally find it not very attractive. The supplier may face more competition after conforming to the standard, since the homogeneity of products and services increases and price competition will become more intensive. Nevertheless it is better to have a share of a large market, especially with strong preferences for network effects, than a monopoly on a small niche market.

In some of the standards races that have taken place in network technologies (such as personal computers, audio recording media, video cassette recorder formats, and so on), the winning technology is not necessarily the 'best' technology from the perspective of technological performance (David 1985; Weiss and Sirbu 1990; Shapiro and Varian 1999). The owner of the winning technology is often the one who has been successful in building a wide network of followers and of complementary products from third party producers (for example software) that conform to his technological specifications.[5]

Standards emerging from such processes are often not standards in a formal sense. They are not defined by a committee in a more or less formal multi-stage process. They represent rather proprietary designs that reach a position of market dominance – and hence earn the title of *de facto* (or informal) standards. When a compatibility standard is a proprietary design rather than an open public document, the content of which can be used by all interested parties, then the owner of the proprietary design is able to misuse his monopoly power. From the perspective of static efficiency, it is generally preferable if standards are public and open rather than proprietary, even though it cannot be denied that some industries have grown based on proprietary standards. However, proprietary standards represent strong incentives for companies to develop new technologies superior to existing technologies. Therefore, from the perspective of dynamic efficiency, open public standards may not be the first best solution.

Besides, in physical networks, compatibility and interface standards are also important for product or service systems containing numerous different components, since they allow multiple proprietary component designs to coexist, enabling innovation at the component level (see also Section 4.2). Furthermore, the consumer of the product system can select the particular components that optimize his system design. Compatibility and interface standards also allow the substitution of more advanced components – supplied not only by one producer – as they become available over time, thereby greatly reducing the risk of obsolescence of the entire system.

The economic effects of compatibility standards will be analyzed in depth in Section 4.2 in the context of technical change and Section 4.3 regarding market structure and competition. Therefore, some of the aspects already discussed will be repeated and integrated in the respective context.

3.3   MINIMUM QUALITY AND SAFETY STANDARDS

In the standard neoclassical model of product markets, it is assumed that one homogeneous product is supplied and that customers are fully informed about the characteristics of this product. Reality is much more complex. First, the product variety has increased, especially with the increasing income level in industrialized countries. Second, and more important for standard issues, consumers are often not fully informed about the characteristics of the product and suffer from so-called information asymmetries. These increase if the characteristics of a product can only be observed by using it, in the case of experience goods. Finally, the quality of credence goods, like safety systems or drugs, is also influenced by external factors not under the control of the supplier.

The consequence of these kinds of information asymmetries are adverse selection or moral hazard. Regarding the adequacy of minimum quality or safety standards, we concentrate on adverse selection. In his pioneering paper, Akerlof (1970) showed that in such circumstances the information asymmetries between buyers and sellers could lead to adverse selection and severe market failures. If buyers cannot distinguish high quality from low quality before purchase, then it is difficult for the high quality seller to sustain a price premium. In the absence of this premium, and if the high quality seller's costs exceed those of the low quality seller, then the former may not be able to survive. Suppliers of bad quality crowd out good quality sellers from the market by undercutting them. The market for good quality breaks down, trade in this segment does not take place and reduces the surpluses both of the producers and the consumers of high quality goods.

There are several solutions to this problem. The demand side can try to reduce the information asymmetry by screening the quality of the supplied goods and services. The suppliers may signal the quality of their goods and services to the customers by building a long-term reputation or by guaranteeing a certain level of product quality. Besides market solutions, governmental interventions are also effective instruments. Leland (1979) showed that *minimum quality or quality discrimination standards* can solve the phenomenon of adverse selection. He finds that minimal quality standards are especially advantageous in markets with greater sensitivity to quality variations, low elasticity of demand, low marginal cost of providing quality and low value placed on low quality service. If these quality standards exist and are well accepted, then the buyer can confidently distinguish high quality from low quality before purchase, and then the high quality seller can sustain a price for his superior product. Standards are not the only way of overcoming the problem of adverse selection, since signaling strategies of the supply side[6] and screening activities of the demand side are also effective. However, standards can be more effective because they may represent a public good, which can be used both by the suppliers and users without additional cost. Kindleberger (1983) argues in the same direction and emphasizes that standards do not only reduce transaction costs, but also enable economies of scale with an increasing number of transactions.

And even if these minimum quality standards are not 'public', they may have the character of a 'club good', which may benefit a limited number of club members, who have co-operatively defined the standard. However, if professional groups or industry confederations define the minimum levels of quality standards, then there is a tendency to set the levels higher, in order to generate extra profits by restricting the total supply and increasing the price (Leland 1979).

Besides enabling a market for high quality products, minimum quality or quality discrimination standards can reduce what economists call *transaction costs* and *search costs* (Hudson and Jones 1996, 2001). If a standard narrows down the spectrum of product characteristics, then it reduces consumer uncertainty. Consequently, there is less need for the consumer to spend time and money evaluating the product before purchase. In commodity markets, traders must be able to buy and sell large volumes without even viewing their goods, and this is only possible if there is complete confidence about the characteristics of what is being traded. Therefore, a clearly defined standard is needed and it must be certified that all products traded meet the specifications of the standard.

As we have seen in the previous section, compatibility standards are able to generate positive externalities and therefore help both producers and customers. Standards can also protect third parties from negative externalities, both generated in the production and consumption of goods – as for example in the case of environmental standards (Baumol and Oates 1971; Skea 1995). Here, maximum amounts for the production or the emission of the materials causing the negative externalities are defined by governmental regulatory bodies in performance standards, or product standards specify requirements which try to reduce the negative impacts for the environment generated during the use or the consumption of the product. However, environmental standards are less efficient in comparison to other instruments, such as taxes, emission trading permits or liability rules.[7]

3.4 VARIETY-REDUCING STANDARDS

Standards limit a product to a certain range or number of characteristics such as size and quality. The majority of standards perform this function. One famous example is the series of paper format standards (for example DIN A4). *Variety reduction* performs two different functions.

First, it leads to economies of scale by reducing the number of variations of a product or even technology. Focusing on just one standardized model allows, first, the mass sourcing of input materials; second, mass production, and third even advantages via mass distribution. All three aspects together lead consequently to lower costs per unit.

The trade-off already discussed in the seminal contribution of Dixit and Stiglitz (1977) operating here is between choice and cost or price.[8] Variety reduction standards emerge in a first stage mostly within a company, since the economies of scale can in principle be obtained by concentrating on a limited model range. This is the best-known function of variety-reduction standards. However, variety reduction may cause adoption costs or a utility loss for the user, because the distance between his most preferred specification and the supplied specification increases.[9]

Consequently, the reduction of a broad range of technical alternatives and preferences down to just a few or even one variety lowers the willingness to pay of the consumers. Therefore, the company has to consider this trade-off and the reaction of its competitors in its price and product decisions.

The model developed by Farrell and Saloner (1986) tries to make explicit the tension between standardization and variety from the perspective of the users, if network externalities are relevant for the utility of the users and their willingness to pay. In their simple model, the users are divided into two groups, who have different preferences regarding the two technologies offered. The authors are able to prove first that multiple equilibria are possible, which include also incompatible, non-standardized solutions. Second, the equilibria do not have to be optimal from a social welfare perspective.

There is, however, a second and an even more important role for variety reduction, and this operates for the benefit of the producer as much as for the customer. Standards can also reduce the risks faced by suppliers – even if this also means they face more competition (Swann 1985). The availability and use of standards often shape future technological trajectories, and these are instrumental in the development and growth of new markets (Dosi 1982). In the early stages of a market for a new technology, standards can play an important role in achieving focus and cohesion amongst the pioneering companies, since sometimes technologies get locked into a pre-paradigmatic stage because suppliers and users are too dispersed and there is no focus or critical mass in developing a market for that technology. The variety-reducing standard can help to achieve that focus, and hence help the market to take off.

However, the variety-reducing function is the most difficult category to analyze because of its ability to either enhance or inhibit innovation. Variety reduction typically enables economies of scale to be achieved, but larger production volumes tend to promote more capital-intensive process technologies. This common evolutionary pattern of a technology over a number of product life cycles usually reduces the number of suppliers and increases their average size. Such trends may or may not reduce competition, but often progressively exclude small, potentially innovative firms from entry due to increased minimum efficient scale thresholds.

3.5   INFORMATION AND MEASUREMENT STANDARDS

Standards of information and product description are usually treated as a distinct category from the above (Tassey 2000), but for many purposes it is sufficient to treat these as a hybrid of the above three categories. Swann (2000) refers to the example of different grades of petrol. These are standards of product description that also offer the other three features. Most car drivers are confident that one type of petrol star is *compatible* with another, and so they can fill up their cars at gasoline stations belonging to different petrol companies. Equally, these grades satisfy certain quality standards. And of course there are major economies of scale in distribution from the limited range of petrol grades.

Those near-market measurements that are carried out to confirm that a product is what it is supposed to be would appear to have much in common with this type of *product description* standard. The producer can confirm that the product to be sold is indeed what he expects it to be, and that reduces the risks (of compensation or litigation) to him and also the risks to the user. In principle, the consumer can buy with confidence and without the need to carry out his own independent test that the product is what it is supposed to be. As such, this sort of certified measurement can help to reduce transaction costs, and hence make markets work better.

In the area of science and technology, standards help to provide evaluated scientific and engineering information in the form of publications, electronic data bases, terminology, and test and measurement methods for describing, quantifying, and evaluating product attributes (Tassey 2000). In technologically advanced manufacturing industries, a range of measurement and test method standards provide information which, if universally accepted, greatly reduces transaction costs between buyer and seller.[10]

3.6 SUMMARY

Table 3.1 summarizes these four different purposes of standardization and highlights their positive and negative effects. The usefulness of this distinction lies in what it contributes to our understanding of the variety of standard types and their sometimes ambivalent effects. This background information is necessary for interpreting the results of the forthcoming empirical analyses.

Table 3.1   General effects of standards[11]

	Positive effects	Negative effects
Compatibility/ interface	• Network externalities • Avoiding lock-ins • Increased variety of systems products	• Monopoly
Minimum quality/ safety	• Correction for adverse selection • Reduced transaction costs • Correction for negative externalities	• Regulatory capture 'Raising rival's costs'
Variety reduction	• Economies of scale • Building focus and critical mass	• Reduced choice • Market concentration
Information standards	• Facilitates trade • Reduced transaction costs	• Regulatory capture

In addition to the arguments and theories addressed in the previous sections, two concepts in this table may need further explanation. The concept of 'regulatory capture' is the idea that some producers may lobby so skillfully that they persuade the standardization development organizations to define regulations in the interest of the producers rather than in the interest of the customer (as originally intended). In that context the concept of 'raising rivals' costs' (Salop and Sheffman 1983) may also be highly relevant. Some high cost and high quality producers may find it in their interest to lobby for an unnecessarily high minimum quality standard, because that will in effect exclude their lower cost, lower quality rivals from the market.

NOTES

[1]   Hesser and Inklaar (1997) provide on overview of the role of standards and standardization in various disciplines.

[2]   This chapter refers to the surveys of Swann (2000), Tassey (2000) and Kleinemeyer (1998).

[3]   De Vries (1999) makes a considerable effort to approach a systematic classification of standards.

[4]   It is unlikely that individual utility is a linear function of network size. And a linear relationship only emerges under quite strict assumptions (Swann 1998). Nevertheless, utility would normally be a monotonic increasing function of network size – unless consumers desire exclusivity in the networks to which they belong in the sense of luxury goods.

[5]   Some papers have attempted to measure network externalities and demonstrate their important role in the standardization process: Gandal (1994), Greenstein (1993), Shurmer (1993), Shurmer and Swann (1995), Swann (1987, 1990). See also the references on empirical evidences of price effects of network effects in Section 15.3.

[6]   In the food sector, the adequate and reliable labeling of ingredients is one instrument of signaling quality (cf. Michaelis 1990b).

[7]   See the evaluation of instruments to solve negative externalities in Fritsch et al. (1996).

[8]   Compatibility standards are able to solve the conflict between variety and economies of scale in the case of system products consisting of various components, since they increase the number of combinations between components and allow the specialization of suppliers and therefore economies of scales.

[9]   It may also be argued that standardization by reducing the product and service assortment also makes it easier for customers to compare price–performance ratios. In the extreme case, the dimension of competition is reduced to price alone. This drives the market structure away from that of monopolistic competition and incentives to extract rents by excessive product differentiation.

[10]   On the economics of measurement standards, see Tassey (1982) and Swann (1999).

[11]   Modified after Swann (2000).

4. Specific Economic Impacts of Standards

4.1 INTRODUCTION

Based on the classification of standards according to their economic impacts, we proceed with more in-depth surveys regarding four different dimensions. First, and most important for the other three dimensions, the ambivalent role of standards for technical change is discussed. Very closely related to this first aspect is the question whether standards foster or prohibit competition. In the international context, standards have also rather contradictory impacts on trade flows, because they can serve as trade-facilitating instruments, but can also be misused as non-tariff barriers to trade. Finally, we close this chapter with a brief discussion about the role of standards as one type of codified technical knowledge for economic growth.

In all of the following four sections, we will come back to the impact dimensions used for the general classification of standards. Therefore, some arguments of Chapter 3 will be repeated and set into the context of the respective issue. Furthermore, the demarcation of the four aspects is often not clear-cut, because market structure is relevant both for competition and technical change and innovation. Consequently, similar arguments will be presented in more than one of the following sections.

4.2 STANDARDIZATION AND TECHNICAL CHANGE[1]

Technical change, or rather the innovations accompanying it, are the guarantee for our economic prosperity.[2] However it is not enough that our researchers and inventors produce lots of new ideas. In order to trigger off significant, positive economic effects, these product and process innovations must be successfully positioned in the market and diffused. Diffusion, however, can be fostered by a functioning standardization system. Existing standards may also present hurdles for new technologies and products, because they compete with existing technologies and products which are more familiar to the users, and in which additional human and physical capital has been invested. This ambivalence will be the focus of the analysis.

In this section, we shall proceed as follows: in Section 4.2.1 a short overview of possible connections between standardization and technical change follows. For the microeconomic analysis of the influence of standardization on technical change, the complex problem of network externalities and the related role of compatibility standards are especially relevant, which will be dealt with in more depth in Section 4.2.2. Finally, in Section 4.2.3 three co-ordination mechanisms are examined for their economic efficiency regarding technical change.

4.2.1 Overview

Technical standards can exert influence on technical change in various ways. A central aspect is the argument of reducing variety or diversity, that is standards limit the diversity of products and thus the consumers' possible choices and exclude special individual wishes.[3] If a standard sets the exact composition of a product regarding quality, form or interface, then alternative (deviating) product designs can only be procured under considerable additional expenditure of cost and time. A basic driving force for technical change is questioned: the existence of variety is a vital pre-condition for the selection possibilities in the market.[4] This selection is possibly brought forward to an earlier point in time by standardization. Standards have a further excluding function: consumers become dependent on the enterprises utilizing specific standards. This applies especially to the use of so-called company-specific standards, which make individual system components offered by one company compatible between each other and thus limit the exchange of single components to the range offered only by the supplier of the system. The phenomenon of variety reduction is represented graphically in Figure 4.1.[5]

Four different cases are depicted in this figure: A describes the number of product variations which are available at a given time in an enterprise (for example due to product modifications and different specifications). B describes the number of product variations at the point in time when all existing national (respectively European or international) standards have been put in place. The adoption of own, company-specific standards is represented by curve C, the readjustment takes place here successively. D shows the average growth of product variations if no standardization (neither company-specific nor industry-specific) has taken place.

This graphic illustrates, albeit in a very simplified fashion, that standardization negatively influences the variety of products.[6] It can similarly be used to illustrate the reduction of product variations in entire sectors or economies. Through this reduction of variety, it also negatively influences technical change by preventing possible product variations which could provide the basis for the development of new products. According to Saviotti (1991), this also has negative impacts on the economic growth and thus the economic development of a country (see also Section 4.5).

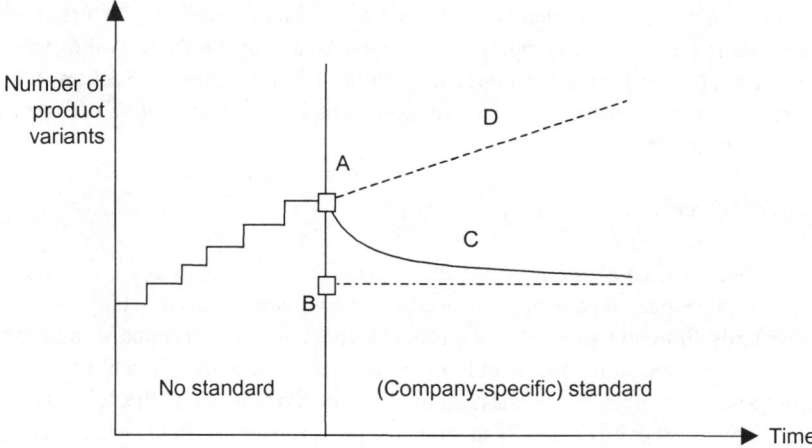

Figure 4.1   Development of product variety within a company[7]

This negative effect is counterbalanced by the argument of advantages through specialization, namely, the decrease in the number of system elements because of variety-reducing standards and the increase in the combination possibilities of single elements through interface compatibility.[8] This opens up possibilities for mass production, reduces the costs and thus the prices, and enlarges the potential circle of consumers. Then there is the additional advantage of a concentration of research and development (R&D) on a manageable number of product options, which involves a lesser market risk. The costs of searching information are decreased from the perspective of the demand side (Foss 1996). Further, standards which start from quality or processes can also influence the work sequences in R&D; whether the effects are positive or negative on the innovation process depends on the degree of obedience to the standards and the mode of implementation (Senden and Wöckel 1997).

A consequence of inflexible and 'false' standardization can be the 'cementing of the state of technology'.[9] What is meant by this is cutting down on other variations in the interest of standardization and the low incentive to change to another standard.[10] This incentive is again lessened by the fact that the broader the standard is defined, the greater is also the probability that own products fulfill the requirements of the standard. On introducing a new, better technology, a so-called cannibalism effect would occur, which means the replacement of the existing standard and the suppression of own products (Heß 1993).

Bauer (1980) points out that technical standards, respectively their drafts, contain information about the state-of-the-art of technology and in addition the state of science, and provide – if publicly accessible as formal standards – a good basis, like patent documents, for researchers to generate new ideas. A free know-how transfer takes place between the authors especially of formal standards and their users. By these means, information flows and cost savings in the innovation process are generated, almost for free. According to Thiard and Pfau (1991), the same advanced effect can also be achieved by collaboration in standardization committees and bodies, in which informal contacts can be established with engineers and other experts who are working in the same technological area.

The problems of variety reduction can be alleviated by standards which do not determine the exact content or design, but only certain characteristics or the performance of a product.[11] In this connection the problem of the subjectivity of the definition of quality can be solved by quality standards, which do not determine the quality of certain products but only define fundamental points of the quality (Liphard 1998). Another positive effect of standards can be derived from this, namely, that product and process innovations which fulfill the minimum requirements of the currently valid formal standards (for example quality and safety standards), are facing a lesser market risk on principle.[12] Because for the consumers the information asymmetries on the part of the manufacturer are less with standardized products and processes as regards product characteristics and quality and thus their trust and willingness to pay more. This gives the innovator a greater probability of success when introducing his new product to the market.

The argument of reducing transaction costs can be applied in any case. The transaction cost approach applies in institutional economics as the reason for developing the 'institution standard' and corresponds to the target of economic efficiency.[13] Variety-reducing standards allowing economies of scale are almost exclusively founded on this argument. On the other hand, quality, environmental and safety standards meet more societal goals so, for example, environmental standards prevent damage to the environment as a socially desirable goal. Common to all standards is that they codify technical knowledge like patents, which may be the base for further technical progress.[14] This differentiation in the kinds of standards reveals the basic problem of a general examination of the impacts of standards on technical change: compatibility standards work differently from variety-reducing standards, these again have another impact than quality, environmental or safety standards. Table 4.1 sums up the above results in an overview.

Table 4.1   Overview of the influence of standards on technical change

	Positive effects	Negative effects
Compatibility/ interface	• More possibilities of combining system elements, forming network bridges	• Impeding the transition from old to new technology
Minimum quality/ safety	• Reducing information asymmetries • Greater probability of market acceptance of new products	• Lock-in of technology status quo
Variety reduction	• Cost reduction, which fosters the accomplishment of critical masses of new products	• Reduction of variety
Information standards	• Information about the status of technology; source for new technological innovation (i.e. idea generation)	

Moreover, a differentiation must be made between process and product standards. Willgerodt and Molsberger (1978) examined the influence on product and process innovations and came to the conclusion that basically process standards (for example safety standards or environmental standards in manufacturing) do not impede product innovations[15] and product standards (for example compatibility standards or safety standards in consumption) do not hinder process innovations. Product standards, however, hamper product innovations, as they establish preferences of customers (or reinforce preferences for standardized products with the customers) and these can only be surpassed by a technologically vastly improved innovation. In order to simplify the further argumentation, only the relationship between product standards and product innovations will be dealt with in this section, if the other two standardization or innovation types are not explicitly referred to.

4.2.2   Microeconomic Theory: Network Externalities

The microeconomic theory of standardization deals to a large extent with the explanation and evaluation of network externalities in view of their economic effect. One aspect of this economic effect – technical change – will be critically scrutinized from the microeconomic standpoint, in particular from the demand side, and its significance for standardization investigated.

The problem of external effects can be described by the phenomenon of the possibility of compensation not regulated by the market or price. One or more economic units act and influence the costs or benefits of other economic units, without them having to pay monetary compensation, or respectively be paid. In the case of net(work) externalities, positive external effects appear in consumption because the benefit from the use of a network good depends directly on whether other economic units also use this good and thus increase the advantage of networking with other economic units. 'Network goods' are therefore goods or services which are privately owned, but which only develop their full usefulness through technical, economic or application-oriented connections to the same network goods of other owners.[16] Local and global network goods must be differentiated; the former limit their external-ities to a geographically restricted area, the latter are spatially independent. Local network goods often develop their external effects in conjunction with close communication and interaction in connection with learning and impart-ing knowledge or are tied to limited locations for cultural or logistical rea-sons. Cowan and Miller (1998) cite the example of video cassettes, which are locally available and must only be locally playable.

Other examples are the telephone, the Internet, and also video recorders and typewriter keyboards. Joining the network in these examples is done by new telephone or Internet connections, buying a video recorder of a certain system or taking part in a typewriting course. The network effect is unmis-takable with the first examples. Only when enough other participants have a telephone or Internet connection can the own network good be adequately utilized, for the direct benefit lies in the communication with other network members. These network goods are global, as one of their fundamental char-acteristics is interlinkage over great distances. In the case of the video re-corder or the typewriter keyboard, the network effect is indirect and is estab-lished via the market of complementary products and services.[17] In order to use a video recorder it is vital to have compatible cassettes to hand. The more one system becomes accepted, that is the more recorders have been sold, the greater the probability that compatible cassettes are sold and traded on the market. A similar case is the example of the standardized typewriter key-board.[18] The compatibility is produced by the ability of the user to operate the keyboard (in the best case) with the ten-finger system. Here the network ef-fect describes the possibility to be able to write quickly, not only on one's own machine, but others also, and thus for example to be able to change jobs. Both network goods are regionally restricted if the typewriting expert or the owner of the video recorder is immobile and hence requires the right cassettes or the proper keyboard only in his vicinity.

In a theoretical model the network effect can be explained for local and also global network goods by a rising utility curve with an increasing number of participants – assuming homogeneous preferences and no overcrowding in the network.[19] Joining the network will accordingly be more likely to be contemplated the more other members it already has. With overcrowding, for example in the case of overloaded telephone lines, this would no longer apply, as here a further membership would create additional costs for the participants. The decision depends basically on the costs of joining – including the opportunity costs from not joining another network. In the individual case the personal (private) benefits and costs are taken into consideration, but not however the benefit to the network community through the increase of the number of members.[20] Thus the social benefit is not adequately considered; under the premise of same costs the danger of the (collective) non-membership is given, as to begin with a too small number of participants (in the extremist case none) keep the network small and so the costs per member are greater than the utility/benefit which by pure network goods tends towards zero (Liebowitz and Margolis 1994). Only when a so-called 'critical mass' is achieved, is the net benefit positive. Figure 4.2 elucidates the problem of critical mass.

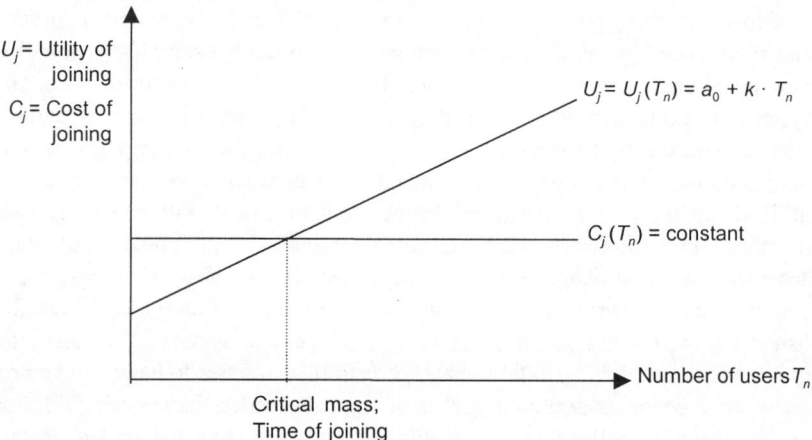

Figure 4.2   Critical mass on joining an economic unit in the new technology T_n with network externalities

It becomes apparent hereby that the new technology can only be accepted if a sufficient number of economic units has already accepted it, even if this technology is basically advantageous and efficient. Assuming the same preferences, the new technology will not succeed unless individual economic units are more prepared to take risks, are more innovative in their consumption behavior, or their costs for membership are less. If the critical mass is reached, then it is also advantageous for the economic units with the above benefits and costs function to adopt the new technology, thus further increasing the number of members and making membership worthwhile for economic units with even less benefits or more unfavorable cost structures. This *bandwagon effect*[21] means that the technology will achieve maximum possible diffusion.

The diffusion of technologies with global network externalities is a fundamental argument in favor of introducing standards. Standards which apply to a new technology and contain compatibility aspects reduce the risk of not achieving critical mass and thus increase the probability of the new technology being accepted. On the other hand, Cowan and Miller (1998) point out that standardization for local network externalities is not necessary, or even detrimental, as the standardization of local differences is not practical. The following thoughts should therefore exclusively take global network goods into account.

The range of possible compatibility standards extends from purely company-specific standards over branch-specific and formal standards to network bridges. But safety and quality standards also contribute to the diffusion of the new technology, as they increase the utility for the users of the new technology by creating greater transparency about product characteristics. The costs are reduced if they contain the risk of unsafe product characteristics. In both cases standards push the utility curve upwards, respectively push the cost curve downwards and thus reduce the critical mass.

The problem is somewhat different if the new technology is supposed to be substituted for an older and less efficient technology, also with network externalities. Here the opportunity costs from giving up the old technology are added to the costs of joining the new technology, including the network externalities of the old technology. In Figure 4.3, both technologies are made comparable, in that participation in the one or the other technology is alternative, that is a new technology T_{new} is adapted only by consumers of the old technology T_{old} by a change, and thus the entire participant number n_{total} is determined for both technologies ($n_{total} = n_{old} + n_{new}$ = constant). This participant number is located at the ordinate. The superiority of the new technology is represented by the higher utility curve U_{inew}[22], the critical mass is achieved at the point where both opposite curves U_{iold} and U_{inew} intersect (on the premise, that no entrance costs arise[23]).

A change from old to new technology is therefore only attractive if a sufficiently large number, the critical mass, undertake the change; as shown in the previous example, this will not happen if all have identical utility functions.[24] Farrell and Saloner speak on this dilemma allegorically about penguins adrift on an ice floe, wanting to catch fish in the water, but unable to decide to jump for fear of possible predatory fish swimming around.[25] In this allegorical example, standardization is analogous to a simultaneous jump by all the penguins. The introduction of a standard is a signal for the users that a sufficiently large number are pursuing this technology and the network externalities are large enough.[26] Moreover, standards can achieve a regulated co-ordination of the adaptation of network goods. They show the direction in which the technical change is heading and thus solve the problem of – to borrow a further picture from Farrell and Saloner (1987) – horses tethered together which could not budge from the spot because of failing co-ordination.

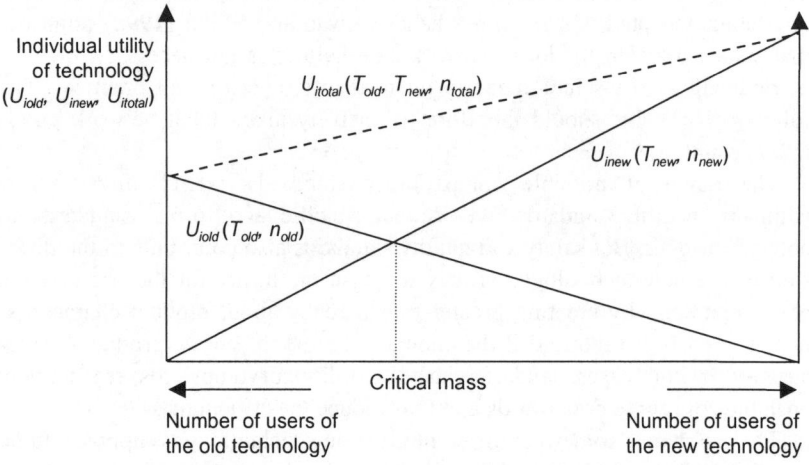

Figure 4.3   Critical mass in the transition from old (T_{old}) to new technology (T_{new})

Compatibility standards have a different effect in the case that the old and the new technology can be made compatible by means of a network bridge or a converter.[27] Thus the transition of especially innovative individuals to the new technology does not cause any negative network externalities for those who at first stick to the old technology, and in the long term when the old technology is completely abandoned the maximal benefit is achieved (in the graphic, addition of both utility curves to U_{itotal}). A completely new technology without network-bridging potential, however, will always be positioned behind the compatible technologies in these assumptions, which impedes the enforcement of radical innovations (as in the above example) and so a more rapid technical change.[28] The individual system elements as well as the systems can be rendered compatible through standardization. Technologies where this is not possible will either achieve the network effect through standardizing their own system elements or will not become sufficiently widespread and be unsuccessful in the market (technological dead-end).

In a study by Glanz (1990) the up till now – purely static – view was infused with more dynamism in a similar graphical form as Figure 4.3. The network benefit was deemed an additional benefit like the other product advantages such as quality, safety and the probability considered that a standard will assert itself. A system made compatible by a formal standard is compared to a company-specific standard and the cases of diverging technological development over time and also different preferences (for a formal standard or for the prior company-specific system) compared. The result, which standard asserts itself, is determined by the parameters preferences, network utility, the respective basic utility and by the course of the technological development of a system.[29]

Following up the 'penguin problem', Choi (1997) designs a model of asymmetrical incomplete information, in which it is assumed that less knowledge is available on the utility of the new technology than the old and for this reason the old technology is preferred – even for anticipated greater utility – so that no transition will take place. Farrell and Saloner (1986) took up another aspect of the change from old to new technology. They differentiate between the adoption of a new technology by new users (for example in the case in which the costs of switching over are too high for the old users and they will not change over) and the adoption of a new technology by the existing old users (as in the above example). In two models they examine the economic consequences of these constellations and arrive at the result that in the first case externalities arise through the fact that firstly the new users of the new technology reduce the possible utility growth of the old users, and secondly increase the utility for the following new users. In the second case the utility of the old users is decreased by those changing and the changeover to the new technology is made more attractive.

Assuming that these externalities will not be internalized, Farrell and Saloner derive from these models the economic inefficiencies of *excess momentum* and *excess inertia*. Under excess momentum – the inefficient adoption of the new technology – the users of the old technology must bear costs because the new technology was chosen by the new users, respectively those old users who changed over. Excess inertia – the inefficient appendage to the old technology – is caused if the new users or those old users who potentially could change over have to bear the brunt of pioneering (innovation) costs (for example initially too low network effects), which are not reimbursed by the following users and beneficiaries of these positive external effects and therefore higher than the utilities of new users adopting the new technology. Excess momentum can therefore lead to an inefficiently early change from the economic viewpoint; excess inertia to a late change. Which effect dominates is not ascertainable a priori, but depends on the size of the installed base of the old users, the quality differential between the old and the new technology and the formulation of expectations. The adoption of, or change to, the new technology can for example be influenced by price reductions or advance announcements as competitive strategy on the part of its suppliers.[30]

The economic inefficiencies described by Farrell and Saloner play a role in the timing of the change from old to new technology and are thus of decisive significance for the replacement of an old standard by a new one. Not every new technology is necessarily better than the old one and is not so mature from the start that it cannot be improved. Too early a change would not only impair the technological development, but also the economic efficiency, if the utility from an improved old technology is greater with the same number of participants. A danger exists in the development into a *blind giant*, which means the further development of the actually inferior new technology. A society in which 'novelty' *per se* is valued, that is many innovators determine the demand behavior, is in danger of adopting every new technology without reflection and follows up even inferior alternatives. It behaves – to take another example from the animal kingdom – like the lemmings and plunges into the technological abyss.[31] Another explanation for this pattern of 'blindly' following is given by Choi (1997); here *herd behavior* is named as the release motive. Everyone adopts the technology which the majority of other users are employing, as no one wants to adopt a technology that nobody else is using, not wanting to be 'stranded' on too small a beach. 'Blind' must therefore in this context be understood as technological blindness, although herd behavior is rational for each individual.

The opposite is graphically described in the 'penguin problem'. All users shrink from a change as the opportunity costs are too high, although if behavior is co-ordinated, the utility exceeds these costs. Society remains a prisoner of the old technology due to the *lock-in effect* and restricted to the *narrow windows*, that is the possibility of a further technological development is limited to the further development of the inferior old technology.[32] Farrell and Saloner (1985) emphasize a further bias to this behavior by arguing that every change is irreversible, that is only sticking to the old technology keeps both options open. For this reason the change will take place as late as possible, possibly at an economically inefficient point in time.

Another consequence can be the inefficient, parallel existence of two or more incompatible technologies, that is the positive network externalities are not being fully utilized.[33] Farrell and Saloner (1985), attempt to prove that this constellation is rather improbable, given the same positive expectations of utility, in that they derive the theory by means of model building in game theory, that the participating suppliers will reach agreement on a standard in any case and the transition – when it comes – will take place simultaneously.[34] But if an 'old' technology is still relatively young and not all economic units have adopted this technology, a change to a new technology would be likelier to be accepted by latecomers, but not by the adapters of the old technology, as they have to bear the opportunity costs or the costs of joining the 'old' technology and therefore their net utility can be negative. They are referred to as *angry orphans*. They would remain embedded in the old technology ('stranded') and would lose not only the external network benefit of a large number of participants, but also the net advantage of adopting of a new, better technology. The possible consequence would be that the – technologically and economically desirable – technical change is hampered and sabotaged by these *angry orphans*.[35] The same phenomenon would be seen in the case of a (too) late change from one to several already existing technologies.

The previous demand-based reflections have shown that if network externalities exist, the agreement on a common 'network' with respect to technology is the optimal solution. This common network can be achieved by the limitation to one supplier with a company-specific technology or by means of compatibility agreements among various suppliers. In a supplier-oriented view of standardization, the impacts on market structure are described.

In the case of an agreement on the specific technology of one manufacturer, this would make this manufacturer a monopolist for this particular good and would – according to the monopolist theory – lead him to offer too small a number at too high a price, and thus cause welfare losses as opposed to positive network effects. Lack of competitive pressure would furthermore reduce the incentive for the monopolist to seek further innovations, as assumed in the case of Microsoft. Possible competitors would be scared off this market, since although they could achieve a technological improvement with their new technology, this improvement must compensate for and outdo the positive network externalities of the already established supplier. For enterprises with a large market share, this incompatibility with products of other manufacturers can become a competitive strategy, as by this method the market must not be shared with other suppliers.[36] The change in the technology area concerned is determined by the strategy of a single enterprise, not by market and technological optimization.[37] *Sponsored technologies* have a similar effect which, according to Katz and Shapiro (1986) are characterized by property rights of the market actors to the corresponding industry standards. By asserting the own company-specific standard as the industry standard, for example by lowering the price by cross-subsidies from other business fields, a monopolistic position can be reached after eliminating the competitors.

By uniting several suppliers to support a common industry standard, a typical polypoly would emerge, with the possibility for small suppliers to also enter the market. *Non-sponsored technologies* in the form of formal standards are accessible for all market participants and are realized by all of them. In this situation there is an incentive for single 'strong' competitors[38] to raise their own specific technological basis via price reduction and product incompatibility (*sponsored technologies*), respectively, assert their own industry standard, eliminate competition and assume a market-dominating position. If this strategy succeeds, the above-mentioned monopoly situation comes about.[39]

In the previous sections technological change was dealt with exclusively from the market aspect, which basically reflects the discussion carried on in the 1980s and 1990s. It was implicitly assumed thereby that the innovation efforts are given by the enterprises exogenously and only the breakthrough of the new technologies by market demand was influenced by the network externalities. This point of view is extended by Kristiansen and Thum (1997), in that they regard taking advantage of network externalities as a strategic innovation decision, that is as a pre-condition to carry out R&D activities in a technology area. This economic approach is based on the thesis that interdependencies between various market niches (for the sake of simplification, to be called A and B) can emerge through compatibilities and in the case of network externalities, the opening up and extension of a new or existing market A has positive external effects for another market B. The markets observed offer different products and are only linked by network effects.[40]

From the existence of positive external effects the authors derive a suboptimal and too low R&D activity in a theoretical model, as these external effects are not internalizable and not the only innovating enterprise profits from the positive demand effects on the compatible market B. So the decision to innovate is only influenced by the profit consideration for the own company. In the extreme case no R&D will be carried out, as the envisaged profits are too low. Another consequence from this situation can, however, also be sub-optimal, because of exaggerated R&D activity. This can make sense for company strategy if the quality of technology A, improved by the R&D activities, forces the competitors in market A to lower their prices. This price reduction increases the demand for the product in market A and thus the positive network externalities for the other market B. It can therefore make sense to develop a technology further although the enterprise might not want to become active in market A. This happens only in order to take full advantage of the network externalities in market B, in which it has a large market share.[41]

Besides the decision about the amount and utilization of R&D expenditure, the existence of network externalities also affects the timing of technology development and introduction.[42] Kristiansen (1998) shows that through network externalities incentives arise for enterprises to introduce new, incompatible technologies prematurely and immaturely to the market. On the other hand, the early agreement on a common standard leads to an economically efficient delay in market introduction.

Another example of innovation behavior distorted by network externalities can arise on the supplier side because of the penguin effect. Potential innovators pay more attention to observing competitors than to developing new technologies, since by the network effect an own technology can be outdone by a superior other technology (or one which utilizes the network effects better). This *'sharing the blame'* effect (no enterprise wants to become active on their own in a new network technology area) leads to a purely reactive innovation behavior which is neither technologically nor economically optimal.[43] According to Farrell and Saloner (1985), further inefficiency can be caused by the *bandwagon strategy*, in which a company does not place its hopes on the technology for which it possesses the most competence or the greatest know-how, or which is the technologically superior, but on the technology for which the greatest possible diffusion is expected.

4.2.3 Co-ordination Mechanisms and their Impact on Technical Change

The remarks up to now have dealt (using the micro-economic tool box) with the question, as to what extent formal standards and industry standards exert influence on the technical change and which fundamental problems standardization must overcome. The question of how the standards are arrived at and which institutions and other economic actors play a role herein has not been touched.[44] But according to Thum (1994), there are also aspects in the impacts on technical change which can be traced back to different co-ordination mechanisms in the evolving of a standard. There are three basic possibilities of co-ordination for standardization: governmental regulation (laws and directives), co-ordination by voluntary committees (formal standards) and the market co-ordination (industry standards).[45]

The market mechanism can produce an optimal solution, if perfect competition and different product or standardization preferences dominate. If no standardization would take place, this would correspond to the preferences and would thus be efficient. If standardization takes place due to identical or majority preferences, then it must be examined if this has taken place to an optimal extent. The danger of an exaggerated (over-)standardization is given if a certain leeway in setting prices is possible (no perfect competition). A supplier could subsidize his own product, increase its market share and thus lay the foundation for an industry standard which corresponds to the characteristics of his own product (sponsored standards). In this form of industry standardization, the technical change would be guided in a direction predetermined by one enterprise, and in the rarest cases does this option produce the technologically and economically optimal technical change.

Inadequate (under-)standardization appears if costs from the standardization process are borne exclusively by the standardizers and the positive external effects (for example network externalities, but also positive environmental and health impacts) cannot be fully internalized. The positive effects of standardization, also on technical change, would not be completely realized. Generally, it can be stated that markets with network externalities tend towards inertia and prevent the adoption of new technologies through the lock-in effect. The standardization of new products is impeded by the above-mentioned co-ordination problem, and also by positive network externalities. The problems of network externalities described above speak against a purely market solution, although it is argued by the school of New Institutional Economics that standards are efficiency-enhancing because of their capability to decrease transaction costs and that therefore standardization should not be influenced by external forces, like governmental institutions (Wey 1999).

An optimal solution is possible using a co-operative committee solution, if a common interest to agree on a formal standard exists on the part of the actors involved, and these actors adequately represent the demand and supply sides. Further, the standardization committee must be well-balanced regarding power structure, as well as technological, economic and social competence. In bodies such as the standard development organizations (SDOs), an attempt is made to fulfill these requirements by making standardization work in principle always accessible to 'interested parties' who unite these competencies. Over-standardization emerges when standardization committees are not evaluated by the economic and technological efficiency of their released formal standards, but by quantitative output. An exaggerated propensity towards standardization can lead to a too early or wrong technology being established by standards. Moreover, the possibility exists that non-participating suppliers are deliberately disadvantaged by the standardization committee, either through one-sided standardization or the strong interest congruency of a few powerful participants influences formal standards in a fashion inadequate to technical change.[46]

There is a tendency towards under-standardization, if differing standardization preferences on the part of the participating actors emerge, or single suppliers pull out and no or too little standardization takes place. David and Monroe (1994) examine, on a game theory basis, the probability that an agreement will be reached and come to the conclusion that under certain conditions, depending on the strategies of the participants, the role of the mediator and information asymmetries, the agreement will not take place in time. Economic losses through non-utilization of network effects, rationalization and economies of scale, as well as technologically unsatisfactory standards, can be the result. It can be generally stated that the problem of network externalities can be better solved with this co-operative solution than through the market mechanism, with the danger inherent in the strategic behavior of the participating actors and thus the tendency towards a technical change strongly influenced by interest groups.[47]

Regulation by governmental institutions results in an optimal solution, if the standardization succeeds in internalizing external effects. This target can only be aspired to from the government side, as the governmental bodies should represent the interests of all the members of an economy and possess the necessary authority to issue binding regulations and instigate punitive measures in the case of non-observance. A sub-optimal standardization takes place when the state cannot adequately foresee the technological development and a technical regulation is introduced too early or too late. The problem of narrow windows and blind giants crops up again, that is a restriction of the future technological development to the further development of existing technologies without the chance of switching to a new technology (blockade by angry orphans) or to the further development of a new technology with only small technological potential.[48]

Although co-ordination against excess inertia or excess momentum may be a welfare improving activity of the governmental bodies (Adams 1996), it can be said that the governmental solution is the least promising one, as it cannot be assumed that the governmental decision-makers are well informed about the market chances of the technological developments and thus capable of making optimal decisions. As a consequence of this inadequate competence for standardization, governmental institutions are given the responsibility of creating the framework conditions for standardization, but the standardization process itself should by conducted by private committees.[49] Table 4.2 sums up the results of this section.

The preceding reflections have shown how ambivalent the effects of the standards can be for technical change. As compatibility standards they can make positive network externalities possible, encourage improving innovations and product differentiation, internalize positive and negative external effects from consumption and production, and strengthen (innovation) competition. On the other hand, compatibility standards can hinder the radical transition to totally new technologies without interfaces to old technologies and provoke economically and technologically sub-optimal strategic innovation behavior. Quality and safety standards, by reducing risk in new technologies, however, tend to have a positive impact by reducing the critical mass to be reached and thus support the change to a new technology in the case of network externalities by the utility increase. Variety-reducing standards reduce the diversity of system elements and products, which in the first case can lead, in combination with compatibility standards, to a greater number of combination possibilities of elements and thus to an extended product assortment, and in the second case to the possibility of limited choice for the consumers.

Strategic behavior and differing interests of the participants in the standardization and innovation process play a not insignificant role in the question of whether standards come into being, and thus in technological development. In the case of network externalities a new standard can be undermined by latecomers and powerful suppliers, or the excessive enthusiasm for standardization of bureaucrats and competitive politicians can produce sub-optimal standards which lead technical change in the wrong direction or into a dead-end street. Standards which lead to the compatibility of various systems can cause strategic innovation and standardization behavior which on the other hand can lead to inefficiencies in the economy as a whole. A false standard or one introduced too early or too late lessens the potentially positive effects of the standards on technical change and can even transform them into negative ones. Whether the traditional standardization process within standards development organizations prevents a timely standardization which meets the demands of researchers and developers and leads to technologically sub-optimal standards, will be investigated in Chapter 16 of Part D.

Table 4.2   Co-ordinating mechanisms

	Over-standardization*	Under-standardization*
Market (industry standards)	• Leeway for price setting, cross-subsidies • Result: sponsored standards	• Positive externalities via non-internalized costs of standardization • Co-ordination problem • Lock-in effect
Committees (SDOs) (formal standards)	• Incentive for participants to produce too many standards • Interests of individuals influence standards	• Interests of individuals hinder standardization
Governmental institutions (technical regulations)	• False estimation of technological development • Blind giants	• False estimation of technological development • Narrow windows • Angry orphans

Notes: *Over-standardization should be equated in this connection also with premature or inadequate standardization and under-standardization with too late standardization.

4.3   STANDARDS AND COMPETITION

After the discussion of the relationship between standardization and technical change or innovation, special focus is given to the economic impacts of standards, especially compatibility standards, on competition based on the seminal papers by David and Greenstein (1990) and David and Steinmueller (1994), although some issues of competition and innovation have already been discussed in Section 4.2. The interrelationship between standardization and market structure will only be taken into account in the empirical part of the study when determining the driving forces for standardization. The impact of standards and standardization processes on the market structure will not be further empirically investigated. However, this dimension represents an important background, especially for the role of standards in dynamic frameworks also taking into account innovation and will therefore be briefly outlined.

Both pro-competitive and anti-competitive effects can be generated by one single standardization activity and its output. These positive and negative impacts may be at work in the same market at the same time. Moreover, the establishment of a standard will generate effects on competition that differ in the life cycle of a product. Finally, in the long run both market structures and standardization activities are endogenous phenomena.[50] Consequently, generalizations about the effects of standardization on competition can be derived only to a limited extent.

The consequences of standardization for competition also depend on the timing of the development of the industry and the maturity of its technology. The impact of standards are therefore influenced by the dynamics and the speed of technological change. Incremental changes in rather mature markets or radical innovations in emerging markets represent rather different framework conditions both for the development of standards, and for their implementation and impact.

Taking these various caveats and shortcomings into account, David and Steinmueller (1994) present an overview of the pro- and anti-competitive impacts of compatibility standards.[51] From a general perspective, the informational attributes of standards being close to public goods tend to be favorable for competitive conditions (Tassey 2000). When agents or companies are not endowed with the same access to resources and existing assets that are complementary to the public good, differences may be created that will lead towards the dominance of one or a small number of suppliers. The most popular example is the case of Microsoft (Liebowitz and Margolis 1999). Having acquired market power, this dominant group may be able to influence standard-setting in a way which may secure or even expand their dominant position.[52] In summary, where the benefits of standardization cannot be appropriated in the same way by all the companies involved, and if the access to such standards by rival firms may be restricted, standards may become a strong strategic instrument to reach, secure and expand market dominance.

4.3.1   Pro-competitive Effects of Standards

The publication of the technical information contained in any type of standard in the course of standardization that reduces the asymmetries in the costs of access to information will tend to level the playing field among competing companies. This is the effect of public information reducing unequal starting positions.

Compatibility standards reduce entry barriers in markets for network products and services or multi-component products, because companies can concentrate on reducing costs or increasing the quality of those attributes not specified by the standard. When companies do not have to invest in the capital facilities and R&D of producing a complete, integrated system, not only is entry into the market easier, but also price competition will lead to prices of the single components near to their marginal costs. In addition, David and Greenstein (1990) argue that in complex product and service markets, where conformity with a performance standard for the inter-operability of systems is not transparent to the consumers, the certification of conformity by independent testing institutions presents a dimension of quality competition among suppliers which has positive impacts on consumers' surplus.

The production or publication of compatibility standards or the achievement of interoperability by means of converters and gateway technologies (Farrell and Saloner 1992) that make use of information about sub-system interfaces, reduces the incentive of suppliers to create switching costs for their customers and to extract rents on replacement purchases by customers who have had to acquire complementary assets that are more durable than those provided by the supplier. Lower switching costs will increase the threat of losing customers even in network industries, and therefore will increase the competitive pressure.

The production and publication of interface specifications that are non-proprietary, a requirement of voluntary SDOs (see Chapter 13 for details), tends to restrict the opportunities for the 'leveraging' of monopoly power through interface manipulation. The latter is a special form of 'tying' which depends upon the technical interrelation between the system component that makes use of the supplier's proprietary technology and other non-proprietary components.

Finally, the introduction of compatibility standards represents a requirement or institutional pre-condition for the effective deregulation of network industries that previously were organized as regulated natural monopolies or as state-owned monopolies. Technical compatibility standards are critical tools in moving network industries towards more decentralized, less hierarchically-controlled system technologies, wherein numerous suppliers compete in the provision of services and producing network components, such as in telecommunication industries.

The previous list of impacts of compatibility standards with positive effects for competition made it obvious that as long as equal access, user rights and transparency are ensured, standards either provide fairer framework conditions for an existing field of competition or elevate the competitors to a more advanced and intensive level of competition in a dynamic setting.

4.3.2 Anti-competitive Effects of Standards

The anti-competitive effects of standard-setting activities are based on the possibility that one or very few companies internalize the benefits of standards or control the specification of standards, so as to extract higher rents or even to impose higher costs on their rivals. Particularly if existing property rights on technological specification can be leveraged by proprietary standards, competition may suffer. The consequences for competition are the following, according to David and Steinmueller (1994).

The procurement standards of large clients, such as governmental bodies, may favor existing major producers that are thought to have sufficient resources to assure supply or meet other reliability criteria. The resulting standards may dominate the whole market, even though their technical characteristics do not fit the needs of other customers. This outcome is especially likely where network externalities make compatibility between publicly and privately owned assets a requirement among private customers, for example the telecommunication infrastructure is purchased by a state-owned public monopoly and the private users have to buy compatible telephones, although they have other preferences.

In addition, regulatory bodies or standards development organizations that publish standards are vulnerable to capture by large domestic producer interests, as the latter possess the technical expertise necessary to specify standards; one form of this is the use of national standard-setting by domestic producers as a non-tariff barrier against products of foreign competitors. These anti-competitive consequences are not peculiar to the case of compatibility standards, but can also arise in the case of minimum quality, safety or environmental standards. This aspect is therefore relevant for the role of standards on foreign trade (see Section 4.4 for more details).

SDOs, especially those technical committees working at the European or international level, are subject to being dominated by representatives of major vendors, rather than users or minor suppliers. Even where the consensus rule is applied and professional engineering considerations are given most weight, the dominant incumbent firms in the industry have advantages when it comes to undertaking background R&D and sending expert personnel to participate in the meetings of the standards committees.

Whereas the previous argument also applies to other types of standards, the following aspect is especially relevant for compatibility standards. Coalitions of producers can use the formal standards writing process to publish product specifications that impose cost burdens upon current rivals (Salop and Scheffman 1983). They may even circumvent possible future competition by issuing anticipatory standards that have the effect of channeling technological development into areas where they have control of complementary or basic technology through intellectual property rights, like patents, or cost advantages based on already realized learning effects.

Restrictions on innovation by rivals tend to be more binding where detailed descriptive standards for interfaces between network components rather than standards for performance are issued. However, it has to be emphasized that an exceptional constellation exists with respect to the effects on innovation in complementary input markets. There performance standards are less informative for specialized suppliers who try to adapt their product specifications insofar that conformity with a given performance standard could be implemented in various ways, which may increase the design costs especially.

Where de facto, market-driven processes of standardization are involved, installed base effects, which induce or even force consumers to follow choices of former buyers, may protect a dominant firm that will thereby receive a greater market power in comparison to a constellation of just having proprietary control over one among a broad set of alternative technologies or products.

The perceived strategic value of rapidly establishing a predominant installed base position in markets where network externalities are important is likely to induce predatory price-setting by the standard's owner. Because the positive network externalities tend to foster the demand for the good or service, the firms with the expectation to become dominant in the market may reduce their prices not just based on the expectation to take over the sales volume of the competitors but also the volumes caused by market growth. These pricing strategies may be used to keep out entrants and to establish a monopoly position (Klemperer 1987). Church and Gandal (1993) show in their model that proprietary interface standards can also create barriers to entry on the markets for complementary products.

The negative impact of standards for competition are mostly caused by a biased endowment with resources available for the standardization process itself. Therefore, even when the consensus rule is applied, dominant large companies are able to manipulate the outcome of the process, the specification of a standard, into a direction which leads to skewed distribution of benefits or costs in favor of their own interests.

In summary, standards possess both a very broad variety of impact dimensions and a high degree of ambivalence concerning competition. Reflecting the discussion of their impacts it became obvious that the more open and less proprietary they are – at least under a static framework – the more beneficial they are for social welfare.[53] For the development of new markets proprietary standards represent (like patents) often necessary incentives for companies to invest significantly in R&D, especially in order to introduce radical innovation into the market. Obviously, the general tension between static and dynamic efficiency can also be observed regarding the economic impact of standards.

4.4   STANDARDS AND FOREIGN TRADE[54]

Some of the arguments on the impact of standards on competition can be simply extended to the competition between companies located in different countries. However, besides this microeconomic perspective foreign trade is also influenced by the competitiveness of countries as a whole, which depends on the efficiency of the national economic system and even the national system of innovation. Therefore, we complement the microeconomic perspective by macroeconomic viewpoints, including the aspect of intra-industry trade.[55]

4.4.1   Standards and Competitive Advantage

In the increasingly important literature it is alleged that the development of (world) market shares is not determined by price competitiveness, but by quality and service competition. Simultaneously, the similarities in competitiveness in enterprises from the same economies indicate that non-price competitiveness should be analyzed in the context of the relevant national economic system.

The following economic ideas (Verspagen and Wakelin 1997) underlying the neo-Schumperterian approach are of most relevance to the scope of this section. First, absolute differences rather than the Ricardian comparative differences are seen as the motivation for economic dynamics. In the context of international trade, absolute differences in product quality or price between different producers determine competitiveness. Firms which score above their competitors will see their market share increase, and firms with low competitiveness will lose market share (Amable and Verspagen 1995).

Secondly, technology is seen as an endogenous phenomenon. One characteristic of technology at the microeconomic level is that it has important private as well as public aspects. As a result, the benefit of innovation can be at least partly appropriated. Assuming that diffusion occurs more easily within a country than between countries, technological differences between countries are assumed to be persistent to a certain extent, that is no country can completely rely on imitation to catch up with the technological frontier. Therefore, technology gaps result from a process of knowledge accumulation and not from different endowments.

Finally, the importance of the role of institutions in the development of technological change is emphasized. Institutional differences between countries may lead to, or relate to, technology gaps and therefore have an influence on growth and trade. Such national institutional characteristics include legal methods to protect intellectual property, such as the patent system, or official platforms for uniform product or process specifications like the national standardization development organizations. More generally, the history of the country, its past technological achievements and the characteristics of its institutions strongly influence its present potential for innovation.

There have been a number of attempts to assess the importance of differences in technology empirically. The survey of the empirical literature by Wakelin (1997) confirms the role of technology in affecting trade performance, especially its long-term impact. In general, the empirical literature underlines the importance of non-price factors, such as quality and innovation, in influencing the trade performance of developed countries. For the United Kingdom, Greenhalgh (1990) and Anderton (1999) find impressive empirical evidence.

Whereas innovations protected by intellectual property rights like patents have unambiguous positive impacts on trade performance (Maskus and Penubarti 1998), because they restrict others from using the technologies covered, formal technical standards are in general public goods. Therefore, in a dynamic theory for the location of production – the product cycle theory – after the innovation process the product or the process innovation becomes more elaborated and mature, leading to de facto or de jure standardization processes. Whereas de facto industry standards can be protected by intellectual property rights, the technical specifications described by de jure standards published by national standardization bodies can be used principally by every producer worldwide. Therefore, goods which rely on these kinds of standards may be also produced in regions which are not the original source of innovation. However, because of high adaptation costs, outsiders with no influence on the standardization process may face considerable disadvantages in using the specifications of the standard (Antonelli 1994; Matutes and Regibeau 1996). Furthermore, the content of the standard can only be efficiently used in another country when there is an absorbing capacity with corresponding technical knowledge. Consequently, formal technical standards, especially national ones, are also an indicator for the innovative potential of a country.

Besides the national R&D capacities and the national innovation system in general, the national systems of product and process standards represent a significant element for the innovativeness and the competitiveness of a country, for the national standardization system can increase the perception of the quality of domestic products, not only at home but also abroad, and thus increase their competitiveness. Standards and technical rules can also lead to a trade advantage, in that they boost the quality of national products or through the generation of scale economies which make price advantages possible. This applies for the implementation of both international and originally national standards, whereby the impact should be strong for the latter, because they facilitate the product differentiation from foreign competitors. However, if standards represent the preferences of consumers and there are differences between national preferences, the positive effect of national standards from the technology perspective may be offset by demand side frictions. Secondly, the importance of cross-border network effects makes international or at least compatible national standards more favorable, also from the supply side (Gandal and Shy 2001).[56]

4.4.2   Standards, Trade Deterrence and Competitive Disadvantage

A more pessimistic and traditional view assumes that national standards and technical rules can be deterrents to trade and competition either intentionally, to avoid negative environmental effects (Barrett 1994; Fischer and Serra 2000) or to be used as instruments for strategic industrial policies (Brander and Spencer 1981, 1985), or unintentionally, depending on their specific design.[57] In the simplest case they have a symmetric effect on exports and imports, because domestic products manufactured according to the guidelines of idiosyncratic national standards can be incompatible with the standards and consumer preferences applicable in export markets and this leads to sales problems. Regarding imports, the idiosyncratic national standards are difficult to maintain for goods imported from abroad, and thus form import barriers.[58] In reality, however, we can assume slightly asymmetric impacts, because the effects of national standards are more serious for importers than for exporters, who have to comply with the customer wishes of the export market. If this is the case, national standards cause a divergence between domestic and world-market prices, which allow domestic producers to raise prices without losing their competitiveness. Consequently, the quantity demanded, imports and consumer surplus decrease, whereas domestic production and producer surplus increase (Nunnenkamp 1985). In addition, several further negative impacts on the allocation efficiency may occur due to higher risks and information costs for importers.

The 'competitive disadvantage' hypothesis, coming as it does from the political discussions, states that the national standards can also lead to a competitive handicap for the domestic producers, because the production costs and the length of the registration procedure are increased by conforming to them.[59] In the case of telecommunications prior to deregulation, it is accepted as proved that a strong orientation to the national standards (the 'standard technology' of the monopolist customer) de facto decreases exportability, because the costs of conversion may be too high in late stages of product development.

4.4.3   Standards and Intra-industry Trade

Besides the inter-industry trade, the intra-industry trade becomes more important the more highly industrialized the considered countries are, because it increases the diversity of product specification in one product class (= horizontal differentiation) and results in higher utility levels (Greenaway and Milner 1986).

The qualitative and the quantitative impact of standards depends both on their regional dimension and on their functional character. Concerning the first distinction, the third theoretical perspective is principally directed towards supra-regional, international standards, because common and harmonized standards can be conducive to trade and demolish trade barriers, by creating a consensus on sizes, weights and quality characteristics.[60] In the more recent literature on the integration of economic areas, the trade-promoting effects of common standards and technical rules have been proved (Flam 1992). But even the publication alone of a standard or technical rule on a national level codifies local technical knowledge and preferences, which in the long term can be more easily anticipated by foreign competitors, so that their import efforts are facilitated.

From the viewpoint of intra-industry trade, international standards are seen as especially beneficial, because they facilitate the specialization in product variations which is characteristic for intra-industry trade. A difference can be made, however, between the generally trade-promoting effects and their specific effects on scale economies and product diversity, which boosts the intra-industry trade. Here we must differentiate between the effects on the one hand of compatibility and quality standards (David and Greenstein 1990), and on the other hand of variety reduction standards.[61] Whereas compatibility and quality standards are basically not only export-promoting but also favorable to intra-industry trade because of the implementation of economies of scale,[62] the implications of standards which lead to a reduction in product diversity are ambivalent, because their positive effects on foreign trade in general are obviously contradicted by their negative effect on the product diversity, which is based on the demand for greater variety from intra-industry trade. However, based on a harmonized standard, many product variations can easily be built on, which makes the negative impact on product diversity ambivalent.

In summary, it appears that the impact dimensions of the idiosyncratic national and the assumed international standards on foreign trade are not completely unambiguous and are rather contradictory. The positive implications of international standards on foreign trade are certainly stronger than those of the idiosyncratic national standards.[63] However, the existence of the latter is preferable to a situation without any standard. Besides, the substitution of idiosyncratic national standards by international ones basically promotes trade, but can lead to a reduction of product diversity, and thus slow down the incentives for intra-industry trade. Although each standard has some kind of variety-reducing effect, the trade-fostering impacts of the compatibility or minimum quality standards weigh more in quantitative terms.

4.5 STANDARDS AND GROWTH

A rather new aspect, which has not been dealt with in depth in the economic literature, is the role of technical standards for economic growth, although the importance of technological activities as an essential determinant of the economic performance of industrialized economies is generally acknowledged today.[64] It is undisputed in the meantime that technical standards are very important for the fast diffusion of new technologies, but this covers only one macroeconomic impact dimension of standards. Furthermore, the standardization process itself is a platform especially for the exchange of knowledge relevant for the implementation of new technologies among its participants. This dialogue enables the generation of new incremental and more application-related technological know-how instead of radical breakthroughs in basic research. In the following section, we do not concentrate on the aspects of tacit knowledge exchanged and created by the participants of standardization processes, but discuss the impact of the already used categories of standards on growth. In addition, the three aspects of technical change, competition and trade also have an impact on growth in general, for example standards promoting competition are also beneficial for growth, because they improve the allocation of resources. Consequently, some arguments already mentioned might be repeated and related to the macroeconomic growth aspect.

All types of technical standards codify technological know-how, therefore the fourth category of information standards is not discussed separately. Codified technological know-how has a public good character in the sense of non-rivalry in use and can therefore be easily distributed among different companies[65] or the whole industry (Cohendet and Steinmueller 2000). Besides the common dimension of non-rivalry in use we divide codified knowledge into two subsets distinguished by the degree to which property rights and excludability are attached to them. Whereas innovations protected by intellectual property rights like patents restrict others from using the technologies covered, technical standards are in general public goods and a form of technical infrastructure (Tassey 2000). From a theoretical perspective, public infrastructures are able to increase the productivity of the other input factors labor and capital, and from empirical studies (for example Schlag 1997) we learnt that public infrastructures positively influence economic growth. The faster and greater the diffusion of private technological know-how by technical standards, the greater also the pool of this publicly available information and the stronger its impact on growth.

Discussing the role of compatibility and interface standards leads us also to the infrastructure argument, because our transport systems, the networks supplying us with water, gas and electricity, and finally the telecommunication infrastructures depend crucially on this type of standard. Since all the service sectors based on these physical networks are both the most dynamic sectors in highly industrialized countries and also enrich development in the manufacturing sector (Jungmittag and Welfens 1996), overall growth can be positively influenced by efficient compatibility standards. However, if central compatibility or interface standards are in the ownership of a single company or a small group of companies, their monopolistic behavior can decrease consumer surplus, inhibit innovation and therefore hamper economic growth in the short and long run.[66]

The category of minimum quality and safety standards is growth-enhancing, because these standards reduce transaction costs especially in respect of markets for complex and 'risky', but also innovative and high quality, products and services. Consequently, they have a positive impact on the development of new markets and high quality segments of existing markets. These markets are decisive sources for growth, especially for highly industrialized countries and their tendency towards saturated markets for durable consumer products. Finally, safety standards are means to restrict negative externalities damaging health and the environment. The resources saved and not used for fixing the damages can be allocated more efficiently, which enhances growth. Negative for growth is the misuse of minimum quality or safety standards by small groups of suppliers, which try to manipulate the specification of these standards in a way that raises their rivals' costs and allows them to behave like monopolists. Furthermore, the restriction of a product or service range by this category of standards can be too strong in the sense that the number of customers with a demand for low quality and low safety products is relatively high compared to all customers. Then the loss of their consumer surplus is higher than the gain of consumer surplus for customers with stronger preferences for quality and safety features.

Regarding the previous categories of standards, the growth-enhancing effects dominate the growth-hampering effects. For variety-reducing standards the situation is rather ambivalent. On the one hand, variety reduction allows the exploitation of economies of scale (Dixit and Stiglitz 1977). Furthermore, variety reduction is a necessary condition for the development of new technologies and markets, because a dominant path or trajectory for technologies or markets has to be found in order to reach critical masses, which make both an investment attractive for entering companies and use attractive for customers. Both aspects are certainly positive for the growth of the economy as a whole. On the other hand, variety reduction restricts the choices for customers. Following Romer's (1990) seminal contribution to endogenous growth models, Grossman and Helpman (1991) assume in their endogenous growth models that R&D activities aim to reduce either production costs or to extend the product range. Growth in this special framework is equal to the number of product variants. In addition, Saviotti (1996) postulates that growth in variety is a necessary requirement for long-term development, but also a phenomenon for growing economies.[67] Besides the tension between the exploitation of economies of scale and variety, the reduction of the latter may facilitate the concentration within a market to a smaller number of suppliers, who are able to realize higher economies of scale on the one hand, and to misuse their market power on the other hand. Weighing all arguments against each other, it remains open whether variety-reducing standards are growth-fostering or -hindering.

Finally, we have to come back to the role of standards for trade. Assuming that standards are fostering trade instead of hampering trade, we can postulate an indirect positive impact on growth according to Grossman and Helpman (1991), since trade is an additional channel for the worldwide diffusion of technological knowledge stimulating innovation and therefore growth. Furthermore, trade of technologies reduces the duplication of R&D effort and consequently increases the productivity of resources employed in the R&D sector, which is also enhancing growth.

Taking all categories of standards and all economic explanations together, one can assume that standards are likely to be positive for economic growth, especially if the standardization process is open and transparent for all interested groups and the standards themselves not proprietary but accessible and usable for all producers and customers.

NOTES

[1] This section refers in parts to chapter 2 of Blind and Grupp (2000).
[2] We use technical change in the sense of technical innovation. Therefore the use of the term innovation implies a technical dimension and excludes innovations in the service sector

without a technical dimension and organizational innovations.
[3]   Cf. Wölker (1996, pp. 68–70). In this connection, the non-standardizing of teletext and videotext is presented as positive (see Besen and Johnson 1986 on this).
[4]   This assumption is fundamental for the evolutionary innovation theory which, in analogy to nature, regards successful innovations on the part of enterprises as selection mechanisms for the economy and thus explains technical change (cf. Nelson and Winter 1982, pp. 234–45). Metcalfe and Miles (1994) analyze the role of technical standards for the development and the diffusion of technology under an evolutionary paradigm.
[5]   Although SDOs merely concentrate on process and interface standards and avoid direct product standardization, the former types of standards also have negative impacts on the number of product variations.
[6]   See also ISO (1982, pp. 15 f.) on the problem of diversity.
[7]   Depicted according to ISO (1982, p. 16).
[8]   Cf. Wölker (1996, p. 68). On the definition of compatibility see also Blind and Bühring (1996, pp. 543–5) or Voelzkow (1996, pp. 140–42). An economic analysis of interface compatibility is to be found in Matutes and Regibeau (1987, pp. 23–8). Beitz (1986, p. 86) speaks of limiting 'creativity', but fostering 'virtuosity'.
[9]   Hereby is meant the lock-in of a once attained level of technical development with a steering of the development (canalization) (see Lukes 1968, pp. 172, 175).
[10]   In order to counter this danger, in DIN Norm 820 the rule was introduced that standards must be checked at the latest every five years and corrections because of abortive developments made possible.
[11]   This is also a principle of the New Approach in European standardization (cf. Europäische Kommission 1990, 1998b). Referring to the production processes, Marino (1998) derives regarding information asymmetries a greater efficiency of process standards than product standards regarding items of equipment.
[12]   In contrast, governmental regulations are binding for suppliers.
[13]   See also Reimers (1995) and Voelzkow (1996, pp. 133–42) on the transaction cost approach.
[14]   Within the standardization process itself, tacit knowledge is exchanged between the participants.
[15]   However, in chemistry new production processes are decisive for the development of new chemical products. The same structural relationship is true for other new technologies, like biotechnology.
[16]   Compare the textbook of Shy (2001) for a comprehensive overview on the economics of network industries and the seminal contribution by Economides (1996) on the economics of networks.
[17]   The direct utility effect is drawn directly from the network good, the indirect one from the possibility of finding compatible goods or service, like repair and maintenance on the market.
[18]   Cf. the famous QWERTY discussion starting in David (1985).
[19]   In Steyer (1997, p. 207), in a similar illustration rising marginal utility is assumed, in Blind and Bühring (1997, p. 519), implicitly sinking marginal utility. Both assumptions are theoretically explicable; here however – for simplicity's sake – we should assume constant marginal utility, that is a linear correlation between number of participants and utility curve.
[20]   Cf. on this the portrayal of Blind and Bühring (1996, pp. 518–21), respectively Blankart and Knieps (1993b, p. 44).
[21]   The *bandwagon effect* (or omnibus effect) describes the phenomenon of the radical change of individual decisions in favor of a network good, when the critical mass has been achieved.
[22]   The graphic is adapted from the illustration in Steyer (1997, p. 207). In this illustration an increasing marginal utility is postulated. Another assumption would not change the basic statement, so in Figure 4.3 for the sake of simplicity we should assume linear utility curves. The indices *o* stand for old, *n* for new, *total* for whole and *i* for the user under consideration. In this model, and in the following reflections, the assumption of perfect information about the benefit/utility of the technologies should be made.

[23] Entrance costs could be represented in this graphic by a parallel shift of U_{inew} downwards, that is the entrance costs (or better, costs of change) would be subtracted from the gross benefit. The new utility curve can in the case of a vastly superior technology always be higher than U_{iold}, so that the above graphic can also apply to the case with entrance costs.

[24] Cf. also Hemenway (1975) on this subject.

[25] See Farrell and Saloner (1987, pp. 13 ff.), or also Heß (1993, pp. 20–22). The core of this allegory is the uncertainty about the consequences of jumping in; is a shark lurking there and eats the first penguin, or is there no danger and the fish can be caught, that is a superior situation to the starting position.

[26] Thus the probability of being eaten, that is to suffer a loss of utility due to the new technology is extremely small. See also Choi (1997, pp. 421 ff.). Choi also points out that standardization prevents the economic units with incomplete information from seeking further information about the new technology before the change, thus bringing about a sub-optimal technical change.

[27] See Blankart and Knieps (1993a, p. 47), or Steyer (1997, p. 207).

[28] Molsberger (1977, p. 400) points out that with a lack of compatibility the radical innovations have a greater probability of success, as only here is the utility growth great enough to counterbalance the disadvantage of lack of standardization. Marginally improving innovations would have no chance of success in this case.

[29] Blankart and Knieps (1992, p. 80) describe the two most significant utility dimensions of network goods technologies as technology effect and network effect.

[30] This aspect of influencing the technical change by competitive strategy instruments of the enterprises is the real emphasis of Farrell and Saloner's work and should not be treated in depth here. It is only important to know that these externalities can have different impacts on technical change (cf. on this Farrell and Saloner 1986, pp. 954 ff.). For an overview of this subject complex, see also Tirole (1989, pp. 406–8).

[31] See Farrell and Saloner (1987, pp. 16 ff.).

[32] Cf. Arthur (1989, pp. 116–31).

[33] Woeckener (1996, pp. 257–9), speaks of network fragmentation.

[34] Cf. Farrell and Saloner (1985, p. 72–4). In the model however – unlike the penguin problem and Choi's observations (1997) – the assumptions of complete information and infinitely rapid adaptation are met.

[35] See David (1987, p. 232).

[36] Cf. Katz and Shapiro (1985, pp. 424–40).

[37] Cf. Farrell and Saloner (1985, p. 82).

[38] Tirole (1989, p. 409) interprets this expression 'strong' or 'large' used by Katz and Shapiro (1986) as the majority of consumers of the products offered by the enterprise and thus as a great market power.

[39] Woeckener (1998) designs a model in which by differences in quality even this monopolist position or incompatibility lead to a higher welfare.

[40] In the work of Kristiansen and Thum (1997, p. 56) the World Wide Web is cited as an example of these interdependencies; on the one side firms are offered the software to make a website, on the other side the consumers have to buy the necessary web browser software to surf the Internet. Both products are only linked by the network effect.

[41] See Kristiansen and Thum (1997, p. 57).

[42] Jensen and Thursby (1996) analyze the rather academic impact of anticipatory product standards on R&D and very often find time-inconsistent constellations.

[43] Cf. Choi (1997, p. 411) and Scharfstein and Stein (1990).

[44] Cf. Farrell and Saloner (1988), Kleinemeyer (1998) and recently Belleflamme (2002) on this general issue.

[45] This section refers basically to the work by Fredebeul-Krein (1997).

[46] Böhm et al. (1998, pp. 42 ff.), suggest, in contrast to current practice, restricting the duration and number of the memberships, as well as appointing representative actors to the committees.

[47] Cf. also Steffensen (1997a, pp. 152–67) and Böhm et al. (1998) on this subject. Lim (2002) describes in an abstract manner pre-standardization in ICT as a negotiation process, whereas

Chiesa et al. (2002) highlight the important role of standard development organization in the mediation process between different interests illustrated by two case studies.

[48] Cf. Thum (1994, pp. 487 ff.).

[49] Cf. Helbig and Volkert (1998, p. 5).

[50] Besen and Farrell (1994) present an overview over the different competitive constellations and their impact on standardization strategies. They show that winners of standard wars exploiting their dominant positions realize only average profits because of the high investments necessary for winning the standards war.

[51] Cf. also the brief discussion in Thum (1995), who notes the limited literature on standards as barriers to entry.

[52] Katz and Shapiro (1985) had already pointed out that dominant firms are against open compatibility standards.

[53] Gandal (2002) discusses different policy implications of compatibility standards. All of the discussed competition-related issues are connected with technical progress or innovation, which confirm their close relationship also discussed in Section 4.2.

[54] This section refers to Blind and Jungmittag (2001).

[55] This categorization follows Swann et al. (1996).

[56] Barrett and Yang (2001) discuss constellations under which it is rational for companies not to adapt international standards.

[57] Ganslandt and Markusen (2001) differentiate in a two-good, two-factor, two-country general-equilibrium model these negative impacts by country size and find that small countries will lose 'standard wars'.

[58] The Cassis-de-Dijon decision intended to implement the principle of origin, which does not allow a discrimination of imports, and aimed for the mutual recognition of national product requirements (cf. Sykes 1995, pp. 97–100). On the mutual recognition of food standards in the EU see Michaelis and Borrmann (1990) and Michaelis (1990a).

[59] In Germany, domestic producers of beer obey the purity law ('Reinheitsgebot'), whereas foreign breweries do not have to follow this rule. This leads to a discrimination against national suppliers. To compensate for this competitive disadvantage, the German Brewers' Association started a vigorous campaign to discredit foreign beers made with additives and chemicals (Egan 2001).

[60] The welfare impact of an international harmonization of quality standards depends on their level, because there is a trade-off between increasing consumer rents and reducing producer rents (Boom 1995).

[61] The information dimension is common to all three types of standards and therefore not discussed separately.

[62] Compatibility standards, which increase the number of combinations of complementary products, also cause a higher variety of system goods, which is positive for intra-industry trade (see also Ganslandt and Markusen 2001). In the same paradigm, they show that standards leading to an increased substitutability between products are ambivalent for trade, since they decrease product variety, but increase competitive pressure.

[63] Casella (2001) raises in her model the question whether the uncontrolled production of international standards through private coalitions of very few large companies may lead to anti-trust problems. If this is the case, then all kinds of harmonized international standards may be negative for the development for intra-industry trade.

[64] Standards can also be interpreted as institutions. Institutional economists postulate a close relationship between institutional development and economic growth (cf. the general survey in Frenkel and Hemmer 1999, who do not consider standards).

[65] Benezech et al. (2001) analyze the role of ISO 9000 for knowledge codification within companies.

[66] The most popular example is the case of Microsoft accused of misusing its de facto standard on office software.

[67] Funke and Strulik (2000) explain the increasing variety of goods within an endogenous growth model by the output of R&D efforts.

PART B

The Supply Side of Regulation:
Definitions, Legal Implications and
Institutional Framework of Standards[1]

5. Introduction to Part B

Whereas the measurement of technical change can already look back on a longer tradition and tried and tested indicators (Grupp 1997), the quantitative approaches to measuring technical standardization are still in their infancy.[2] Accordingly, the systematic classification of indicators in this field is still under-developed. The premature stage of using standards as indicators makes this chapter about the institutional framework necessary.

Only since the middle of the 1990s have technical standards been used as indicators for the emergence and diffusion of new technologies (Hawkins 1996). Many SDOs possess extensive electronic databases that contain a broad range of information concerning both the standardization process and its output, the standard document. These could be sources of indicators of the 'state-of-the-art' in any particular technological field. Other indicators about science and technology, already well established, are R&D investments, patents and scientific publications.

The examination of the output of *formal de jure* standards edited by national and international SDOs turns out to be more feasible than the input into their development. Empirical examinations of data that collect the outcome of standardization processes help to develop an indicator complementary to the existing and well-established indicators, which can be located closer to the implementation of technology in products. As Hawkins (1996, p. 1) postulates 'the collected body of [formal] technical standards as developed by national or international SDOs is a codification of the global technology infrastructure'. However, the validity and reliability of this indicator has to be proved by analyzing the institutional framework, including the relationships to other forms of regulation in a wider sense.

Part B is structured as follows. In Chapter 6 the question of definitions is tackled, as a demarcation of general technical rules – the comprehensive term for formal standards – must be arrived at, not only for company and industry or branch standards, but also regarding government regulations and laws, in order to pre-empt misunderstandings. Then in Chapter 7 the legal implications of formal technical standards will be described, in order to underpin their economic effects. Chapter 8 presents the institutional framework of technical standardization on a national, European and international level. Both of these sections serve in the final analysis to legitimize the data basis used. The relevant characteristics of the used database PERINORM edited by European SDOs are introduced in the following chapters of Part C and Part D, when the other data sources used are also presented and discussed.

NOTES

[1]   Part B draws mainly on part 2 of Blind and Grupp (2000).
[2]   Cf. Hawkins (1996) and Swann et al. (1996) on this subject.

6. Definitions

6.1 TECHNICAL RULES IN THE CONTEXT OF THE GENERAL CONCEPT OF REGULATION IN THE ECONOMY AND SOCIETY

Over-regulation, 'a flood of regulations' and 'legal jungle' are catchwords often used, especially in the current discussion about Germany's and Europe's attractiveness as a location for industry.[1] How to differentiate technical rules and standards in this multiplicity of terms, and which interdependencies exist within the mentioned concepts, is the task of this first step towards definitions.[2]

From the perspective of economics, regulation basically applies to market access, prices, qualities and conditions such as, for example the obligation to accept contracts (OECD 1997e). Regulative measures are direct requirements and prohibitions, which aim at individual economic actions (for example permission or prohibition to invest, production requirements, regulations concerning pricing or tariffs, admission regulations for professions, exceptions to competition laws). Apart from protective and safety regulations – which apply equally for all market participants – regulation does not rate as system-conform in the free market system. Regulations may distort competition, hinder investments and hamper economic growth.

Ideally, the efficiency of the allocation of scarce resources is ensured by the market forces in the different submarkets, for example labor, capital, and product markets. It is assumed for imperfect (regulated) markets that they are hampered in their efficiency. However, the market power of monopolies, ruinous competition, external effects and the need for public goods provide the welfare-theoretical argument for state regulative interventions, which should restore efficiency and compensate welfare losses. This neo-classically influenced argumentation is extended by the discussion about the New Political Economy and the normative theory of state regulation to include important elements: the regulative process is regarded from its origin as in part a politically motivated decision-making and authoritative process. In this way it is possible to integrate the institutionally founded positions of power of politicians and the rationality of politically motivated decisions 'in the interests of all'. This approach also tries to find an explanation for regulated public enterprises or regulated markets.

6.1.1 Regulation Instruments

At the center of the debate on regulation are instruments such as price, market access and quality rules. The effects of the different rulings of individual branches or sectors were discussed in detail regarding their allocative, technical and qualitative (in-)efficiency (cf. Deregulierungskommission 1991). We differentiate in this first step towards definition between: (1) socially regulative measures; (2) competitive policy regulations; and (3) industrial policy measures:

(1) Since the 1960s, *social or behavioral regulations* have gained significance. Their justification stems from (not completely to be anticipated) negative external effects or negative aspects of technology used in the production process and in the consumption of goods and services produced and the strategies necessary to ward off these dangers and protect the general public (cf. Roßnagel 1993). At the center of the social regulation are qualitative aspects like environmental protection, product safety and safety at the place of work, which are often not limited to one branch of industry, but apply across branches. They are expressed in:

- EU guidelines;
- laws;
- legal decrees;
- administrative regulations;
- technical rules and standards.

(2) The regulation of the remaining public enterprises, so-called exceptional areas (for example public utilities and transport services), can be regarded as a conscious policy intervention on the part of the state in the increasing industrialization and the development of large industrial systems. But it is not only public enterprises which are subject to regulations; price, profit and market access regulations are also quite usual in branches with a competitive structure.

Competitive policy regulations – restricting market access and price regulations – are only 'legitimized' in limited areas in exceptional cases permitted by the German Trade Regulation Act. The freedom of trade allows everyone the fundamental right to carry on a trade. The license to trade can depend on subjective and objective reasons, for example the educational background and training for doctors, membership qualifications of professional associations and chambers, or safety criteria for the operation of new machines.

Competitive policy regulations include various instruments, such as:

- pricing regulations;
- market access regulations;
- regulation of company co-operations;
- regulation of products and processes; as well as
- regulation of intellectual property rights (for example patenting).

The regulation of prices is justified by the prevention of an exorbitant price level and price discrimination in monopoly-type structures. In competitive sectors, the setting of minimum prices is credited with shielding the enterprises from 'too much competition'.[3]

The regulation of products and processes are subsumed under general heading of quality regulation, such as:

- restrictions for a qualitative product policy;
- the obligation of public enterprises to furnish services to all;
- rulings on product classification;
- labeling of goods; as well as
- the avoidance of external effects through the production process.

By fixing technical standards, safety regulations or other quality criteria, the number of suppliers, the production methods and the production result can be controlled or restricted. Once again, the question of the steering and effectiveness perspective crops up again. Product regulations can directly affect the quality, on the other hand they can also lead indirectly to quantitative effects. Quality regulations thus have equally allocative and distributive effects.

(3) Equally, industrial policy measures also exercise a corrective influence, whose origin is rather to be sought in distributive policy and not in allocative goals. They can also be understood as regulation in the wider sense, such as:

- tariff or non-tariff restrictions to trade;
- financial burdens and relief for individual branches or regions; as well as
- real transfers (for example in the framework of support programs for industry or innovation) and decisions in fiscal policy.

The spectrum of regulative interventions in the social and economic sphere is broad, and the instruments utilized are numerous. State authorities are only in part the origin and executor of regulations; significant areas are regulated by the social and economic actors themselves. According to a classification by Hoffmann-Riem (1996), all regulation of societal and economic activity can be located on a *sliding scale* between mandatory government regulation, government regulation with self-regulative elements, government regulation of the self-regulation and private self-regulation (see Table 6.1).[4]

Table 6.1   Regulation in the economy and society

	Government versus private regulation			
Economic and social dimension	government mandatory regulation	government regulation with self-regulative elements	government-regulated self-regulation	private self-regulation
Social regulation	e.g. fixing emission limits	e.g. Packaging ordinance	*e.g. building standards for the disabled in DIN*	*e.g. Total Quality Management (TQM)*
Regulation of economic competition	e.g. government screened local monopoly	e.g. Recycling Management Law	e.g. status of architects and tax advisors	e.g. agreements, cartels
Industrial policy regulation	e.g. Teletel introduction in France; Electricity Supply Law	e.g. Transrapid maglev train	*e.g. compatibility standardization in DIN (paper standards)*	*e.g. company internal and cross-company standardization*

Hoffmann-Riem (1996) characterized these four types of regulation as follows:

- One extreme pole of his scale is formed by the classical *government regulation*. By issuing rules and prohibitions the state attempts to attain the desired goals. 'Compulsory licensing (*Erlaubnispflichten*), prohibition reservations (*Untersagungsvorbehalte*), reservations (*Untersagungen*), sanctions and the like are manifestations of regulative law.'
- The second type is a *government* or *sovereign regulation* including

self-regulative elements. In this case those to be regulated, 'in a sense the objects of the regulation', are drawn into the problem-solving process; scope is deliberately left in the decision-making in order to stimulate the willingness to collaborate in the search for the best and efficient solutions.[5] As an example, the author quotes the well-known compensation solutions in the Immission Protection Law, according to which the enterprises may choose where they introduce certain environmentally friendly improvements.

- The author calls the third type of regulation *sovereignly regulated societal self-regulation*; here the task fulfillment is basically left up to self-regulation by the private actors. From the government side, however, a structural framework is supplied, 'which should above all ensure, that the self-regulation does not only obey the rule of the strongest'.[6] As an example the author names the 'Open Network Provisions', among others, in the area of liberalized telecommunications markets, which ensure certain minimum rules by being obliged to furnish services to all.

- At the other end of the scale, purely *private self-regulation* can be found. In this case the general legal system among others of the German Civil Code apply; no government steering attempts are desired or expected here.

In the following chapter and the study as a whole, the broad notion of regulation with its various dimensions will not be further pursued, but will be limited in the next step the narrower terms of the technical rules and standards, whose scope is given in italics in Table 6.1, before finally the DIN definitions of technical standards are presented.[7]

6.2   ECONOMIC DEFINITIONS OF STANDARDS

In the framework of the national, regional and international technical standardization, there is a great number of terms which are not uniformly and exactly defined, especially among the scientific disciplines.

If the somewhat imprecise division into social, competitive and industrial policy regulations is set aside, as on the level of general technical rules, safety and quality standards cannot always be differentiated in their function from compatibility standards, then government regulated standardization is to be distinguished from private standardization activities according to economic criteria.

According to the generally accepted David Greenstein categorization of 1990, four categories of technical rules or standards can be distinguished from an economic perspective in the sense of their authorship:[8]

(1) '*Unsponsored standards*':
     This concerns a set of specifications which are well documented, but for which no author is identifiable. The best example of this is the key layout of the typewriter keyboard QWERTY, the origins of which are still disputed today.[9] Historical-evolutionary processes accompanied by increasing returns to scale and positive feedback are finally responsible for the emergence and the specification of such a standard.

(2) '*Sponsored standards*':
     In contrast to the 'unsponsored standards', in the case of so-called 'sponsored standards' individuals or groups posses proprietary rights to a set of technical specifications (for example IBM or Microsoft standards). The owners of this form of standards can determine their specifications and use them in a sophisticated market strategy in addition to pricing options. In a situation with several actors, this can lead to incompatible solutions and thus to an under-standardization, although this is usually not optimal from the economic viewpoint.

(3) *Standards from voluntary standardization development organizations (committee solution)*:
     Whereas the sponsored and unsponsored standard types can be classified as private self-regulation and are also called *de facto* standards, the elaboration of formal standards in voluntary standardization institutions falls under the category of government regulated self-regulation. As will be shown in Chapter 7 in detail, interested groups agree on common formal standards in processes mediated by private standardization institutions, such as, for example the British Standards Institution (BSI) or the German Institute for Standardization (DIN). However, the standards elaborated by private enterprise consortia do not come under this classification but under either 'unsponsored' or 'sponsored' standards. These are of ever-increasing importance.[10]

(4) *Mandatory standards or regulations of government regulatory authorities*:
     To complete the David Greenstein categorization, the standardization by government regulatory bodies must be mentioned, which actually come under the state imperative or self-regulating regulation.[11] Concrete reasons for a direct government regulation by means of norms and standards were given in Section 6.1. From a theoretical economic perspective it should be added that with strong externalities and low possibilities of asserting property rights in view of the positive economic effects of standards, the private incentive to standardize is sub-optimal, so that the state must develop technical standards itself (for example standards for measuring and testing). For this reason many government rulings in the areas of environment, health, traffic, trade etc. contain references to technical

standards.[12] The significant indirect impacts of the formal standards of voluntary standardization institutions on laws and ordinances will be dealt with explicitly in Section 7.2.3.

6.3 OFFICIAL DEFINITIONS OF STANDARDS

After the first general and subsequent narrowing down of the term standardization from an economic perspective, the DIN definitions of technical rules or standards must be cited, in order to arrive at a common working basis for the following analysis.

According to DIN-EN 45020 (edition 1994, p. 12, in DIN 1995, p. 276), the standard for general technical terms and their definition concerning standardization, a normative document is a 'document, which lays down rules, guidelines or characteristics for activities or the results thereof', while a standard (norm) is defined as a:

> ... document, which has been elaborated consensually and accepted by an acknowledged institution and which lays down for general and recurrent application rules, guidelines or characteristics for activities or the results thereof, whereby an optimal degree of regulation in a given connection is striven for.

Through the strong influence of English literature in economic analysis, norm and the English translation of 'standard' are treated as synonyms.

Technical product specifications or company-specific standards are to be differentiated from the concept of the given standard definition, as these apply only within the company. However, they can assert themselves in the market place and develop into *company and industry standards*. They can also gain acceptance in the branch-internal regulation process and are then referred to as *branch standards. Technical product specifications or company-specific standards* also find their way into the standardization process according to DIN 820 at the DIN institute and are then published as formal DIN standards.

The object of analysis in the study will basically be the 'technical rules', a collective term which encompasses not only formal standards but also the entire collected rules of all acknowledged private law rule-makers. However, we will use the term (formal) standard, since it is compatible with the terminology in the economic literature. As in certain cases, for example the environment, laws and ordinances with technical specifications play a significant role, the data basis and thus the object of analysis will be correspondingly extended, as will be shown.

6.4 INTERIM FACET: ARGUMENTS FOR THE RESTRICTION OF THE STUDY TO FORMAL STANDARDS

After the definitory demarcation and limitation of the term formal standards, economically meaningful and pragmatic-empirical arguments for restricting the object of the study to formal standards are outlined, which will then be dealt with in more detail in the following sections.

The following economic reasons speak in favor of limiting the study to formal standards:

- Formal standards have no direct legal effect, it is true, but in Chapter 7 their far-reaching, indirect legal implications are made clear. Purely company and industrial standards do not have these implications and the economic effects resulting thereof.
- In Chapter 8 the co-ordinated system of national, European and international standardization is depicted. On the horizontal-national, but also vertical level it prevents that standardization is duplicated simultaneously, resulting in competing standards. Further standards according to EN 45020 or DIN 820 must have a certain uniform form and content. This structuring makes a uniform counting of the standards documents possible, not only within a sector but also between areas – a necessary pre-condition for the development of indicators based on standards.
- Company and industrial standards often become acknowledged branch standards or formal standards.[13] The co-ordination between branch standards and DIN standards, for example in the case of the chemical industry, is guaranteed in that the VCI, or its working group 'Standardization' makes its results known to the DIN committee and the same persons serve on both committees.[14]
- Whereas product specifications or company-specific standards can *de facto* affect one, a few or many enterprises in a branch, general formal standards are potentially applicable for all enterprises concerned, so that an equal impact can be assumed, while the former requires a more differentiated analysis of the impacts which cannot be achieved practically.
- Formal standards have the character of public goods and thus are basically the responsibility of the state and its institutions, while the company standards, protected by ownership rights, offer sufficient financial incentives for private initiatives, rather fear the competition-restricting effect of re-active regulations and demand no action from public institutions.

Besides the economic arguments, a number of pragmatic empirical reasons can be advanced to restrict the object of analysis as formulated above:

- Formal standards are, at least for Germany and other European countries, on principle classified according to the international standards classification ICS, indexed and provided with a publication date, or in some cases a withdrawal date, and made available for a selection of certain countries.[15] Against the background of the analysis of interdependencies between standardization and technical change, it should be noted that patents as indicators of technical change are also to be classified, and arranged timewise according to application or priority year and to regions according to the applicant's address. Thus a congruent data basis and corresponding time series can be made, which will serve as the basis for empirical analyses.
- Company standards are on the one hand not directly accessible, as here partly explicit trade secrets or informal, company-internal, but also cross-company agreements are concerned. On the other hand, they are not uniformly documented and not classified or indexed.[16]
- Company standards are partly protected as patents by ownership laws; which patents, however, have the character of a standard is not to be determined by patent searches. Therefore unidentifiable intersections between the standards and the patent databases would emerge and impede the quantitative analysis in an uncontrollable manner.[17]
- In the analysis of the empirical correlation between technical change and standardization, the intersection problem mentioned is not solved. Patents will flow not only into the indicator for technical change, but also into the standards database, when patented product specifications enter into formal standards. This procedure is justified by the fact that the monopolized ownership rights and the private good character of such a patent are practically withdrawn and it becomes quasi a public good or club good.[18] The line of demarcation between innovations and standards is drawn by the formalities of the patent application and so operationalized.

NOTES

[1] The Council of Experts (*Sachverständigenrat*) in its report 'The Lean State' (1997, pp. 15–26), gives the highest priority to the reduction of the 'flood of regulations'.
[2] On the following see also Kuhlmann et al. (1997). De Vries (1997b) provides an overview of official definitions and the practical application of the term 'Standardization'.
[3] Cf. Kruse (1989, p. 11) and von Weizsäcker (1982, p. 330).
[4] The regulations shown in the matrix serve merely as an illustration. Most rulings cannot be classified to one single field of the matrix; the transitions between the fields must be understood as not clear-cut.
[5] Cf. Hoffmann-Riem (1996, p. 16).
[6] Hoffmann-Riem (1996, p. 17).
[7] Fundamentally, the regulation of economic competition can be regarded in close connection with industrial policy regulation, so that the corresponding boxes of this row could also be

shaded.
[8] Gabel carries out a similar systematization (1993, pp. 12 ff.). We have used this classification already in Section 4.2.3, since it is helpful as a background for the analysis of the relationship between standardization and technical change.
[9] Cf. David (1985) on this.
[10] See Weiss and Cargill (1992) on a classification of these consortia.
[11] Thum (1994) calls this category the bureaucratic solution.
[12] Cf. Mansell (1995).
[13] See on this David and Greenstein (1990, p. 25), and the DIN Business Report (1995–96, p. 21). At the end of 1997 DIN agreed to the elaboration of publicly available specifications (PAS), 'a document, which contains technical requirements of a product, a process or a service, expresses its connection to DIN standards and is available to everyone in Beuth Verlag GmbH', is a further form of publication for technical rules. Due to today's rapid innovations there are technical specifications on the market which were worked out by consortia and are publicly accessible, whereby 'public' does not only mean the availability of the documents, but in particular the 'know-how access' to these specifications. On a European level, the recent CEN Workshop Agreements (CWA) introduced a new form of documents to shorten the standardization process.
[14] Cf. Graßmuck and Heller (1986).
[15] Cf. the description and discussion of the ICS in Nohr (1997).
[16] The admission of publicly available specifications (PAS) as a further publication form for technical rules reduced the outlined documentation problem for future investigations.
[17] See Section 7.2.1 on the legal dimension of this interface.
[18] Cf. on this also Antonelli (1994) and especially Chapter 11.

7. Legal Implications of Formal Standards

7.1 INTRODUCTION

Formal technical standards direct their focus mainly on the technical specification of products and processes, but despite their fundamentally voluntary character they do not, however, exist in a legal vacuum but are also fraught with legal implications. In order to provide an overview of the legal interfaces of standardization with various legal fields, the standardization process is divided into the phases of standards production, publication and application, and the adjoining relevant legal fields are analyzed.[1]

7.2 PRODUCTION, DISTRIBUTION AND APPLICATION OF FORMAL STANDARDS AND THE LEGAL IMPLICATIONS

In order to analyze the interdependencies between standardization, technical change and international trade streams, the legal dimensions of their distribution and application must be examined, besides the legal framework conditions for the production of formal standards, because they may have repercussions for their part on the standardization activities, and thereby also on the connections between technical change and standardization which are to be examined.

7.2.1 The Standardization Process

The national standardization organizations have as a rule the status of a registered non-profit society or an association. The Society and Association Law contains guidelines which regulate the structure, the elections and the arrangements for dissolution, but also the competencies of the various organs and thus are also to be found in the statutes of CEN or DIN. The legal relevance of these statutes lies therein, that they regulate the relations between the society members and so create their own legal sphere.

Besides the internal regulations, the external relations of the standardization organizations are determined by competition law and public law. Competition law is relevant inasmuch as in standardization processes market-dominating enterprises make decisions which are relevant to, or especially restrict, competition. Not only national but also European law can ensure that on the one hand the standardization process is made practically accessible to concerned actors (SMEs and consumers) and on the other hand certain standards which would distort competition are controlled and prevented.

Besides the competitive law implications of the outputs of standardization processes, on the input side conflicts exist between the right to intellectual property, in the form of patents protected by ownership law and the unprotected, uncopyrighted standards.[2] The owners of patents are willing for their innovative technology to enter the standardization process and so to support its diffusion, refuse however for reasons of strategic competitiveness to sell licenses under appropriate conditions to competitors. An attempt by ETSI to make licensing of technologies mandatory for the patent holders failed and was withdrawn in favor of voluntary solutions.

7.2.2   The Publication and Distribution of Standards

As many of the national standardization institutions are primarily financed by the sales of standards documents, the law governing copyright usage is of supreme importance for them.[3] Whereas the actual copyright remains with the individual originators, they leave the right to use the copyright up to the standardization institution. If the standardization institute has developed the standard on its own, that is exclusively with own fully employed staff, then it also holds the copyright. This procedure is not in contradiction with the (ideally) typical process of drawing up standards, which allows the public possibilities to raise objections and suggest corrections. For no watering down of the authorship of the standards takes place.

Standardization at the European level is faced with a similar problem,[4] because before the final passing a number of internal, and also external institutions can contribute commentaries, which will enter into the final version. Further, these standardization institutions are partly represented by persons who are not employed by them but by companies.

This legal position is clarified by the special regulations of copyright law, as challenges arise for copyright in standards through being referred to in areas of public administrative law. In some national copyright laws the copyright for standards is lifted if government rules and regulations refer to them, because government ordinances and laws are not protected by copyright law.

Contract law is also touched on, through the sale of standards documents.

Here the Product Liability Law is in the middle of the legal discussion, whereby according to contract law, standardization institutions can in reality be held responsible only up to a point for the content of standardization documents, and individual members of standardization committees are explicitly protected from product liability claims.[5]

7.2.3 The Application of Formal Standards

The legal implications of the application of standards are most serious in comparison with their elaboration and especially their diffusion. Besides the already mentioned direct references in ordinances and regulation guidelines, the application of formal standards exercise considerable effects on Competition and Contract Law and on questions of liability. It must be mentioned in this connection, that safety and quality standards play a considerably greater role in this context than compatibility standards. In many cases the compatibility standard contains safety and quality specifications, so that they are also surrounded by (liability) law interfaces.

Due to the variety of interfaces, the application areas of formal standards and their impacts on the different legal fields are classified as follows:[6]

- the application of formal standards in associations;
- the role of formal standards in laws and ordinances;
- the significance of formal standards in the New Concept of European Standardization.

The application of formal standards in associations
If an association recommends its members to use certain standards, for example safety and quality standards, then this has no initial legal significance. For the members are absolutely free to follow or reject this recommendation. For reasons of competition or by strong implicit obligation to the association, such voluntary guidelines soon become obligatory formal standards for association members, but also for other economic entities active in the same field. This indirect compulsion to observe certain formal standards is also effective, if enterprises want to be awarded with certain test and quality marks. For the societies and associations which award such marks, for example the Committee for Terms and Conditions of Sale and Quality Control, require as a rule from their members or their members' products the observance of a number of basically voluntary formal standards. In cases of conflict, the judiciary will refer to these originally voluntary recommendations in questions regarding the observance of due diligence for orientation.[7]

The role of formal standards in laws and ordinances
Although technical standards are fundamentally not legal standards, they can attain legally binding status if they are referred to in legislation.[8] Various possibilities regarding the connection between legal requirements and formal standards exist, which can be categorized as follows:[9]

- *Incorporation*: The standard is incorporated word for word in the text of the law or its annex. It is thus an integral part of the legal provision and becomes a binding legal rule. The disadvantage of the incorporation is its inflexibility with regard to technical changes.
- *Static reference (Starre Verweisung)*: In the legal provision a specific version – as a rule the most recent – of the formal standards is referred to.
- *Dynamic reference (Gleitende Verweisung)*: A standard in the respectively valid version is referred to, so that the reference can also apply to the future versions of the standard. The great flexibility of this solution conflicts with the fact that standardization institutions can curtail the competencies of the legislative.[10]
- Blanket Clause (all-purpose clause) method: The legal provision requires the observation of 'generally acknowledged rules of technology', the consideration of the 'status of science and technology', application of the 'best available techniques' or uses similarly indefinite legal terms or 'all-purpose clauses', which can be distinguished from each other as follows: in a blanket reference the indefinite terms do not concretely refer to particular, valid technical versions and by a static reference to a specific version; for dynamic references, the opposite is true. Here too the advantages and disadvantages of static and dynamic references already mentioned above apply.

The consequence of the reference to formal technical standards in rulings of public law is that many laws and ordinances, such as for example those of the Construction and Trade Regulation Acts, but also the Technical Plant and Equipment Act[11] require consideration or compliance thereto.

In private law,[12] in implementing contract law, in cases of faulty contract performance, compliance with the relevant technical standards is taken as a measure of due diligence and the decision made accordingly.[13] Also the right arising from tort, from which product liability law is derived, is based on the compliance with technical standards as a source of knowledge, but is not a sufficient condition for proper conduct of a producer. Conversely, the application of technical standards does not necessarily infer certain product qualities.[14]

The legal significance of formal standards in the new approach to European standardization

As already described in the previous chapter on the institutional framework conditions of technical standardization in Germany, the national work on standardization is in the meantime strongly integrated in European standardization. This development is not without consequences for the legal significance of technical standards.

At the European level, a specific co-operation has been established between the public regulation and the private standardization process in the framework of the New Approach. Because of the difficulties in passing EU guidelines with detailed technical regulations aimed at breaking down technical obstacles to trade, the Council of Ministers decided in 1985 that the 'harmonization of legal requirements to determine the basic safety requirements (or other requirements in the interest of public welfare) be limited to the scope of guidelines'. The committees responsible for standardization (CEN and CENELEC) will have the task of elaborating the corresponding technical specifications. The national standardization institutions are then obligated to adopt the European standards and to withdraw the corresponding national standards (see Figure 7.1).[15]

Although these technical specifications – in contrast to the legal provisions – do not have an obligatory character, but remain voluntary standards, they are for practical purposes binding for enterprises and producers. For the producer must prove that products which are not manufactured according to the technical specifications fulfill the essential requirements of the EU guidelines. Further, the Member States are obliged in government procurement to keep to the observance of the European product standards.[16]

All in all, the European standards have a high legally binding effect by comparison with the national standards, which will gain in importance due to their superiority regarding the national standardization systems of the Member States.[17]

7.3 THE SIGNIFICANCE OF THE LEGAL DIMENSIONS OF FORMAL STANDARDS FOR THE DATA BASIS AND STANDARDS INDICATORS CLASSIFICATION

The overview of the interfaces of technical standardization and formal standards with the relevant legal fields has made it clear that, especially in the area of application of technical standards, it can no longer be assumed that manufacturers or service providers will only voluntarily adopt them. Both the direct and indirect references to technical standards in legal requirements and administrative regulations, and their role in the competitive process, make a direct application or a close approximation quasi obligatory in the last analysis.[18] Due to the completion of the European Single Market and the necessary legal harmonization of technical regulations, the voluntary character of formal standards will continue to lose significance.[19]

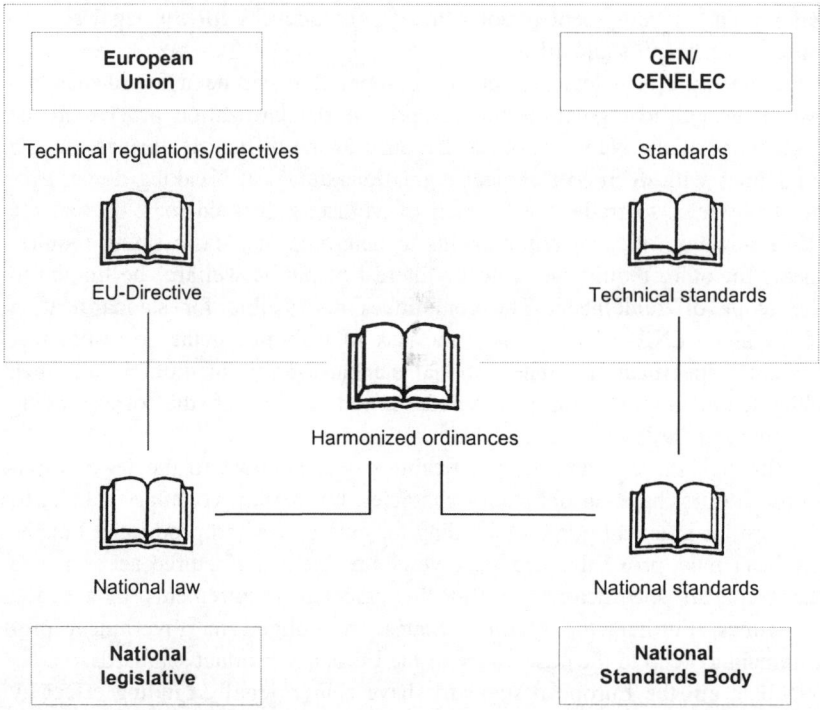

Figure 7.1   Standardization and regulation in the European Union[20]

To what extent this development will influence the distinction between the depicted elaboration of formal standards and the informal, non-binding company and industry standardization, and what consequences will result herefrom for the data basis, can only be taken into consideration by means of a direct survey of the enterprises involved.

As formal standards – as shown – effectively have an obligatory character for the most part, they are no longer to be strictly separated in their sphere of influence from legal guidelines and ordinances with technical content. This will have consequences for the data basis, for it does not contradict an equal treatment for formal standards and legal requirements. For example, in the environmental field the share of legal requirements is very high in comparison with formal standards, while in many other areas legal requirements do not play a role. This different composition should not present a fundamental problem for the generation of the data basis and the classification of standards indicators building on it.

NOTES

[1]   Cf. Stuurman (1997) on this approach to categorization.
[2]   Cf. Thiard and Pfau (1991, pp. 16 ff.) and Chapter 11.
[3]   A further legal aspect, which should not be gone into here, is the market-dominating position of DIN (cf. on this subject Bartsch 1987).
[4]   See DIN (1996a, pp. 17 ff.).
[5]   Cf. Stuurman (1995, p. 519).
[6]   See Stuurman (1997) and Hesser et al. (1994) among others.
[7]   The so-called prima facie evidence is accepted, if proceeding according to DIN standards, and the burden of proof is shifted.
[8]   According to information from DIN, 10 per cent of the German stock of standards in embodied in laws.
[9]   See Neisen (1977, p. 442).
[10]   Cf. Voelzkow (1996) on the role of private institutions in technology steering.
[11]   Cf. Hesser et al. (1994, pp. 9 ff.).
[12]   See Köhler (1985) on the significance of formal standards for liability law.
[13]   Whereas the principle of liability for default practiced in Germany is oriented to compliance with standards, according to the principle of liability based on causation irrespective of fault which is predominant in the USA, the consideration of formal standards is of limited significance (cf. Stuurman 1995, p. 521).
[14]   Cf. Stuurman (1995, p. 520).
[15]   Cf. Europäische Kommission (1998b) on the efficiency of European standardization in the framework of the New Approach.
[16]   See among others Berg (1998, p. 8).
[17]   In addition to the New Approach, the conditions for reliable conformity assessment are defined by the European Council Resolution of 1989 on the Global Approach (cf. European Commission 2000a on the implementation of the New Approach and Global Approach).
[18]   Another, not standardized, building technique can be used, for example, if the Institute for Building Technology proves that this other technique is the equivalent of the standard.
[19]   European standardization is alleged to be democratically deficient as regards the under-representation of certain groups, which can in practical terms only be partly corrected (see Voelzkow 1996, chapter 8.2) for a discussion of this. Compared with this, Korinek (1996, p. 60) argues that on the one hand, European standards also have a voluntary character and that on the other hand, many guidelines of the Community Law also contain references to the standards. And thus the legitimization to such declarations of commitment lies in the democratic legitimization of the declarer of the commitment.
[20]   According to Hesser et al. (1994, p. 21).

8. Institutional Framework for the Elaboration of Formal Standards

8.1 INTRODUCTION

The institutional structure of technical rule-making and standardization is extremely complex, not only within a regional level but also between regional levels. As in other countries, in Germany a great number of organizations are involved with the elaboration and diffusion of technical rules, in the form of standards, guidelines or worksheets.

In this chapter the most important organizations with standardization competencies in the area of technology and their role at the national (Section 8.2.1), European (Section 8.2.2) and international levels (Section 8.2.3) are introduced. Subsequently, in Section 8.2.4 the vertical relationships between the various levels are dealt with.[1] The chapter is to be understood against the background of examining the institutional framework conditions of the various regional levels and the organization of the standardization process in view of the suitability of standards as indicators for the diffusion and implementation of technology and therefore for innovative capacity.[2] For this reason the internal organizational structures of standardization bodies will not be dealt with further,[3] as they should have no direct influence on the focus of standardization work and the number of standards published, because they provide only the platform for the interested groups to perform standardization processes.[4]

8.2 THE SYSTEM OF FORMAL STANDARDIZATION

8.2.1 The System of Formal Standardization in Germany

The institutional structure
The elaboration and diffusion of technical rules and formal standards lies traditionally in Germany in the area of technical, industrial and professional associations, which receive the support of governmental bodies. Of the

approximately 150 private law organizations, the German Institute for Standardization (DIN) is the best-known and most important body, but also the only institution which deals exclusively with technical standardization. The Association of German Engineers (VDI) must be mentioned as another important institution, which devotes itself to the elaboration and diffusion of technical guidelines besides safeguarding the interests of its members. The great number of associations and societies which either represent a specific professional group, the actors in a defined market or groups interested in a particular subject, and which develop or diffuse technical rules and formal standards among other activities are not specifically named here.

The body of rules of the associations which are organized according to criteria of professional or specific interests are formally independent of each other, but are linked together for reasons of simplification and so form a systematic network. The rule-making organizations are not all on one level, but in their entirety form 'a pyramid of many steps, the bricks of which are joined together in manifold ways, whereby the organizations narrowly defined by policy interests form the supporting structure' (Voelzkow 1996, p. 94), on which the relevant umbrella organization for formal standardization – in the case of Germany, DIN – builds.

If one climbs up the pyramid of the standardizing institutions, the heterogeneity of the members in the standardization committees increases. They consist of representatives of various stakeholders and thus form 'committees of functional representation' (Voelzkow 1996, p. 94). The organizations 'of higher order' are therefore not pressure groups in the narrower sense, but provide interest groups with a platform for discussion and decisions. DIN is, therefore, similar to other governmental authorities, an addressee for a multiplicity of already organized interests, which should be led towards a consensus in a transparent process.

Although these organizations with standardization competencies are regarded as voluntary partnerships in a 'pre-government' area, they claim to consider the 'public interest' and to serve the common weal. This state-supporting function is explicitly granted to DIN in the 'Standardization Contract' with the Federal Republic of Germany of 1975, in which it is referred to in §1 paragraph 1 as 'the responsible standardization organization for the area of the Federal Republic' and in §1 paragraph 2 commits itself 'in its work on standardization to consider the public interest' and to make standards applicable for the concretisation of undetermined legal terms. In return, DIN will receive funds from the federal budget to fulfill its tasks.[5]

If the interfaces of the work of DIN with other private-law organizations possessing standardization competencies are considered, then the German Electrotechnical Commission (Deutsche Elektrotechnische Kommission, DKE) can be cited as a joint organization of DIN and the Association of

German Electrical Engineers (Verband Deutscher Elektrotechniker, VDE) as an example of the co-ordination of originally separate standardization institutions with identical areas of responsibility (DKE 1996).

DIN cultivates intensive relationships also with public-law organizations, such as trade associations, which refer explicitly to DIN standards in their safety regulations. The collaboration was even institutionalized in an agreement which grants the trade associations an explicit influence on the national, but also on the European and international standardization processes.

Besides the interfaces with other standardization bodies, the connections between DIN and the so-called interested circles are of importance. These circles are as a rule already organized in associations of various kinds, whereby industry and its associations are the main pillar in the organizational foundation of technical standardization, in that their representatives act as members or sponsors of the technical standardization. In individual cases standardization committees are even moved from DIN and assigned to the appropriate trade or professional associations, such as the Association of German Machine and Plant Builders (Verband deutscher Maschinen- und Anlagenbauer, VDMA). Conversely, the Federation of the Chemical Industry (Verband der Chemischen Industrie, VCI) established a working group on standardization in 1972, in order to elaborate joint company standards for the chemical industry using company-specific standards provided by the member firms, which do not pass through the official standardization process and therefore cannot be considered as general rules of technology. Moreover, this working group also formulates and represents the position of the VCI in national, European and international standardization.

The formal standardization process

Apart from the institutional structure of standardization work, the organization of the actual standardization process has a particular significance for the national stock of standards. A standardization project passes through different phases, which are illustrated in Figure 8.1.

The working group which is responsible for a certain technology area is referred to as the standardization technical committee, which is made up of representative experts of the partners in question – manufacturers, suppliers, trade, users, laboratories, public authorities. In Germany, the standardization committees are directly located at DIN, the national standardization institute.

After the creator, an enterprise or group of enterprises has submitted an application for standardization with a corresponding presentation of the proposal, the documentation and feasibility of the project is examined by the standardization institute before the relevant technical committee decides about accepting the project into the work schedule. After these preparatory steps are taken, the actual standardization work can begin.

After the specification of the standard by the technical committee, a group of experts elaborates a working draft which will be examined. An internal process to establish consensus is followed by a public inquiry among possible interested parties, whose suggestions are incorporated before final publication. The standardization process takes approximately one year and in the case of so-called pre-standards (*Vornormen*), which have a certain experimental character, among other things by dispensing with the public inquiry, can be shortened to a few months.

8.2.2 The System of Formal Standardization in Europe

The institutions
The responsibility for European standardization lies primarily with the European Committee for Standardization (CEN) and the European Committee for Electrotechnical Standardization (CENELEC). The European Telecommunications Standardization Institute (ETSI) was established in 1988 for the standardization in telecommunications.

CEN, CENELEC and ETSI elaborate:

- European standards (EN), whose text must be adopted in the exact wording by the national member organizations;
- European Harmonization Documents (HD), whose technical content must be adopted in the corresponding national standards (whereby certain freedom is permitted);
- European pre-standards (ENV), whose application is intended for areas with a high degree of innovation.

CEN and CENELEC consist of the standardization organizations of the Member States of the European Community and the EFTA countries. ETSI on the other hand is open for all organizations, which are interested in the standardization of telecommunications. They must merely be located in a country which belongs to the area of the Conference of European Post and Telephone Administrations (CEPT). Thus private telecommunications providers are also admitted to standardization work.

In the recent past the European Commission successfully urged CEN and CENELEC to accord sector-specific and rather informal European standardization organizations like AECMA (Association Européene des Constructeurs de Matérial Aérospatial), ECISS (European Committee on Iron and Steel Standardization) and EWOS (European Workshop for Open Systems), the status of so-called associated standardization bodies (ASB).

The formal standardization process

The course of the standardization process on the European level is fundamentally identical to the national standardization process depicted in Figure 8.1. Only the actors and organizations differ. Thus the application for a new standardization project can only be submitted by the member organizations or committees of CEN/CENELEC, by the European Commission, the EFTA Secretariat or European specialist organizations. Members of the Technical Board (TB), organizations to steer the European standardization work, decide after consultation with national standardization institutes about the adoption of the proposal in the work schedule. The real standardization work from the elaboration of a draft, during the enquiry and up to the final reaching of a consensus will be performed by a Technical Committee (TC), which will be appointed. The administrative work of the Technical Committee will be undertaken by individual national standardization institutions, while all the member organizations can have their interests represented in the Technical Committee by delegations. On the national level, so-called 'mirror committees' are set up to ascertain the national position, so that the way to a European standard normally leads via the national standardization organizations. Even the enquiry carried out among the interested groups is carried out by the national institutes. After evaluation of the commentaries and a phase for reaching a consensus in the Technical Committee, a formal vote takes place among the CEN/CENELEC members – unlike the national procedure – whereby the votes of the member states delegations are weighted according to their economic performance.[6] A standard is then accepted when a simple majority of the countries voting and at least 71 per cent of the yes votes (without abstentions) have been cast. The whole process takes between 24 and 75 months on average.[7] A simplified, shortened procedure applies for the elaboration of European pre-standards.[8]

8.2.3   The International System of Standardization

The institutions

On the international level, the International Organization for Standardization (ISO) and the International Electrotechnical Commission (IEC) form the central institutions for technical standardization, whereby the IEC is responsible for electrotechnics and electronics and the ISO for all other standardization questions. Analogous to CEN and CENELEC, ISO and IEC are organizations which are open to all representative national standardization organizations. ISO numbers at present over 85 members, IEC 54.[9] For example, Germany is represented in ISO by DIN and in IEC by DKE. Analogous to ETSI, on the international level the public institution International Telecommunications Union (ITU) exists, which besides other tasks addresses itself to the question of standardization in the field of telecommunication, although more in the form of recommendations, and closely collaborates with the IEC.[10]

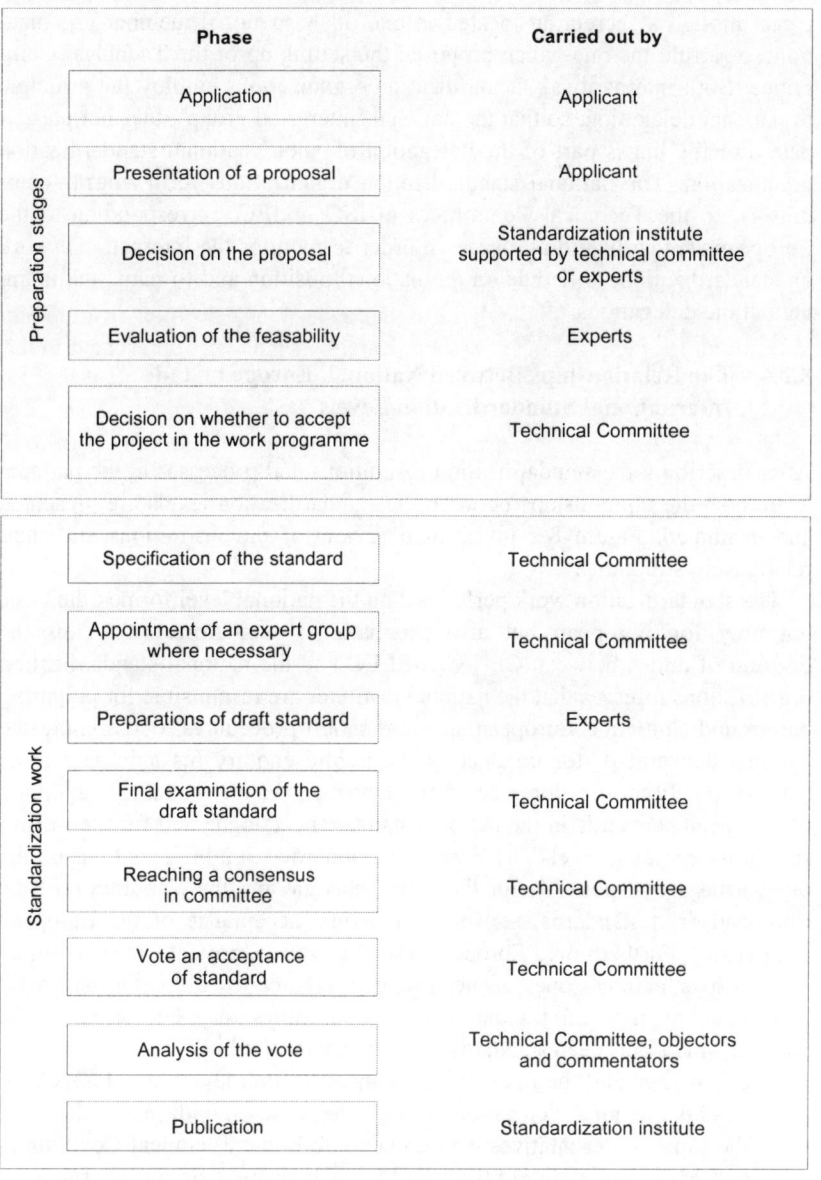

Figure 8.1   The national standardization process[11]

The standardization process

The standardization process on the international plane is basically comparable with the European one. Technical Committees are set up, which are charged with carrying out the standardization procedure. Administration takes place through a secretariat located in one of the national member organizations, as a rule the one which proposed the setting up of the Technical Committee. Both international standardization organizations employ the principle of national delegation, so that the national 'interested group' does not participate directly, but is part of the delegation of 'their' national standardization organization. The national standardization organizations form 'mirror committees' to the Technical Committees of ISO or IEC, corresponding to the European standardization process, in order to monitor the international work on standardization, to decide on the national position and to name the members of the delegations.

8.2.4   The Relationships Between National, European and International Standardization Levels

After describing the standardization institutions and processes at the national level, now the relationships between the standardization levels are presented and examined. Figure 8.2 gives an overview of the institutions and their relationship structures.

The standardization work performed on the national level formed the basis not only for European but also international standardization. Thus the division of duties between CEN/CENELEC and the national standardization organizations foresees that the national institutes are responsible for preparing the ground (initiating European standardization procedures, determining the national consensus), for conducting the public enquiry (as a basis for the national position on a European draft standard), for the mandatory adoption of European standards in the national standards catalogue and for their diffusion at the national level. The European standardization organizations on the other hand are responsible for the work schedule and the elaboration of the European draft standards, as well as the final acceptance of the European standards.[12] Furthermore, European standardization projects have absolute priority over national ones, as according to a so-called obligatory standstill agreement, no national standardization proceedings may be started in the areas in which European standards are to be established.[13]

The European and the international standardization levels are closely connected. As the national delegation principle is in use on both levels, there are often the same representatives who collaborate in the Technical Committees of CEN/CENELEC and ISO/IEC. Further a large part of the international standardization work is borne by the CEN/CENELEC members. In contrast to the European ruling, the ISO and IEC members are not obliged to adopt the international standards as national standards. Although their adoption is actually voluntary, international standards have increasing significance for the national standards, as the European standards organizations transpose the international standards into European standards. Also, in the co-operation agreement between CEN/CENELEC and ISO/IEC the exchange of information, the striving for standards valid worldwide and the prevention of duplicating work are defined as the common goals.

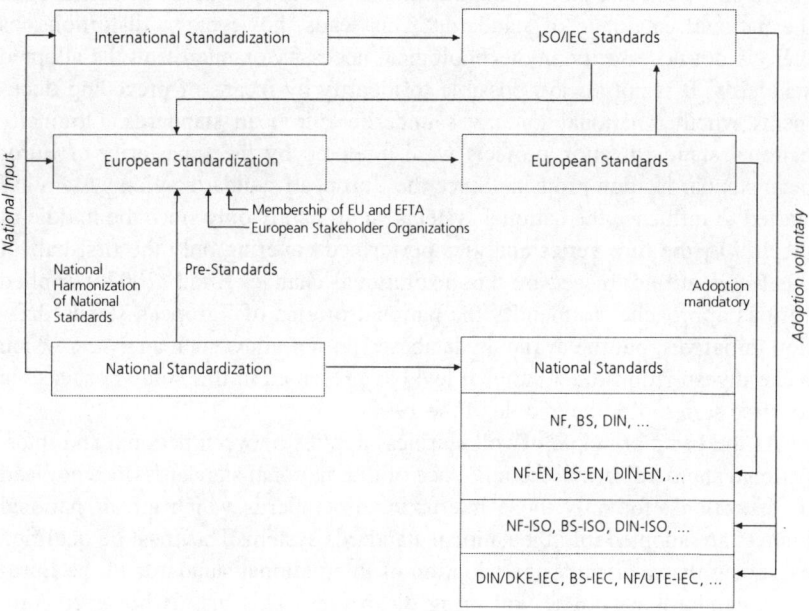

Figure 8.2   Relationships between national, European and international standardization levels[14]

8.3   THE SIGNIFICANCE OF THE INSTITUTIONAL FRAMEWORK CONDITIONS OF FORMAL STANDARDIZATION FOR THE DATA BASIS AND STANDARDS INDICATORS

The institutional framework conditions described above have a fundamental significance for the analysis of a database with the standards indicators derived herefrom.

On a national level, the hierarchical structure and involuntary co-ordinating function of DIN and other national institutions like BSI in the United Kingdom avoids duplication of work in principle.[15] This means that not only their documents, but also for example guidelines and the technical rules of other standard-setting institutions can be taken into consideration, since they are integrated into the database. Thus the potential data basis is significantly widened.[16]

The relationship between national and European standardization is characterized by the fact that a compulsion exists to adopt European standards in the national catalogue of standards. This leads, however, to distortions, as there is not necessarily any technological necessity or interest in the adopted standards. It is not always possible to identify by means of preceding documents whether national initiatives underlie European standards. Originally national standardization projects were hindered by the superiority of European standardization projects. Since the European standardization system has started to influence the national systems significantly only since the middle of the 1990s, the time-series analyses performed covering only the first half of the 1990s are not biased by this institutional change. Blind (2002c) applied various approaches to identify the national origins of European standardization initiatives, but the available databases do not allow such analyses, which make investigations on a national level as presented in this study not feasible for time series after the middle of the 1990s.

As the basic principle of voluntariness applies between national and international standardization, the influence of international standards does not lead to distortions, for only those international standards which are of national interest are adopted into the national standards system. This must be qualified by saying that indirectly, the adoption of international standards in the European standards catalogue will bring distortions. This bias is however comparatively still small in the periods analyzed.

If international comparisons are made, the framework conditions in the EU countries are similar, so that distortions appear in the same direction if a similar national adoption mode for the European documents is assumed. Switzerland is a special case, because it is not obliged to adopt the European standards, or only in the cases when it has agreed to the standard.[17]

NOTES

[1] The chapter is oriented basically on Voelzkow (1996), Thiard and Pfau (1991) and DIN (1996a).

[2] See OECD (1996, pp. 21 ff.), on the problems of indicators to measure technology diffusion and Geroski (2000) for an overview about models of technology diffusion.

[3] Cf. on this Lehr (1992), Voelzkow (1996), and Wölker (1996) among others.

4   From the viewpoint of the bureaucracy theory, an influence on the standardization activities can absolutely be derived from the financing structures.

5   On average 15 per cent of the costs are covered by federal funds (see DIN 1996b, p. 42; DIN 2000a, p. 14).

6   Cf. DIN (1996a, p. 19).

7   Cf. Europäische Kommission (1998b, p. 9).

8   Due to a similar organizational structure, the standardization process in ETSI runs through basically the same phases, whereby the administrative accompaniment of the actual standardization process is not delegated to a national standardization institute (cf. www.etsi.fr). The associated standardization bodies elaborate draft norms with full-time project team members, which are submitted directly to the CEN, so that the standardization process can be speeded up (cf. Woeckener 1997, p. 4).

9   See www.iso.ch and www.iec.ch for the present status.

10 Cf. www.itu.ch for the tasks and organization of ITU.

11 Source: Thiard and Pfau (1991, p. 43).

12 This principle of the indirect participation and the national representation should be increasingly replaced by direct participation of the parties in European standardization organizations (see Woeckener 1997 for an economic critique of this concentration process).

13 European standardization further gains considerable legal influence from being embedded in European regulatory policy in the framework of the New Approach by comparison with national standardization (see Section 7.2.3.3). Further the national regulatory institutions have to make their regulations known and take the principle of reciprocal recognition into account (cf. on this Hildebrandt 1995, pp. 89ff.). It follows that national standards lose in importance from this process.

14 Source: Thiard and Pfau (1991, p. 60).

15 On a national level, DIN has only a focussing function, but for the other national institutions, access to European standardization institutions is only possible via DIN. In this way DIN finally has a formal co-ordinating function.

16 The standardization system in the United States is organized in a much more decentralized way (National Research Council 1995; Toth 1997), which makes the construction of a consistent database much more difficult.

17 Cf. Holler (1996, p. 140).

PART C

The Demand for Standardization at
Standardization Development Organizations

9. Introduction to Part C

The globalization of the economy and the increasing importance of systems technologies, such as information and communication technology, are additional driving forces for the expanding need for technical standards in recent years. From the standpoint of SDOs, the following well-known reasons support the hypothesis that the agreement on common formal standards encourages trade and the exchange of products and services, and the diffusion of new technologies:[1]

- Common standards make for higher and more reliable product quality at a reasonable price.
- Common standards take environmental and safety issues into account.
- The compatibility of different components is improved by standardization.
- The handling of standardized products is easier by comparison with non-standardized products.
- The number of different types of one product is reduced by standardization, so that economies of scale can lower production costs and therefore prices.
- The diffusion process of standardized products is more efficient.
- Maintenance and repair are less difficult with standardized products.

In view of the many advantages of standardized products and processes, standards development organizations were already founded at the beginning of the 20th century, at the national but also international level, mainly by private initiatives of companies (Wölker 1996).

Part C of the study is devoted to analyzing the driving forces for standardization within SDOs and the output of standards. The following paragraphs consider the most relevant aspects discussed in the theoretical literature. Therefore, some arguments already presented in Part A about the impact dimensions of standards will be taken up again, before the question of an adequate database for empirical analyses is tackled.

Issues of standardization have already been economically analyzed and modeled in the 1970s, for example by Hemenway (1975), but the era of economic publications about standardization started in the 1980s with authors like Farrell and Saloner, and Katz and Shapiro. The economic literature about standards and standardization can be divided into two segments: those which concentrate on the microeconomic analysis on the one hand, and those which focus on the importance of standards in international trade and competition on the other.

In order to categorize the huge number of microeconomic analyses which deal with the role of standards in isolated markets, it is useful to differentiate the research approaches in an evolutionary setting according to the stages in the product life cycles. At the innovation stage, the question of the best technical design and the optimal point in time to establish a standard arises. The benefits of an early standardization, advantages of compatibility for example, have to be weighed against the loss of diversity and the danger of a lock-in at an inferior level. At the diffusion stage, standards are very important for the economic success of goods and services with positive network externalities like telecommunication and information networks and services. Finally, in mature markets, compatibility and incompatibility standards are, besides price setting, a parameter both in competition and cooperation between companies.

The aim common to all these research approaches is to define efficiency criteria and to identify market inefficiencies, because the unregulated markets tend to provide too many, too few or inferior standards. From a social perspective, standard wars may lead to incompatible standards and fragmented markets not able to exploit network economies (Shapiro and Varian 1999). Since open standards in particular have the classical features of public goods, there is a tendency towards underprovision because of insufficient private incentives. Finally, systems competition may be a competition of laxity resulting in technically inferior or respectively too low quality standards (Sinn 1996, 1997). The technical change can therefore be belated, as shown by Farrell and Saloner (1985) or can be inefficiently accelerated (Farrell and Saloner 1986). Furthermore, privately-owned standards can be used to increase the entry barriers for newcomers, while open standards facilitate market entrance. A further research area is the standardization process in so-called voluntary standardization committees, which tend to retard the standardization process and to result in socially inefficient standards. Many of the cited research approaches remain only at a theoretical level and are unable to support their theoretical hypothesis by empirical tests. Others rely on case studies. A validation of the theories by statistical or econometric analysis is in general not performed. Exceptions are the studies of Lecraw (1984), Link (1983) and Weiss and Sirbu (1990).

If we depart from the restricted microeconomic approaches of standards and widen our view to include the institutional environment of the standards development organizations, a variety of overlapping interdependencies with other public institutions appear. Standards are also an instrument for public policy in safety and quality issues to internalize negative externalities. This research field focuses on the discussion of effectiveness and efficiency of standards in comparison with other policy tools, like subsidies and taxes or liability rules. It is concentrated especially on the protection of the environment and of the consumers. Furthermore, institutions commissioned with the protection of intellectual property and anti-trust duties are influenced by the rules and the consequences of standardization processes (Farrell 1989).

Finally, leaving the national framework, the international interdependencies have to be integrated into the analytical set-up. National standards are still an important tool for national trade policy in order to protect domestic producers by installing non-tariff trade barriers, especially after the worldwide agreement to reduce tariff trade barriers in the former General Agreement on Tariffs and Trade (GATT).[2] Their negative implications for the volume of international trade were already discussed in depth in the traditional trade theory and proved by some empirical studies. However, the trade-fostering effects of harmonized standards in particular at a supranational level is undisputed from the theoretical point of view, but empirically not yet proven, despite first approaches for example by Swann et al. (1996). Besides the growing international competition in the markets for goods and services too, globalization is also fostering the competition for financial and physical capital and consequently for employment opportunities. Because of the Cassis-de-Dijon judgment on the confirmed origin principle, it is likely that a deregulation competition will start in the field of quality and safety regulation and standardization. This leads finally to a consensus at the lowest level, an inefficient result from an economic point of view (Sinn 1996, 1997). Therefore, the tendency towards harmonized European standards is desirable for an effective and efficient protection of consumer interests, besides the advantages of free trade.

This short survey of the state-of-the-art in economic research on standardization as well as Chapters 3 and 4 have shown that the economic theory about standards and standardization already covers the relevant topics. Deficits are perceptible in the empirical testing of the theoretical models and hypotheses. If there is empirical proof of a theory, it is mainly restricted to specific goods and markets or to case studies. Empirical studies, which rely on a broad sample of standards, are hardly available.[3]

The lack of broad empirical studies concerning standardization is the result of data problems between the different kinds of standards. A taxonomy of four different kinds of technical standards exists, depending on how they are created.[4] First, a firm has property rights to a specific technical specification or a product and is able to promote it as a widely accepted and used industrial standard. In the second case, a group of companies are negotiating in an informal way about the specification of a component of a system or of a whole system, which is necessary for the effective diffusion of their products or services. The agreement about a common specification is also an industry standard, which can either be protected by property rights, like a patent, or is reproducible for free by everybody. When these companies go to SDOs, then all interested groups, including consumers, are allowed to join in the standardization process. Furthermore, the standardization process will only be successfully completed when all participants or the majority agree on the set of characteristics for the standard. The obstacles and costs of this process are outweighed by the fact that the standard is regarded as an official document with almost legal status, especially because liability laws and insurance companies rely on it. Finally, the government itself sets up technical standards, mostly as minimum quality or safety standards, which are legally binding for the corresponding suppliers of goods and services. This regulation process is driven by public needs as perceived by government institutions, and not by private firms. However, the latter may benefit from these legal standards, because they can improve the consumer acceptance of new products and services whose characteristics are not very transparent.

In the analysis, we will focus on the standard documents generated by SDOs for the following reasons.[5] First, the industry standards generated by one or more firms are mostly protected by some sort of property rights and cannot be applied by somebody else. Secondly, there is no database which contains these standards.[6] Legal regulations in general are also excluded, because they are not generated by private company initiatives but by government decision processes.[7] Finally, the analysis is concentrated on the standards which underwent the SDO standardization process.

The scope of Part C is to bridge the gap between the theoretical insights about the driving forces of standardization and the lack of empirical results. Based on hypotheses concerning sector- and company-specific factors of standardization, Part C examines empirically both sector- and company-specific driving forces to standardize the processes and products at official national standardization bodies. It is structured as follows. After the elaboration of the hypotheses (Chapter 10), the results of these regression analyses are presented in Chapter 11 based on the German and international sectoral data (Chapter 11). In Chapter 12, a special focus is on the impact of intellectual property rights (IPR) on standardization activities, which already includes firm-specific data. Determinants for participation in standardization processes at firm level are empirically tested in Chapter 13. In the final Chapter 14, the diffusion of quality standards among innovative service companies is analyzed.

NOTES

1   See www.ansi.org (25 June 1996) and Chapters 3 and 4.
2   The GATT always dealt with trade in goods, and it still does. It has been amended and incorporated into the new World Trade Organization (WTO) agreements. The updated GATT lives alongside the new General Agreement on Trade in Services (GATS) and Agreement on Trade-Related Aspects of Intellectual Property Rights (TRIPS). The WTO brings the three together within a single organization, a single set of rules and a single system for resolving disputes.
3   An exception is the analytical work of Swann et al. (1996).
4   Compare the taxonomy of standards by David and Greenstein (1990) or Toth (1997) already presented in Sections 4.2.3 and 6.2. The taxonomy introduced in Chapter 2 categorizes standards by their function, for example compatibility, variety-reducing and quality or safety standards.
5   See Section 6.4 for more details.
6   In Europe, the national standardization bodies like AFNOR, DIN and BSI have been offering companies the opportunity to publish their industrial standards as publicly available specifications (PAS) only since 1997. Therefore there is no adequate sample size available yet.
7   However, technical regulations which rely on SDO standards are included. In general, their share of the total amount of documents is below 5 per cent, and can therefore be disregarded.

10. Theoretical Hypotheses Concerning the Driving Forces of Standardization Activities[1]

10.1 INTRODUCTION

In this chapter, theoretical hypotheses concerning the driving forces of standardization activities are presented, derived both from the need or demand for standards, but also from the characteristics of the standardization process – the supply side. We differentiate between the sectoral and the firm level. First, standards are the result of a complex negotiation process involving several actors besides companies. These other actors represent important framework conditions for standardization processes. On the one hand, the input of research organizations is necessary to catch recent developments in science and technology. On the other hand, various non-governmental organizations, like environmental groups or unions, try to represent their answers to the challenges of new technologies in standardization processes. And the tasks of the companies in standardization processes is to bridge the gap between the impulses of science and technology with the degree of acceptance at the user side. Second, the individual company follows a strategy regarding standardization which is determined by company-specific framework conditions, but the interaction of numerous companies may lead to different outcomes, depending on the composition of standardization committees and the technological framework conditions. Finally, regarding the sectoral level we try to explain the annual output of standards, whereas at the company level the question is what are the determinants for companies to join or not to join standardization processes in their specific frameworks, since it is not possible to attribute the successful publication of a specific formal standard to a single company.

The remainder of this chapter discusses first, the driving forces at sectoral level; second, the determinants at company level, and concludes with a brief comparative summary.

10.2 DRIVING FORCES AT SECTORAL LEVEL

The standardization process can be regarded as the extension of the competitive product development process.[2] After the decision concerning the R&D budget[3] has been taken, the firm has to decide, in a second step, about the protection of its product innovation by going through patenting or other formal processes to obtain formal intellectual property rights. Finally, the firm has to decide on the number of product and process innovations it is going to propose for a standardization process. The expected benefits of a standardized product are advantages in its diffusion and therefore a higher anticipated demand. The costs include the actual financial cost of a standardization process, including the opportunity costs of a delayed marketing of the product.[4] Finally, the company has to publish their R&D results, which turns private knowledge into public knowledge, first available to the participants of the standardization process[5] and later to all buyers of the documents. The knowledge spill-overs will be higher when there is no protection of the R&D results at all. However, the patent protection cannot prevent other companies using the technology, but it may at least control the knowledge spill-overs, because the company has to license the patent for a reasonable amount to the public.[6] Based on these considerations, the following hypotheses are postulated.

Because standardization is a part of the R&D process, the higher the R&D intensity of a sector, the higher the annual standardization output will be. Therefore, both the input indicator R&D expenditure and the output indicator patent applications should positively explain the annual output of standards. However, because of the spill-over problem of the standardization process, the sectoral propensity to standardize should be explained better by the number of patent applications compared to the R&D expenditure.

Moreover, due to the IPR problem, the standardization process can be prolonged or can even fail because patent holders are not willing to license their IPR.[7] Therefore, sectors with a very high number of patents tend to standardize less.[8]

Apart from the role of technical standards in the R&D process, the standardization process is an important strategic measure besides price and quality in the competition with other suppliers. Standardization is a negotiation process between different producers. The basic tenet of the policy decision-making theory is that only consensus (unanimity) can lead to a Pareto-optimal result. However, the process of reaching this settlement is very long-drawn-out and costly, especially when many heterogeneous interests have to be taken into consideration. For this reason, a Pareto-optimal result is not achieved in many cases. Consensus often leads to strategic behavior, where preferences are falsely represented. The result can be a decision which favors one interest group and places the group as a whole at a disadvantage. Therefore the majority rule is often chosen instead of consensus and explicitly accepting Pareto-inferior solutions, in order to be able to ignore individual interests. The advantage is that decisions can be reached more quickly, especially in large groups with heterogeneous interests. The optimal majority size depends on the relationship between loss of benefits of the minority and the costs of coming to a decision, and varies from case to case.[9]

If the characteristic of a zero sum game is assumed for the standardization process, coalitions will be formed which just reach the necessary majority, because in this way the smallest number of coalition members can share the 'prize'. On the other hand, with a large majority or consensus, the 'prize' for each individual would be practically zero, or even zero. Indeed, minorities have a strong negotiating position in a constellation of two large groups which have narrowly failed to gain the necessary majority. Stable voting blocks will not be obtained for a majority ruling, if so-called cyclical preference structures exist. In that case, the final result depends decisively on the order of voting or the day's agenda.[10] To sum up, we conclude that neither the consensus rule nor majority rule are necessarily satisfactory. These are voting rules which give moderate positions a better chance than extreme positions, although the latter may present the better solutions.

As the SDOs mainly use the consensus ruling, the more participants there are, the more difficult and long-drawn-out the negotiating process becomes. A link to the sectoral market structure can be established and the hypothesis derived that, in sectors with a high market concentration, the standardization process is simpler and swifter because of the lower number of actors, so that in the end a greater number of standards can be produced.[11] Another argument for this hypothesis is based on the fact that as a public good is produced in a standardization process, there is – as already mentioned above – the temptation to appropriate the knowledge spill-overs of the other participating enterprises, as a free rider, without providing any significant own input. If all actors were to pursue this strategy in the final analysis, then the standardization process would come to grief. The more actors are involved, the greater the probability of failure, because in larger groups the peer group pressure to become actively and constructively involved in the standardization process is smaller. Therefore, it is easier to exclude this free-rider behavior in sectors with high concentration indexes and fewer actors, and to carry out standardization processes successfully.

Whereas the two arguments above are valid for negotiation processes in general, it has to be considered that, as shown in the introduction, there are different kinds of standards and consequently standardization procedures. Besides the official standardization process accessible to all interested groups and firms and ending in a de jure standard, industry consortia with restricted access can specify so-called de facto standards, like the famous IBM standard. The market power of companies and therefore the market concentration are determining factors for their decision to choose the official or the informal way to reach a standard. Whereas in markets with many small- and medium-sized companies the probability of establishing a de facto industry standard successfully is low, it is much higher in markets with a small number of companies with a very high market share. Therefore, in the latter case the way to reach a de facto standard is more likely to be chosen, with the consequence of a decreased propensity to standardize at a national standardization organization, especially in very highly concentrated markets.

Taking the average company size as an explanatory factor leads in a similar direction to the discussion about company concentration. As considerable (personnel) costs are involved in the standardization process (Berger and Clement 1990) and the standards department of a company represents a fixed cost block,[12] large companies are more likely to be able to afford such a department and become involved in the standardization process. In addition, large companies may have a stronger position in standardization processes with conflicting interests. Finally, they benefit more from a successful standardization process than small companies, since they are able to apply the standard either to a larger volume of produced goods, to larger production processes or to larger organizational units.

Apart from these supply-oriented considerations, there are also demand-oriented ones, for the more fragmented a market is, the greater the benefit for the many small companies to have standard, uniform technical rules. This means that sectors with small market concentration have a greater need for standards than branches with very few enterprises.[13] To sum up, from a theoretical point of view, it is not unambiguous whether the company concentration presents a positive or negative explanatory factor for the standardization intensity of the sectors.

Besides these supply-oriented explanatory factors, the various objects of standards supply a further, sector-specific explanation. Standards have on the one hand the function of guaranteeing the compatibility of interfaces. On the other hand, they should ensure the quality and the safety of products and processes. The third function, of reducing variety, is often merely an indirect effect of these two functions and should not be further regarded.[14] Compatibility standards, especially for the production process, are therefore of particular importance for a sector if the production is very capital-intensive. Apart from building the original manufacturing facilities, the continuous investments for maintenance and extensions depend on correspondingly compatible interfaces in order to avoid high adaptation costs. This leads to the hypothesis that capital-intensive branches of industry produce more standards than those with a lower capital intensity.

Quality and safety standards for processes and products are of immense importance for the production process, if the labor force is involved to a great extent. This means that the need for standards is higher in labor-intensive branches than in capital-intensive ones. Together with the above-mentioned theoretical considerations, no clear-cut hypothesis on the significance of the capital, respectively labor, intensity for the sector-specific standards output can be advanced for the total need for standards.

In the present globalized markets, standardization cannot be considered separately from international economic relations, especially in highly indus-trialized countries. In order to remain internationally competitive, firms in these countries must hold their own market segments with innovative pro-ducts in the quality competition, as other countries can produce more cheaply due to lower production costs. In the meantime, the significance of the na-tional innovation potential for the international competitiveness of the highly industrialized countries has been proved both theoretically and empirically,[15] however, the role of standards has not been clarified yet.[16] As standardization must be regarded as a part of the innovation process, so too should standards be regarded as indicators for the national innovation potential. Sectors which have a high export rate and are thus internationally competitive, should standardize more strongly than those with low export rates. Besides this reasoning, which derives from the R&D process and the resulting indicators, a second line of argument points in the same direction and proceeds from the economic functions of standards, because standards fulfill a general function of informing not only consumers, but also producers. As the information asymmetries are especially large between foreign demand and domestic pro-ducers and exporters, export-intensive sectors have a particularly strong incentive to generate national standards first of all, which may possibly pre-vail in the European or international standardization organizations. Besides this information function, many standards also have an interface or compati-bility function to fulfill. If exporters successfully sell standardized products abroad, they generate, particularly for network products, network externalities which are increasing with their installed base. This in turn leads to a lock-in in the selected technology or product type. This means that the foreign cus-tomers only ask for the products which have interface compatibility with their already installed base, or are prepared to accept a change to other product models only in the case of greatly changed preferences or substantially improved alternatives. It is true that standardization makes the interface or product specification more accessible for all competitors, but the domestic producers still have a temporary time and adaptation advantage and thus also a price advantage. Branches with high export rates standardize more, in order to export domestic standards abroad via standardized products and thus ensure market shares at least temporarily.

Whereas a high export rate can be seen as a success indicator for a sector, low export rates may indicate a limited international competitiveness. In order to compensate for the lack of success in the international markets, one strategy consists in sealing off the domestic market. One instrument to this end is the generation of informal non-tariff trade barriers in the form of national standards and technical regulations. For despite the increased transparency of national specifics afforded foreign competitors by the standards, the domestic producers still have gained a time, and usually also a price, advantage by collaborating in the national standardization process. For this reason, another theory maintains that branches with low export rates standardize more in order to protect their domestic markets from foreign competition. Although, due to the numerous other instruments to protect domestic markets shares, the role of technical standards is rather limited and probably empirically not significant.

In direct connection with this theory, the role of standards for imports must be mentioned. Although the import rate, as opposed to the export rate, is not defined or calculated by the Statistical Offices, along the lines of the above argument the theory can be advanced that branches with high import rates standardize more, in order to protect the domestic market by national standards. The theory of standards as non-tariff trade barriers, however, is opposed to their function as information channels. If it is assumed that a multiplicity of standards makes it easier for foreign competitors to appropriate the national preferences and technical specifications and to adapt their products accordingly, then it should be true for branches with high import rates that they standardize less, in order to make the technical requirements of the domestic market less transparent for the foreign competition. This means that the sign of the sectoral import rate is ambivalent for the explanation of the standards output.

10.3 DRIVING FORCES AT FIRM LEVEL

Whereas we have discussed in the previous chapter determinants for the quantitative output of standards per sector, this chapter is devoted to discussing the main explanatory variables for companies' standardization activities. The decisions about joining a standardization process and therefore of committing resources is made at the firm level, as well as the determination of resources for R&D and innovation (cf. for the latter Wakelin 1998a). However, in contrast to the benefits of innovations which the firm primarily can appropriate – despite numerous problems – caused by reduced production costs, penetrating new markets and exploiting monopoly rents, new standards published and distributed by SDOs are at first glance a public good.[17] Despite the general possibility for everybody to buy a standard at a reasonable price, only the core of companies which have the relevant technological know-how can use the new technical specification effectively and efficiently.[18] Therefore, the participants in the standardization process may have advantages compared to outsiders, due to their early involvement in the development of the standard and the accompanying process of knowledge exchange and creation. Salop and Scheffman (1987) underline this argument in that the establishment of product standards may be a strategy by which firms could disadvantage rivals by raising their costs. Secondly, only the companies which are in the same branch or are using the same technology may benefit in general from a new standard. Consequently, the theoretical approaches of Farrell and Saloner (1985) dealing with innovation and standardization use game theory models with only two companies. Therefore, despite the explicit technological spill-overs of standards, which justify an analysis at branch or even at macroeconomic level, the single company is also an appropriate unit of analysis, especially when considering the relationships between innovation, standardization and export behavior.

Concerning the firms' performance in R&D and its impact on standardization, two contradictory trains of thought have to be considered. Firstly and obviously, the standardization process is a continuation of the development phase of internal R&D. Therefore, companies which are actively involved in R&D are also more likely to participate in standardization processes in order to continue their previous activities and to reach marketable products or process technologies compatible with those of other companies (Farrell and Saloner 1985). This argument is closely linked to the discussion on the incentives to form research joint ventures.[19] Whereas joint research ventures are able to save R&D expenditures, especially when they reduce an excessive duplication of efforts, common standardization activities may reduce the costs for the market introduction of a new technology. However, the involvement in standardization processes is accompanied by the danger that the other participants could use the disclosed and unprotected technological knowledge for their purposes. Therefore, in contrast to joint research ventures, which may help to solve the spill-over problem, standardization activities present another opportunity for the original knowledge base to be leaked out to competitors. Therefore, R&D-intensive companies may be more reluctant to join standardization processes and try to market their products alone without relying on standardized input technologies, common interfaces to complementary products of other competitors or even uniform product designs.

Secondly, and on the contrary, companies with low R&D efforts in particular may take advantage of the spill-over problem and compensate for this by entering standardization clubs of R&D-intensive firms and profiting from the technology transfer there.[20] This view is empirically supported by the analysis of Love and Roper (1999) about the substitute relationship between own R&D and technology transfer. Taking the spill-over problem, respectively the free-rider aspect and the cost-saving incentive together, the companies' R&D intensity may be ambivalent or even slightly negative for the likelihood of joining standardization processes.

Closely linked to the discussion about the role of R&D intensity, the impact of labor productivity on the participation in standardization has to be considered. Assuming a positive correlation between R&D intensity and (labor) productivity because of improved production processes and expanded market shares, the same reasoning about a negative relationship between labor productivity and the probability of joining standardization processes can be justified. Those companies with insufficiencies in their productivity have stronger incentives to join standardization processes in order to benefit from the positive knowledge spill-overs of highly productive companies. In contrast to the R&D intensity, no alternative hypothesis of a positive relationship exists on the probability of joining standardization. Consequently, not just an ambivalent, but a clearly negative relationship between labor productivity and activities in standardization is postulated.

When discussing the role of export activities, a two-way causality has to be considered. Due to institutional paradigms of most national standardization development bodies, the participation in standardization processes at the national level also facilitates influence on the standardization at European or international level. Therefore, exporting companies which try to influence supranational standards in order to secure market shares in foreign markets are more likely to participate in standardization both at the national, but also at the European and international level. However, the influence of a single company on the standardization processes decreases with the increasing territorial responsibility of the standards committee. This fact also reduces the strength of the motivation of exporters to join standardization at a supranational level. Nevertheless, companies actively involved in standardization should be more successful in exporting their goods and services due to their influence on the product specification of supranational standards.

However, whereas participation in standardization is certainly a strategy to shape foreign markets according to specifications of own products or technologies, this advantage is shared by all other participants and also – although with a certain time lag – by the companies implementing the standards. Consequently, standardization constrains competition for a very short time by giving the participants a leading position, which is only temporary, since the codification of the technical specifications allows the other competitors to catch up at the latest by the official publication of the standard. Therefore, companies seeking for a continuous market lead ahead of the competitors, which results in a monopolistic position and high market share worldwide indicated by very high export shares, will be reluctant to join standardization processes.

Finally, the participation in standardization processes causes significant costs, for example, for highly qualified personnel and travel expenses. Since these costs have a fixed-cost effect, small- and medium-sized companies have a smaller likelihood to join the standardization process and behave more as free-riders. In addition, larger companies are more able to internalize some of the generated positive externalities of standards compared to smaller companies. Therefore, standardization is less favorable for smaller companies, not only concerning costs, but also relating to benefits.

Furthermore, sector-specific differences have to be taken into account, because so-called network industries with significant network externalities and complementary components rely crucially on the existence of standards (Shy 2001). Companies in these sectors therefore in general have a higher demand for standards, driven by technological reasons. Finally, the sector-specific characteristics discussed in the previous section are important framework conditions which influence companies' decisions to join formal standardization processes.

10.4   SUMMARY

In the previous two sections, we have discussed the driving forces for standardization both at sectoral and at company level. The summarizing section tries to compare the hypotheses derived for the main determining variables. First, company size is a rather ambiguous variable for the sectoral output of formal standards. However, the likelihood of joining standardization processes for individual companies should increase by company size. On the other hand, the sectoral R&D intensity and patent activities, an indicator for the R&D output, should be in a positive relationship with the sectoral output of standards, since emerging new technologies require new or at least updated standards.[21] For companies, it is not clear whether there is a complementary or a substitutive relationship between performing own R&D activities and joining standardization processes. Finally, sectors with strong export activities are more likely to produce standards, which may be transferred to supranational standardization development bodies, because this strategy may foster their export performance. For the single exporting company it is ambivalent to join a standardization committee, since both the chances of asserting its preferred specifications is dependent on other company characteristics and the threat of losing its technological advantage respective to domestic and foreign competitors.

Concluding this brief comparison, it is sensible to perform empirical tests both at the sectoral and the company level, because it is expected – at least from the theoretical perspective – to reach different results for some of the explanatory variables. The sector dummy in the regression equations explaining the behavior of the single company is the connection between the two approaches and represents the complex framework conditions for the standardization decisions of companies.

NOTES

1   Chapters 10 and 11 are based on Blind (2002a).
2   Cf. Weiss (1990, pp. 36 ff.) and Thiard and Pfau (1991).
3   See among others Harhoff (1997, p. 349).
4   See Farrell and Saloner (1985) and Katz and Shapiro (1985).
5   See the club goods argument of Antonelli (1994).
6   See DIN Standard 820 and Chapter 12.
7   Compare Farrell (1989, pp. 43 ff.). Mazzoleni and Nelson (1998) therefore question the economic benefit of strong patent protection in system technologies where developments rely mainly on efficient and fast diffusion by means of standards. The linkage between strategic competitiveness strategies and the technical specification of designs as candidates for standards agreement are clearly visible here, although no in-depth research on the implications has been carried out as yet. See Mansell (1995, pp. 221 f.).
8   For a more detailed discussion including the institutional framework, see Chapter 12.
9   Cf. Goerke and Holler (1995) on majority rules in the standardization process.
10 In a case study, Weiss and Sirbu (1990) cannot prove that the committee chairman could push through his own interests in the standardization process by bringing his influence to bear on the agenda.
11 In his survey of 881 consumer goods categories, Link (1983) established a positive correlation between the market concentration and the probability of producing a standard, in the United States.
12 The average number of employees in standards departments rises degressively with the size of the enterprise (cf. on this subject Adolphi 1997, p. 4).
13 Cf. on this the theoretical considerations of Katz and Shapiro (1985).
14 To this comes the empirical problem of measuring variety (cf. Saviotti 1991).
15 Cf. here among others Wakelin (1997, 1998b) and Verspagen and Wakelin (1997).
16 Swann et al. (1996) have conducted a first empirical study in British industry.
17 In the case of privately owned de facto standards caused by network externalities, the R&D decision will change towards a socially ineffective speed up of R&D (cf. Kristiansen 1998).
18 Therefore, Antonelli (1994) goes even further and characterizes standards as non-pure private goods.
19 See for a review of the various incentives Röller et al. (1998). The other two incentives of product complementarity and similarity of firm size are not discussed here, since standardization focuses always on one technology or product. Röller et al. (1998) find no clear incidence that research joint ventures are more often formed among companies producing complementary products. Finally, the role of firm size is closely connected with cost and spillover aspects, which will be discussed below.
20 The problem of an inadequate absorptive capacity in companies with no or low R&D activities respective to the integration of standard specification in the own production process or in the own product design is of little relevance in contrast to traditional research joint ventures (Kamien and Zang 2000).
21 It has to be mentioned that the patent intensities vary between the sectors. That means that patent intensity at the sectoral level characterizes technology, whereas it represents innovative capacity at the firm level.

11. Test of the Sectoral Model

11.1 INTRODUCTION

The theoretical hypotheses presented in Section 10.2 will be tested empirically in a two-stage process. First, we perform a multivariate regression analysis based on sectoral data for Germany. Second, we extend this approach to a sample of seven OECD countries in order to prove whether the findings for Germany can also be confirmed for other countries and general patterns on the sectoral driving forces for standardization can be identified. This last objective gains in importance since standardization is becoming more and more international and national characteristics lose in importance.

11.2 THE DATA OF THE GERMAN MODEL

The theoretical hypotheses on possible explanatory factors for sectoral standards output will be empirically tested first on the basis of the following 19 industrial branches in Germany (see Table 11.1), for which the data has been compiled and matched from various secondary statistics. Starting from a concordance between standards and patent classifications (Appendix I), further concordances between the ICS and the industrial sector classification WZ93 and the international industrial sector classification ISIC Rev. 2 were elaborated, in order to be able to examine various explanatory factors for the sectoral standardization intensity. [1]

In order to measure the extent of standardization, the average publication of standardization documents in the years 1991 to 1995 will be referred to, because in the standards databank PERINORM, a database is available for Germany and other countries, which reflects the output of the standardization process regarding both the content and time perspective. [2] The database does not only include documents of the German standardization institute DIN, but also of around 150 other German standardization institutes. The standards data also contain the European and international standards which have been adopted in the domestic market. However, on the basis of the database information alone, it is not possible to identify in which country the initiative for an international or European standard originated, so this fact cannot be used as additional information. Nevertheless, Germany is the most active country in Europe as regards standardization, and the sector-specific characteristics are assumed to be very similar in the highly industrialized European countries.

Table 11.1   List of 19 industrial sectors

Mining and extraction of stones and earth
Food industry, tobacco processing
Textile, clothing and leather industry
Wood (without furniture manufacturing)
Paper industry
Coke, petroleum processing, production of breeding materials
Chemical industry (without pharmaceuticals)
Manufacturing of rubber and plastic goods
Glass, ceramics, handling of stones
Basic metal industries
Manufacturing of metal goods
Mechanical engineering
Manufacture of office machines, data processing equipment and systems
Manufacturing of equipment to generate and distribute electricity
Radio, television and communications engineering
Medical, measuring and control technology, optics
Manufacture of automobiles and automobile parts
Other vehicles (shipbuilding, aircraft)
Furniture, jewelry, musical instruments etc.

The various explanatory factors were compiled from the following data sources. On the basis of these 19 industrial branches, the R&D expenditure of the enterprises in 1995 was taken from the R&D Data Report 1997.[3] The output indicator patents is depicted by the patent applications by German inventors at the German Patent Office in the year 1995. The details of enterprise concentration and size and turnover according to branches is taken from the series published by the Federal Office of Statistics.[4] The export rates of manufacturing industry 1995 were taken from the Statistical Yearbook (Statistisches Bundesamt 1997, p. 197). As import rates are not calculated by the Federal Office of Statistics, the import figures for 1994 according to groups of goods from the Statistical Yearbook (Statistisches Bundesamt 1997,

p. 296) were divided by the total turnovers of the corresponding branch of industry. Finally, the statistics of the Deutsches Institut für Wirtschafts-forschung (DIW) (1995) are the source for the capital coefficients, defined as gross fixed capital gross capital stock relative to the volume of gross value added.

11.3   EMPIRICAL RESULTS FOR GERMANY

The average number of publications of technical standards in Germany differentiated by the above 19 sectors are presented in Figure 11.1. These numbers also include the publication of standard drafts, pre-standards and revised standards. Surprisingly, the number of published documents in the glass and ceramic industry is very high. As expected, the network industries of electricity and communication technology are very active in releasing standard documents. They are followed by medical and measurement instruments, where especially safety and quality aspects play an important role. Very small numbers of standards are found in the low-tech and 'old' areas of the mining, wood and paper industries.[5]

For the simple OLS econometric analysis, the average standard output is explained by variables which are suited to test the theoretical hypotheses.

$$ST_i = f(C_0, Pat_i, RD_i, Co_i, Siz_i, Ex_i, Im_i, Ca_i, e_c) \qquad (11.1)$$

The variables are defined as follows:

ST_i = average standard output (1991–95) in industry i;
C_0 = constant;
Pat_i = patent application in industry i;
RD_i = expenditure of enterprises for R&D in industry i;
Co_i = Gini coefficient in industry i;
Siz_i = average turnover of the enterprises in industry i;
Ex_i = export ratio in industry i;
Im_i = import ratio in industry i;
Ca_i = capital coefficient in industry i;
e_c = error term iid $N(0, \sigma_e)$.

These are the patent applications, the R&D expenditures of the enterprises, the Gini coefficient as concentration indicator, the average turnover of the enterprises, the export and the import ratio and the capital coefficient. In the models, the coefficients of variables originating from joined hypotheses are tested together. The results are shown in Table 11.2.

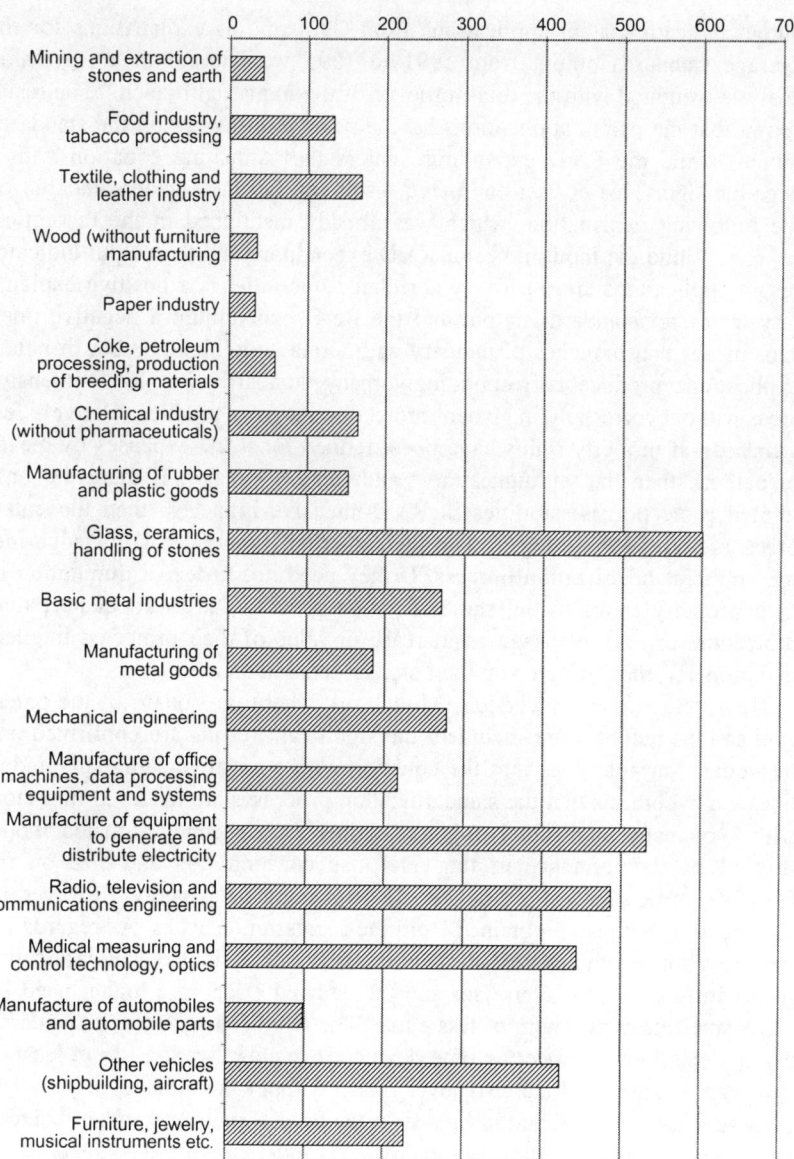

Figure 11.1   Average standard output (1991–95) in Germany measured by the number of standards

In a first OLS model, the two indicators of the R&D process, the input indicator for R&D expenditure 1995 (in million DM) in the enterprises and the output indicator patent applications from Germany, as explanations for the average standards output from 1991 to 1995 were tested in a regression analysis weighted with the total turnover of the industrial branch.[6] The result shows that the patent applications have a positive influence on the standards output, while the R&D expenditure enters the estimating equation with a negative sign. This at first confusing result can be made understandable by the following explanation, which was already mentioned in the theoretical chapter. While the input indicator R&D expenditure and the output indicator patent applications are positively correlated, the latter is a positive explanatory factor for standards output and the R&D expenditure a negative one.[7] This means that branches of industry with a relatively high number of patent applications produce correspondingly many standards, but R&D-intensive areas without corresponding patent protection intensity produce relatively few standards. If property rights have been defined for R&D expenses by means of patents, then the willingness to standardize is greater. If there are only limited protection possibilities in R&D-intensive branches, then the enterprises are not very willing to feed their internally generated knowledge into the public standardization process. If they need not to fear a diminution of their property rights in the standardization process due to adequate patent protection, or even envisage an increase in value of their protected intellectual property, they will step up their standardization activities.

However, in another version (Model 1a), where the square of the patent applications numbers are included, the significant results are confirmed and the square variable goes into the equation with a negative sign. This underlines the hypothesis that the standardization processes are hindered in sectors with very many intellectual property claims. In a second estimated model (Model 2), the influence of the enterprise concentration and size on the branch-specific standards output should be examined separately. However, no significant results were obtained from the regression analysis. As regards the concentration of enterprises, the supply effect of a simpler standardization (procedure) with few enterprises and the demand effect of a higher need for many small firms are more or less equal. The fixed cost effect of a standardization department or working time devoted to standardization is – at least on this highly aggregated analysis level – also without importance, as the company size has no significant influence on the branch-specific standards output. Here also a second variation was estimated (Model 2a) with the square of the concentration index and the average size of the enterprises as additional variables. Whereas the size indicator is still not significant, the sign of the linear concentration variable becomes significantly positive, while the square is negative. This result does not confirm the findings of Link (1983), who can show for 881 product groups in the United States the positive impact of the market concentration after a certain threshold on the likelihood of the existence of a standard. On the contrary, this result shows that standardization activities at national standardization organizations increase with higher market concentrations[8] up to a certain level, whereas in highly concentrated sectors the propensity to standardize decreases. The explanation for this U-shaped relationship can be found in the fact that meanwhile industry consortia, especially of multinational companies, negotiate about technical specification and standards without involving the official standardization bodies to avoid the lengthy standardization process, especially when they control the markets and the technical specifications of the products.[9]

Table 11.2   Explanatory factors for standards output

	Model 1	Model 1a	Model 2	Model 2a	Model 3	Model 4	Model 5
Constants	140.54^b [2.49]	22.03 [0.36]	−152.96 [−0.35]	−5288.69^a [−1.91]	11.62 [0.14]	164.58 [1.258]	−6097.10^b [−2.34]
Patents	0.1529^b [2.65]	0.45^c [3.98]	0.1329^a [2.11]	0.46^c [3.61]	0.1256^b [2.268]	0.1521^b [2.547]	0.34^b [2.44]
(Patents)^2	–	−0.0001^b [−2.35]	–	−0.0001^b [−3.11]	–	–	−0.0001^a [−2.18]
R&D expenditure	−0.0292^b [−2.32]	−0.0247^b [−2.35]	−0.0319^b [−2.28]	−0.02^a [−1.94]	−0.0435^c [−3.07]	−0.0292^b [−2.25]	−0.03^a [−2.10]
Enterprise concentration	–	–	444.37 [0.73]	13107.84^a [1.86]	–	–	15025.84^a [2.25]
(Enterprise concentration)^2	–	–	–	−7989.19^a [−1.81]	–	–	−90092.27^a [−2.14]
Enterprise size	–	–	−0.0811 [−0.96]	0.18 [0.32]	–	–	−0.03 [−0.39]
(Enterprise size)^2	–	–	–	−0.0001 [−0.40]	–	–	–
Export rate	–	–	–	–	7.8673^a [2.06]	–	5.12 [1.43]
Import rate	–	–	–	–	−0.2396 [−0.13]	–	111.62 [0.69]
Capital coefficient	–	–	–	–	–	−9.7485 [−0.21]	−26.96 [−0.55]
R^2(adj)	0.22	0.47	0.17	0.54	0.33	0.17	0.59
F	3.52^a	6.26^c	1.90	3.99^b	3.25^b	2.22	3.85^b

Notes: t-values in brackets; ^{a} = level of significance < 0.10; ^{b} = level of significance < 0.05; ^{c} = level of significance < 0.01.

In a further estimated model (Model 3), the export rate and the 'import rate' should be added as further explanatory factors. Whereas the export rate enters the estimating equation significantly with a positive sign, the import rate provides no additional explanations. This means that export-intensive industries standardize more than less export-intensive ones. Thus, the hypothesis that standardization is a strategy of export-oriented branches of industry, receives a certain empirical support.[10] The import volume per industrial branch, on the other hand, has no empirical explanatory value for the standards output. This indicates that in industrial branches with small import rates standards are not used significantly, or rather successfully, to erect trade barriers, but also do not lead to higher imports through their informational function.

The capital coefficient of the branches of industry (Model 4) also has no significant share in explaining the output of standards, as there are no indications for the hypothesis of an increased standardization tendency in capital-intensive branches. Here the importance of compatibility standards in capital-intensive branches is counterbalanced by the quality and safety standards in labor-intensive branches.[11]

To summarize the results of the empirical tests of the theoretical hypotheses for Germany, the standardization activities of sectors are tested together in Model 5. Finally, they depend very much on their R&D intensities and the propensity to patent. Market concentration has a positive impact on de jure standardization up to a certain degree, whereas in highly concentrated markets companies are inclined to work out de facto standards. The cost aspect of standardization seems to be of little relevance because company size is no explanatory variable. However, export-intensive sectors are still likely to be more active in standardization, but without being significant in the comprehensive model, because of the close correlation with the concentration index.

11.4   DATA SOURCES OF AN INTERNATIONAL POOL MODEL

Because of the low number of observations the theoretical hypotheses on possible explanatory factors for sectoral standards output will be empirically tested on the broader basis of the following 20 industrial branches in Table 11.3 in the seven countries of Table 11.4, for which the data has been compiled and matched from various secondary statistics.[12]

In order to measure the extent of standardization in 20 industries of the seven countries, the stock[13] of standardization documents in the year 1995 will be referred to, because in the standards databank PERINORM the annual output of the standardization process cannot be determined for all of the selected countries in Table 11.4. As in the case of Germany, the database does not only include documents of the main national standardization institute like the BSI in the United Kingdom or AFNOR in France, but also of the other national standardization institutes. The standards data also contain those European and international standards which have been adopted in the domestic market. However, on the basis of the database information alone, it is not possible to identify in which country the initiative for an international or European standard originated, so this fact cannot be used as additional information.[14] Nevertheless, the total stock of standards can be divided in the two sub-groups of idiosyncratic standards and standards which have a reference to an international or European standard.

Table 11.3   List of 20 industrial sectors with ISIC codes

3100	Food, drink & tobacco
3200	Textiles, footwear & leather
3300	Wood, cork & furniture
3400	Paper & printing
351 + 352 – 3522	Industrial chemicals
353 + 354	Petroleum refining
355 + 356	Rubber & plastics products
3600	Stone, clay & glass
3700	Basic metal industries
3810	Fabricated metal products
382 – 3825	Non-electrical machinery
3825	Office machinery & computers
383 – 3832	Electrical machinery
3832	Electronic equipment & components
3841	Shipbuilding
3843	Motor vehicles
3845	Aerospace
3842 + 3844 + 3849	Other transport equipment
3850	Instruments
3900	Other manufacturing

The explanatory factors were compiled from the following data sources. On the basis of these 20 industrial branches, the R&D expenditure (in US$ million) of the enterprises in 1994 were taken from the ANBERD database, published by the OECD (1997a). The output indicator patents is depicted by the sum of patent applications by the inventors of the seven countries at the European Patent Office in the years from 1993 until 1995. The other data concerning the export and import intensities, the labor input and the capital formation compared to total production stem from the OECD STAN Database 1997 (OECD 1997d). As there are no data available on the concentration indices and the average company sizes in the industries of the seven countries, it is assumed that the German data is representative for the situation in all other six countries.[15]

Table 11.4   List of selected countries

- Germany
- Spain
- France
- The United Kingdom
- The Netherlands
- Japan
- The United States of America

11.5   EMPIRICAL RESULTS OF THE INTERNATIONAL POOL MODEL

In order to broaden the statistical basis to test the theoretical hypotheses, the following general pooling model (equation 11.2) over the seven countries and the 20 industries is used and tested, first by the OLS approach applying cross chapter weights[16] (WLS) and secondly by the seemingly unrelated regression method SUR, which is more efficient, if the disturbances of the country equations are correlated, because it takes account of the entire matrix of correlations of all the equations.

$$ST_{ic} = f(a_{0c}, Pat_{ic}, RDI_{ic}, Ex_{ic}, Im_{ic}, Ca_{ic}, Em_{ic}, Co_{ic}, Siz_{ic}, e_{ic}) \quad (11.2)$$

It is assumed that the proposed hypotheses are equally valid in the seven countries selected, in order to obtain a total of at least 136 observations, because of some missing variables.

The variables are defined as follows:

ST_{ic} = stock of standard (either total T or national N or adopted international I) in country c in industry i;

a_{0c} = fixed effect of country c;

Pat_{ic} = patent application of industry i in country c;

RDI_{ic} = expenditure of enterprises for R&D divided by value added in industry i in country c;

Ex_{ic} = export ratio (exports divided by total production) in industry i in country c;

Im_{ic} = import ratio (import divided by total production) in industry i in country c;

Ca_{ic} = capital intensity (gross capital formation divided by value added)in industry i in country c;

Em_{ic} = employment intensity (employed divided by value added) in industry i in country c;

Co_i = Gini coefficient of Germany in industry i;

Siz_i = average turnover of the enterprises in Germany in industry i;

e_{ic} = error term iid $N(0, \sigma_e)$.

The results of the different pool estimations are presented in Table 11.5.

The hypothesis of the standards as part of the R&D process can be affirmed by the empirical results. In particular, the patent applications are very significant in explaining the standard variable, whereas the R&D variable can only explain the total number of standards significantly. When we use both the input indicator R&D expenditure and the output indicator patent application, only the latter remains a significant explanatory variable for the standards, whereas the R&D indicator shows a negative sign.

In order to show the negative impact of too many patents for the standardization process, we add additionally the squared number of patents in the estimated equation. The empirical results underline the theoretical hypothesis that patent protection makes it easier for companies to propose a new standardization project or to participate in an already ongoing standardization process. Sectors with a high R&D intensity and low patent protection tend to standardize less because of the uncontrollable spill-over effects on other participants and competitors. Additionally, too many patents increase the probability that one part of the technical standard is injuring the patent rights of a company which may be not willing to license the patent for a reasonable amount. Then the standardization process will fail.

Whereas the import rate does not have any significant impact on the stock of standards in the respective industries, the export rate has, at least in the SUR model, the expected positive sign. Furthermore, neither the capital nor the employment intensity of the sectors are able to explain standardization output significantly. However, the enterprise concentration is positively correlated with the intensity of standardization up to a certain threshold, above which industries tend to produce less official standard documents and probably prefer informal industry standards. This result corresponds with the initial positive impact of the average company size of the industrial sectors, which turns into a negative influence in industries with very big companies.

Comparing the overall goodness of fit between the three explained variables total stocks, idiosyncratic national stocks and international stocks of standards, the latter are obviously influenced by other factors, especially by the institutional arrangement in Europe to integrate all European standards into the national system of standards. However, due to the increasing importance of European and international standards and the decreasing significance of idiosyncratic national standards, empirical analyses in the future will have to take this development into account.

Table 11.5   Explanatory factors for the different stocks of standards

Method	WLS	WLS	WLS	SUR	SUR	SUR
Dependent Variable	STT_{ic}	STN_{ic}	STI_{ic}	STT_{ic}	STN_{ic}	STI_{ic}
Patents	0.376^c	0.220^c	0.088^b	0.355^c	0.234^c	0.076^b
	[4.363]	[4.092]	[2.164]	[5.021]	[4.490]	[2.352]
(Patents)^2	$-3.59E{-}05^c$	$-1.56E{-}05^b$	$-1.19E{-}05^b$	$-3.09E{-}05^c$	$-1.66E{-}05^b$	$-8.98E{-}06^b$
	[−3.467]	[−2.411]	[−2.431]	[−3.489]	[−2.608]	[−2.301]
R&D intensity	−898.710	-1286.866^c	419.509	-1124.183^c	-1492.401^c	550.197^b
	[−1.525]	[−3.609]	[1.358]	[−2.627]	[−4.763]	[2.200]
Export rate	27.717	50.448	−17.728	93.824	85.333^b	91.229^a
	[0.218]	[0.854]	[−0.193]	[1.303]	[2.063]	[1.945]
Import rate	−1.492	−13.228	26.117	−9.875	−6.509	−48.966
	[−0.017]	[−0.300]	[0.454]	[−0.181]	[−0.206]	[−1.536]
Capital intensity	353.924	199.805	148.026	185.991	16.817	−8.925
	[0.764]	[0.740]	[0.626]	[0.542]	[0.077]	[−0.044]
Employment intensity	−958966.8	1593545	−3590944	−742069.6	1810714	−23777864
	[−0.199]	[0.656]	[−1.238]	[−0.216]	[0.868]	[−1.395]
Enterprise concentration	30464.67^c	12335.75^b	9390.285^a	28193.24^c	17860.63^c	5244.292
	[3.263]	[2.417]	[1.799]	[4.103]	[3.677]	[1.141]
(Enterprise concentration)^2	-19914.78^c	-8373.510^b	-5981.693^a	-18581.45^c	-11980.03^c	−3419.250
	[−3.347]	[−2.567]	[−1.802]	[−4.264]	[−3.890]	[−1.166]
Enterprise size	2.708^c	2.165^c	0.066	2.695^c	2.426^c	0.209
	[4.212]	[6.187]	[0.182]	[5.795]	[7.180]	[0.645]
(Enterprise size)^2	-0.001^c	-0.001^c	$-7.13E{-}05$	-0.001^c	-0.001^c	−0.0001
	[−4.570]	[−6.411]	[−0.459]	[−6.147]	[−7.410]	[−0.779]
R^2(adj)	0.52	0.49	0.37	0.50	0.59	0.30
F	16.56^c	14.83^c	9.60^c	–	–	–

Notes: *t*-values in brackets; ^{a} = level of significance < 0.10; ^{b} = level of significance < 0.05; ^{c} = level of significance < 0.01.

11.6 SUMMARY

In the theoretical discussion a number of driving forces for standardization were elaborated and formulated as hypotheses to test. The econometric analysis both of Germany and of the pool of seven countries showed very similar results. The output of official standards in the industrial branches observed can be explained primarily by the branch-specific patent applications and the R&D expenditure. More patent-intensive branches tend towards a higher number of standards, whereas R&D-intensive sectors with low patenting tendencies are, in principle, more reserved towards standardization. In addition, the export rate was merely a significantly positive explanatory factor, which underlines the importance of standards for the export-intensive branches. In particular the high output of standards in technology- and export-intensive sectors, which play an important role for the economic competitiveness of highly industrialized countries, is a remarkable indication of the significance of standardization for the economy as a whole.[17]

Derived from these results some preliminary recommendations can be formulated for the standardization policy of national standardization organizations. First, the growing importance and speed of R&D and innovations for the competitiveness of Germany in the age of globalization should be taken into account in the standardization strategies by setting priorities on innovative areas and by adjusting the existing stock of standards to keep up with the state-of-the-art of science and technology. In order to keep abreast of the higher speed of technical change, the standardization process needs to be closer to the R&D process. In this context, with the standardization at the R&D stage and pre-standards, a first step is made in the right direction. However, these developments are still questioning the role of IPR especially in standard-intensive system technologies.

Because of the low propensity to standardize in sectors with a small market concentration on the one hand and the higher demand for compatibility in these sectors, strategies should be elaborated which improve the access to standardization in general for the small and medium-sized enterprises of these sectors and facilitate the standardization process. Furthermore, although the de facto standardization cannot be prevented, at least incentives should be provided to make the details of the technical specifications public in order to reduce information asymmetries and wasted R&D invested in incompatible solutions by other companies which are not members of the consortium. A first approach is the new opportunity to publish so-called public available specifications (PAS).

Finally, export-intensive sectors and companies should be supported to start European or international standardization projects, in order to transform or integrate their R&D results into common technical standards and to provide them with at least temporary cost and quality advantages compared to their competitors abroad, without being non-tariff trade barriers.

However, these general recommendations have to be implemented by specific measures. In order to reach effective and efficient solutions, in-depth analyses for the different industries and technologies are needed, which represent a challenge for further research activities.

NOTES

[1] Cf. Appendices II and III.

[2] There is no adequate qualitative and quantitative information available on the input in the standardization processes.

[3] See Wissenschaftsstatistik GmbH (1998, table 22, p. 33). The data on the R&D survey are only available for the years with odd numbers. As the data from 1993 are still classified according to WZ79 and are also not so deeply disaggregated, we referred to the data from 1995.

[4] See Statistisches Bundesamt (1996) Fachserie 4 (series 4.2.3, pp. 42-51). The enterprise concentration is measured by the Gini coefficient (cf. Statistisches Bundesamt 1996, p. 7 for the definition).

[5] The famous DIN A4 standard must not be revised periodically, because of its widespread use and the tremendous follow-up costs for computers, printers and copying machines.

[6] The danger of heteroskedasticity, that is non-constant error terms, of cross-section analysis can be avoided with a weighted estimation (cf. Pindyck and Rubinfeld 1991, pp. 127 ff.).

[7] In a simple correlation analysis it turned out that both patent applications and R&D expenditure are positively correlated with the number of standard output, which is in line with the results of Lecraw (1984).

[8] This confirms the results of Lecraw (1984), who finds for 252 products that standard usage is a function of buyer and seller concentration.

[9] Compare David and Shurmer (1996).

[10] This effect is fostered by the fact that the standard numbers include also the adopted European standards, which are likely to affect especially international and therefore also export-relevant issues.

[11] However, Lecraw (1984) can confirm in his analysis that the importance of product quality for health and safety has an impact on the use of standards.

[12] Compare Appendix III.

[13] The stock of standards is equal to the totality of previous annual standard outputs corrected by the number of withdrawn documents. Additionally, the correlation between patent applications and annual standard outputs has remained almost constant in the last 10 years.

[14] Blind (2002c) proves that there is no systematic way to identify the national origin of European or international standardization processes.

[15] Regressions without these additional data do not change the signs of the other explanatory variables and lead to lower goodnesses of fit.

[16] In order to avoid heteroscedasticity the weights are estimated in a preliminary regression with equal weights and then applied in weighted least squares in the second round.

[17] For the significant contribution of the stock of standards to long-term economic growth in Germany, see Chapter 18.

12. Special Focus: The Impact of Intellectual Property Rights on the Propensity to Standardize at Standardization Development Organizations[1]

12.1 INTRODUCTION

In this chapter the impact of IPR on the propensity to standardize is analyzed in a more detailed and focused manner, based on the empirical data used for the analyses presented in Chapter 11, because the interface between IPR and standardization is a challenge from a theoretical perspective and of high priority for policy makers. Furthermore, it includes a discussion of both the role of different formal and informal strategies to protect intellectual property and of the institutional conditions to handle with IPR in standardization processes. Finally, the observations at sectoral level are supplemented by the experiences of a sample of European companies with their IPR in standardization processes. The chapter closes with first policy recommendations concerning the interrelationship between IPR and standardization or between microeconomic incentive structures and macroeconomic public good concerns.

In January 1998, European manufacturers and network operators, together with Japanese companies, agreed on a uniform standard third generation of the cellular telephone system UMTS, which in one to two years should complete the present GSM standard in the European Standardization Institute ETSI. However, this was being contested by the US firm Qualcomm, which claimed the basic patents of the standard for itself and also decisively co-developed the American CDMA 2000 standard. This topical case is an excellent example of the influence of IPR on standardization. Standardization organizations deal with the problems of integrating protected knowledge in standards as a rule by requiring that the enterprises and individuals involved in the standardization process disclose the patents and copyrights which affect the object of the standard or the technical rule. Should patent protected technologies or protected knowledge become established in a committee standard, many organizations prerequire the right owner to make an advance declaration of willingness to sell licenses at reasonable terms. The problematical part of this procedure is naturally – as the example underlines – that it cannot be guaranteed that the enterprises not involved in the standardization process disclose their industrial property rights before the standardization process. They can perhaps wait until the committee standard has been decided on, published and already widely applied in industry. In this case, the competitors and potential customers have already made considerable investments in the standardized technique. If an enterprise, like Qualcomm, discloses that parts of the standard are protected by their IPR after the investment phase, then the further use of the standard is dependent on paying license fees and the patent holder can appropriate a part of the economic rents connected with competitors' and customers' investments.

The ambivalence of industrial property rights and public available standards for technological development is triggered by the contradiction between static and dynamic efficiency considerations.[2] For the generation of new knowledge, the inventors are awarded exclusive property rights due to dynamic efficiency aspects. The monopolistic effect provides incentives for the production of new knowledge, by enabling the innovators for a limited period of time to sell the innovative products over the marginal cost level of the competitors and thus to achieve adequate compensation for the outlaid R&D expenditure.[3] As, however, the economic benefit of new technologies is based on their wide diffusion and parallel developments are macroeconomically undesirable, the exclusive protection ceases after a certain period and the knowledge is at the disposal of imitating competitors for free in order to respond to static efficiency.[4]

In contrast to the property rights, formal standards are decisive for the diffusion of new technologies.[5] They make information about new technologies available to everyone, for a small fee, and come near to being a classical public good, which is particularly distinguished by non-rivalry in consumption and application.[6] The economic benefit is optimal if all economic units have free access to the public good.

To sum up, it must be said that the economically optimal, strong property rights in the phase of knowledge generation must be relaxed at the beginning in the stage of wide diffusion of innovative technologies. From this it can also be derived that in the standardization process, property rights must be at least coordinated,[7] better moderated for the promotion of the diffusion, in order to enable new standards to be produced.

The discussion of the effects of IPR on standardization has shown that the economic theory already addresses the relevant topics. Deficits are perceptible in the empirical foundation and validation of the theoretical models and hypotheses. If there is empirical proof of the role of IPR at all, it is mainly restricted to specific case studies. Empirical studies, which rely on a broad sample of standards, concerning the influence of IPR on standardization are not available.

The lack of broad empirical studies on the impact of IPR on standardization lies in the data problem. These do not concern property rights like patents, but are due to the different kinds of standards.[8] In our analysis, we will focus on the standard documents generated by the national SDOs, like in the previous analyses.

The scope of this chapter is to bridge the gap between the theoretical insights about the impact of IPR on standardization and the lack of empirical results of an aggregated international cross-section level. Based on theoretical hypotheses concerning the impact of IPR on standardization, this chapter examines empirically, for the set of seven countries and 20 sectors – the database used in the previous chapter – the impact of R&D expenditures and patent protection on the standardization of processes and products at official national standardization bodies. Furthermore from a micro-perspective of European companies their experiences and problems are reported.

This chapter is structured as follows. First, the different protection strategies concerning innovations are discussed and their use in the German manufacturing industries are presented. After the institutional regulations concerning the role of IPR in standardization processes in Section 12.3, theoretical hypotheses are derived (Section 12.4). In Section 12.5, some remarks on the data are made and the results of the regression analyses – similar to the approach in Section 11.5 – are represented. The purpose of these similar insights at sectoral level is to enrich them by survey results about experiences of European companies with the role of IPR in standardization processes. A summary and general recommendations for the standardization policy of SDOs are presented in the final Section 12.7.

12.2   THE MOST IMPORTANT INTELLECTUAL PROPERTY RIGHTS

A number of legal possibilities can be deployed to protect the results of companies' in-house R&D processes. In the following, the most important legal bases in this study for the protection of intellectual property, as well as the legal possibilities to protect industrial secrets – such as patent law, utility or petty patent law or copyright law – will be shortly characterized,[9] before the intensity of their use in Germany is presented.

12.2.1   Patent Law

Recourse can be taken to patent law when an invention results from new knowledge. An invention must have the following characteristics, according to §1 paragraph 1 of the German Patent Law (PatG), in order to achieve patentability:

- The invention must be new.
- The invention must be based on an inventive activity.
- The invention must be suitable for industrial/commercial application.

The obviously central concept *invention* is, however, not specified more closely in the law. The Federal Supreme Court has defined the patent law requirements of an invention as follows: an invention is 'lesson for systematic action using controllable natural forces – outside the reach of human understanding – for the direct cause of a causally assessable success'.

The recourse to natural forces thus limits the applicability of patent law directly to technical innovations.

The granting of a patent is handled differently in different states. Whereas in some countries the date of the invention is all-important, in other states the date of patent application is the basis.[10] All the various forms of patent law have, however, one thing in common, that this right is awarded for a limited period of time. In some states the patent will be renewed only on payment of an increased fee, in order to remove worthless inventions for the inventor from the patent pool, or to internalize the increasing costs of a too small diffusion.

12.2.2   Utility Patents

The utility patent law is, like patent law, geared to technical innovations and is often characterized as 'little patent' or 'petty patent'. It is much quicker to achieve protection by means of an industrial or utility patent, but this lasts for a maximum of ten years (three years initial duration, with a maximum extension possible of a further three years – once – and two years – twice [§23 Utility Patent Law]). The fundamental difference to patent law is that the registration of the utility patent is made without further examination. It is a considerably faster form of protection.

12.2.3   Copyright Law

Copyright law concentrates on the protection of the expression of creative performances, the so-called *works*, which exist in words, images or sound.

The creator of a work is awarded the copyright protection for the period of 70 years after his death. By the 'Second Law to Amend the Copyright Law' of 9 June 1993, the call for a form of protection for computer programs, as raised in the European Directive 91/250, was included in German law. From this date computer programs can be protected under §69 of the Copyright Law.

The Copyright Law tries to differentiate between the 'expression' and the 'idea'. Thus in principle the expression of an idea can be protected, but not the idea itself. This procedure is unproblematic, as long as an idea can be expressed by more than one possibility. This is certainly the case for a great number of creations – especially in those areas which were the original target of the Law of Copyright, such as literature and music. In copyright a further distinction can be made between personal and non-personal rights. The former apply for example to the right of acknowledging authorship or the right of invariation of the work. The latter deals with the commercial utilization of the works; they are therefore called the exploitation rights. Whereas there is no possibility, for example, of selling the right of authorship, every creator is entitled to sell the exploitation rights to his work.

12.2.4 Industrial Secrets and the Law Against Unfair Competition

Besides these laws which directly protect the standardized technology, there is the further possibility of classifying information about the standard as industrial or company secrets in the sense of §§17 and 18 of the Law against Unfair Competition (UWG), and thus render it inaccessible to the general public. In §20 of the UWG, not only the divulging of company secrets but also the incitement to divulge secrets is a punishable offence. Company secrets are understood not only to be information that has been exclusively generated in the company, but which are in principle also generally available, but whose determination and collection is linked with costs.

Besides these legal instruments to protect product or process innovations, other strategies exist. First of all, it is possible to try to score a time advantage over your competitors in the marketing of the products or in the using of innovative processes. Secondly, the complexity of products and processes may also be a strategy to hinder competitors from imitating own innovations. Finally, because important information concerning innovative ideas is tacit knowledge of the employees, companies attempt to make long-term contracts with their innovative labor force.

Figures 12.1 and 12.2 show the intensity of the use of these different legal and informal strategies in 13 German industries. In general, the time advantage and the long-term involvement of the staff are the most favored strategies, both for product and for process innovations.[11] The secrecy and the complexity of products and processes follow. Patents and other protection rights are less important. However, they have – as expected – a higher importance for product innovations than for processes. Between the sectors, they are widely used in the chemical industry and in the non-electrical machinery branch. In the less innovative sectors, such as mining, construction and food, the use of patents to protect innovations is under-represented.

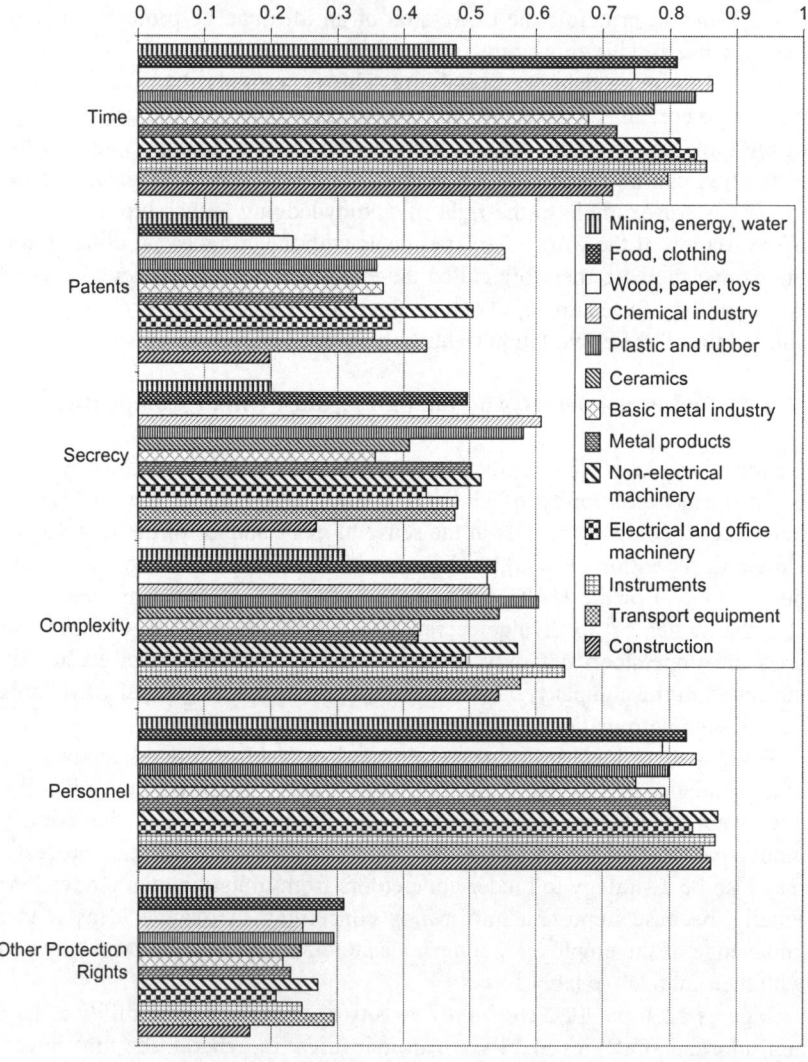

Figure 12.1   Share of companies which use different protection strategies for product innovations in 13 industries[12]

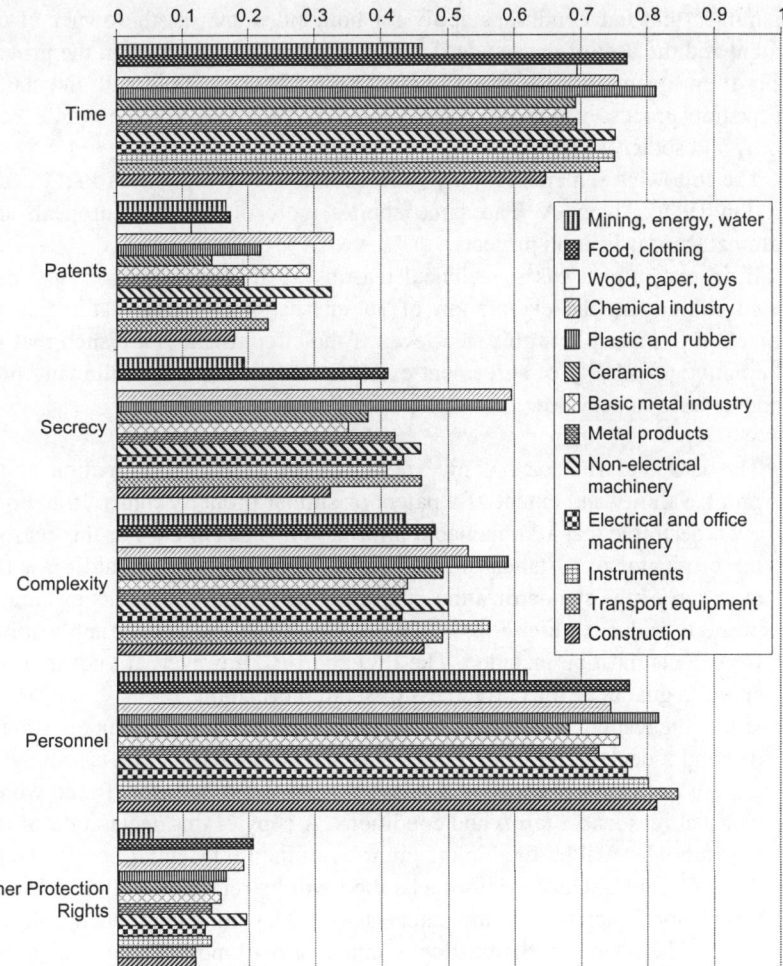

*Figure 12.2   Share of companies which use different protection strategies
for process innovations in 13 industries*

12.3   INSTITUTIONAL FRAMEWORK OF NATIONAL AND INTERNATIONAL STANDARDIZATION BODIES

In principle, the application of patented solutions is possible in the development of standards in SDOs. However, procedures which are practically identical in all standardization organizations must be utilized, in order to guarantee that fair conditions apply for both sides, that is, the owner of the patent and the user of the standard.[13] The question arises, how can the protection of an invention by a granted patent be made compatible with the standardization process which is acknowledged as beneficial for the whole economy and society?[14]

The following remarks roughly reflect the ISO/IEC Directives Part 2, first edition 1989, Annex A. The same applies more or less for European and national standardization projects.

If, in exceptional cases, technical reasons justify the utilization of a patented solution in the elaboration of an international standard, there is no objection in principle to this step, even if the circumstances are such that no alternative possibility of agreement exists. In such a case, the following procedures have to be followed:

1. The ISO and IEC cannot give definitive or detailed information about proof, validity and extent of a patent or similar property rights. It is, however, desirable that all obtainable information is disclosed. For this reason, the originator of a standardization proposal of this kind should draw the attention of the standardization committee to all known patents or similar property rights, whereby the situation worldwide and known applications for patents must be included. The ISO and IEC, however, are not in a position to guarantee the correctness of such information.
2. If the suggestion is accepted on technical grounds, the originator should demand a declaration from each known patent holder, that he is willing to grant a license for patents or similar property rights for users in the whole world at reasonable terms and conditions. A copy of this declaration of the patent holder will be filed in the archives of the ISO Central Secretariat or the IEC Central Office. This declaration will be referred to in the relevant international standard. If the patent holder does not submit such a declaration, then the standardization committee will not continue with the incorporation of the patented solution in the draft standard, unless the competent advisory council agrees.
3. Should it become known after the publication of the international standard that licenses for a patent or similar property rights will not be granted at reasonable terms, then the international standard will be withdrawn by the standardization committee for revision.

The Commission of the European Communities' view of the licensing option was stated earlier in 1992 (European Commission 1992): compulsory licensing would be likely to reduce investment in R&D in affected sectors; non-EU firms would keep their technologies away from the EU market, and low-cost equipment manufacturers outside the EU would benefit from cheap licenses to use indigenously developed technology. However, ETSI's policy requires holders of such rights to disclose them within 180 days after the standard is put into an ETSI work program.[15] If the holder chooses not to license, and no other technical design is found, then a dispute settlement mechanism is provided.

The tension between this procedure and the Commission's desire not to restrict a property right holder's freedom except in exceptional circumstances is evident. Only after a 'relevant market' has been legally specified and the IPR claimant has been found to have prevented the production and marketing of a new product for which there is potential consumer demand, and to have withheld a license in order to secure a monopoly in a derivative market, can a finding be made in favor of a challenging party. Mazzoleni and Nelson (1998) therefore question the economic benefit of strong patent protection in systems technologies where development relies mainly on their efficient and fast diffusion by means of standards.

Not only is the likelihood of a finding in favor of a challenger low because of the difficulties inherent in specifying a relevant market and in demonstrating the presence or absence of potential consumer demand for innovative technologies, the time required to process such disputes can result in de facto monopoly for the right holder during the period in which a technical design remains of high priority in the technical design and pre-standardization process. The link between long-run competition strategies and the technical specification of designs as candidates for standards agreement are clearly visible, although no in-depth research on the implications has been carried out as yet (Mansell 1995, pp. 221 ff.).

12.4 THEORETICAL HYPOTHESIS CONCERNING THE IPR ON THE INTENSITY TO STANDARDIZE AT SDOS

The standardization process can be regarded as the extension of the competitive product development process.[16] After the decision concerning the R&D expenditure is taken, the firm has to decide, in a second step, about the protection of its product innovation by going through the patenting process or by using another of the presented strategies. Finally, the firm has to decide on the number of product and process innovations it is going to propose for a standardization process or to reach a conclusion about participating in ongoing standardization processes.

The expected benefits of a standardized product are advantages in its diffusion and therefore a higher anticipated demand. On the supply side, the participation in the standardization process may reduce the distance and therefore the switching costs between the specifications of the standard and the technical features of the firm's products and processes.[17] Additionally, outsiders of the standardization process face higher adaptation costs and probably a competitive disadvantage.[18]

The costs include the actual financial cost of a standardization process, including the opportunity costs of delayed marketing of the product.[19] Finally, the company has to publish their R&D results, which makes private knowledge public knowledge, first available to the participants of the standardization process, later to all buyers of the documents. The knowledge spillovers will be higher when there is no protection of the R&D results at all. However, due to the outlined institutional framework, the patent protection cannot prevent other companies using the technology, but it may at least control the knowledge spill-overs, because the company has to license the patent for a reasonable amount to the public.

Based on these considerations, the following sector-specific hypotheses concerning innovation and standardization are postulated, which have been discussed in a similar manner in Chapter 10. Because standardization is a part of the R&D process, the higher the R&D intensity of a sector, the higher the annual standardization output will be. Therefore, both the input indicator R&D expenditure and the output indicator patent applications should positively explain the annual output of standards. However, because of the spill-over problem of the standardization process, the sectoral propensity to standardize should be explained better by the number of patent applications compared to the R&D expenditure. Furthermore, the higher the ratio of patents to R&D expenditure, the more the R&D results are protected by property rights and the higher the incentives to participate in standardization processes. However, due to the IPR problem, the standardization process can be prolonged or can even fail because patent holders are not willing to license their IPR.[20] Therefore, sectors with a very high number of patents tend to standardize more slowly because of the negotiations concerning the licensing questions and in the case of unwilling patent holders not at all. However, the total output of standards will be lower.[21] These effects are expected to be stronger for international standard processes because they are likely to be affected by a higher number of potential patent holders. However, the subgroup of idiosyncratic standards should be explained better by the proposed variables compared to the stock of international standards integrated into the national standardization systems, because of the looser link between national R&D and international standardization and the obligation to take over European standards for the member countries of the European Union.

12.5 EMPIRICAL RESULTS AT SECTORAL LEVEL

The theoretical hypotheses on possible explanatory factors for sectoral standards output will be empirically tested on the basis of the same 20 industrial branches as in Chapter 11, for which the data has been compiled and matched from various secondary statistics.[22] In order to measure the extent of standardization, the stock of standardization documents in the year 1995 will be referred to, because in the standards databank PERINORM a database is available for the selected countries in Table 11.4, which reflects the output of the standardization process regarding both the content and time perspective. The explanatory factors were compiled from the following data sources. On the basis of these 20 industrial branches, the R&D expenditure (in US$ million) of the enterprises in 1994 was taken from the ANBERD database, published by the OECD (1997a). The output indicator patents is depicted by the sum of patent applications by the inventors of the seven countries at the European Patent Office in the years from 1993 until 1995.

In order to test the theoretical hypotheses, the following general pooling model over the seven countries and the 20 industries is used and tested by an OLS approach applying cross section weights:[23]

$$ST_{ic} = f(a_{0c}, Pat_{ic}, Pat_{ic}^2, RD_{ic}, Pat_{ic} / RD_{ic}, e_{ic}) \qquad (12.1)$$

It is assumed that the proposed hypotheses are equally valid in the seven countries selected, in order to obtain a total of at least 138 observations, because of two missing R&D variables.

The variables are defined as follows:

ST_{ic} = stock of standard (either total T, or national N or adopted international I) in country c in industry i;
a_{0c} = fixed effect of country c;
Pat_{ic} = patent application of country c in industry i;
RD_{ic} = expenditure of enterprises for R&D in country c in industry i;
e_{ic} = error term.

The results of the different pool estimations are presented in Table 12.1. The hypothesis of the standards as part of the R&D process can be affirmed by the empirical results. Especially, the patent applications are very significant in explaining the standard variable, whereas the R&D variable can only explain significantly the total number of standards. When we use both the input indicator R&D expenditure and the output indicator patent application, only the latter remains a significant explanatory variable for the standards, whereas the R&D indicator shows a negative sign. This result is confirmed by

Table 12.1 *Result of the weighted pooled estimation*

	STT	STN	STI	STT	STN	STI	STT	STN	STI	STT	STN	STI	STT	STN	STI
Pat	0.09c [3.68]	0.08c [3.43]	0.01b [2.33]	—			0.11c [3.54]	0.10c [3.57]	0.01 [1.61]				0.37c [4.17]	0.20c [3.09]	0.09c [2.80]
Pat²							−0.02 [−1.20]	−0.02 [−1.42]	−0.00046 [−0.13]				−0.00003c [−2.98]	−0.00001 [−1.60]	−0.00001c [−2.61]
RD				0.02a [1.66]	0.0156 [1.369]	0.00253 [1.04]							−0.0311a [−1.67]	−0.025a [−1.73]	−0.00158 [−0.25]
(Pat/RD)										26.52 [1.36]	17.63b [2.54]	3.18 [0.51]	—	—	—
R²(adj)	0.16	0.13	0.22	0.08	0.06	0.20	0.15	0.14	0.21	0.05	0.07	0.05	0.24	0.17	0.32
S.E.	708.27	493.16	320.52	750.29	524.28	319.52	713.52	494.08	319.81	756.90	524.25	357.27	662.70	460.00	313.39

Notes: *t*-values in brackets; [a] = level of significance < 0.10; [b] = level of significance < 0.05; [c] = level of significance < 0.01.

the positive coefficient of the patent–R&D ratio of the sectors in a further model.

In order to find statistical evidence for the negative impact of too many patents for the standardization process, we add the squared number of patents in the estimated equation. The empirical results underline the theoretical hypothesis that patent protection makes it easier for companies to propose a new standardization project or to participate in an already ongoing standardization process. Sectors with a high R&D intensity and low patent protection tend to standardize less because of the uncontrollable spill-over effects on other participants and competitors. Additionally, too many patents increase the probability that one part of the technical standard is injuring the patent rights of a company which may be not willing to license the patent for a reasonable amount. Then the standardization process will fail.

12.6   EMPIRICAL RESULTS AT COMPANY LEVEL

Insights from a company survey on the role of IPR for standardization may confirm or question the empirical results at a sectoral level, especially the negative impact of too many patents on standardization. The empirical results presented are based on the survey data deriving from a study on the 'Interaction of standardization and Intellectual Property Rights' carried out by the Fraunhofer Institute for Systems and Innovation Research for the Directorate General Research of the European Commission.[24] The survey focused on problems within the standardization process due to conflicts with IPR, because the technical specifications of a standard may touch the IPR of one or more patent-holders. The answers originate from a survey of a sample of European companies. Companies have been approached, which are members of CEN, participants of the 5th Research Framework Programme and randomly drawn from a commercial database. In total, the data rely on 159 reasonably filled out questionnaires. The sample provides a representative coverage of the European Union, including in addition some respondents from non-EU countries (Iceland, Russia, Norway, Liechtenstein and Switzerland). The participants from the sample show very diverse business activities. This is one reason why it was not straightforward to split the sample into different industries. The sample covers an equal distribution of small, medium and large companies. However, this distribution is not representative for the real distribution of the number of companies in the European Union, which is dominated by small and medium-sized companies.

The following descriptive statistics summarize the problems companies have with IPR in standardization processes. First, Figure 12.3 makes it obvious that more problems are due to others' rights and that patents are the most relevant IPR in standardization processes. Over 30 per cent of the companies indicate that they had problems with own patents and over 40 per cent of them had problems with the patents of others. Particularly large companies, and patent- and R&D-intensive companies more often attest problems with own and foreign patents. Problems with trademarks, copyrights, and trade secrets[25] are less likely in standardization processes. However, a significant share of companies has experienced problems with IPR in standardization processes, which indicates that the negative impact of too many patents found on the sectoral level is reflected at the micro level.

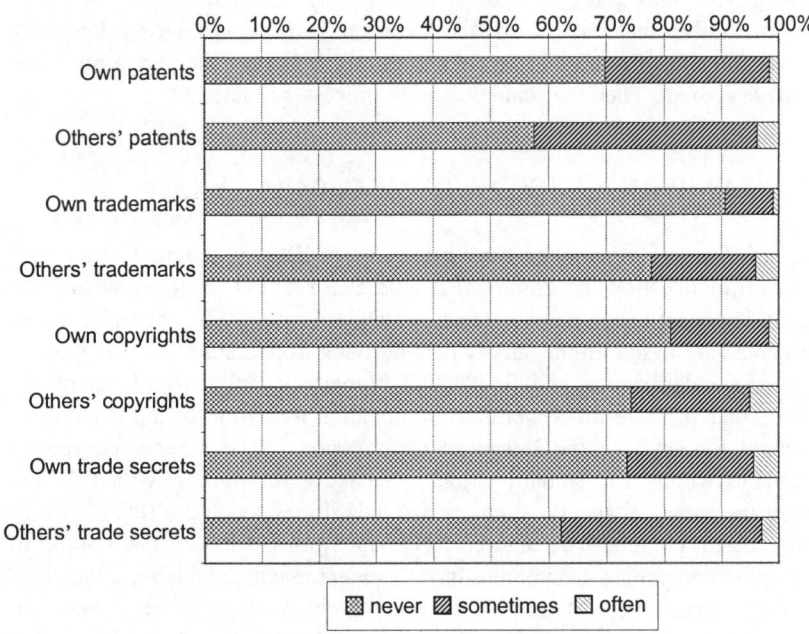

*Figure 12.3   Companies' problems with IPR in standardization processes
             (share of companies in per cent)*

In order to elucidate the kind of problems, various options concerning own and foreign IPR are possible (Figure 12.4). In general, the likelihood that the licensing conditions were not accepted, that the own technology was circum-vented, and that the own IPR was infringed, has been equally high, at around 30 per cent. The technology of companies with low R&D intensity has even been circumvented in over 40 per cent of the cases. The same is true for large companies. Furthermore, over 40 per cent of the large companies indicate that their licensing conditions have not been accepted and more than 35 per cent of patent-intensive companies have experienced infringements of their IPR.

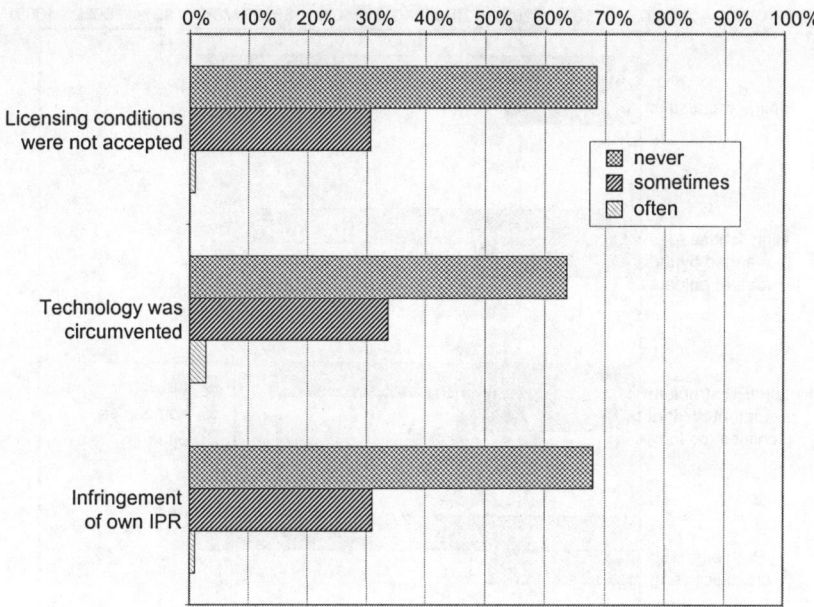

*Figure 12.4   Kind of problems with own IPR in standardization processes
(share of companies in per cent)*

Similar to problems with own IPR, problems with foreign IPR may arise in standardization processes (Figure 12.5). In total, 30 per cent of the companies indicate an experience with infringement suits, too high license fees, unclear IPR structures and problems with cross-licensing. However, companies with a high R&D intensity are in general more likely to get involved in conflicts with foreign IPR. Large companies compared to small companies face more infringement suits, higher license fees and less clear IPR structure. Furthermore, these problems are sector-specific and especially crucial in the radio, television and electrical engineering sectors.

We have seen manifold problems with own and foreign IPR in standardization processes. However, some solutions to overcome the conflicts do exist. Nevertheless, almost 50 per cent of the companies indicate that they have never found a solution to their conflicts. This underlines that there is a real problem with IPR in standardization processes. In particular, over 55 per cent of companies with high R&D intensities and medium-sized companies were not successful in reaching a solution for their IPR-related problems. The further observation at the micro level is in line with the negative impact of too many patents at the sectoral level assuming a positive relationship between R&D intensity and patenting.

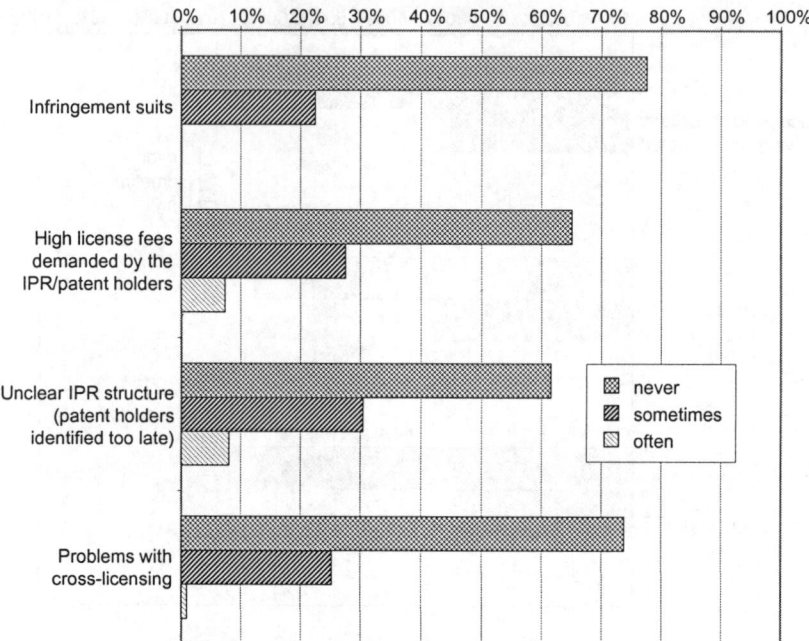

*Figure 12.5   Kind of problems with foreign IPR in standardization
            processes (share of companies in per cent)*

Figure 12.6 illustrates that the purchase of licenses and the circumvention of
protected technologies are the most popular strategies of solving the IPR-
related problems. Joining a patent pool, and the aggressive strategies of
mergers with, and acquisitions of, the patent-holding companies are less
preferred measures. Less R&D-intensive companies favor more strongly to
purchase a license instead of developing an own solution. Patent pools are
more likely in the radio, television, and electrical engineering sectors, due to
the need to rely on different components in one standard.[26]

We have already seen that almost 50 per cent of the companies have not
solved the conflict with IPR in the standardization process. Figure 12.7 makes
it obvious that too high costs for licenses is the most important reason. The
other reasons, like the failures to circumvent the technology and to create a
patent pool, have an importance below average. The merger with or the acquisition of the patent-holding companies are only very rare solutions. For
small and for R&D-intensive companies all other reasons have a higher importance. This result confirms that IPR-related problems are most crucial for
the R&D-intensive and the small companies. Consequently, all policy conclusions have to consider the special needs of these target groups.

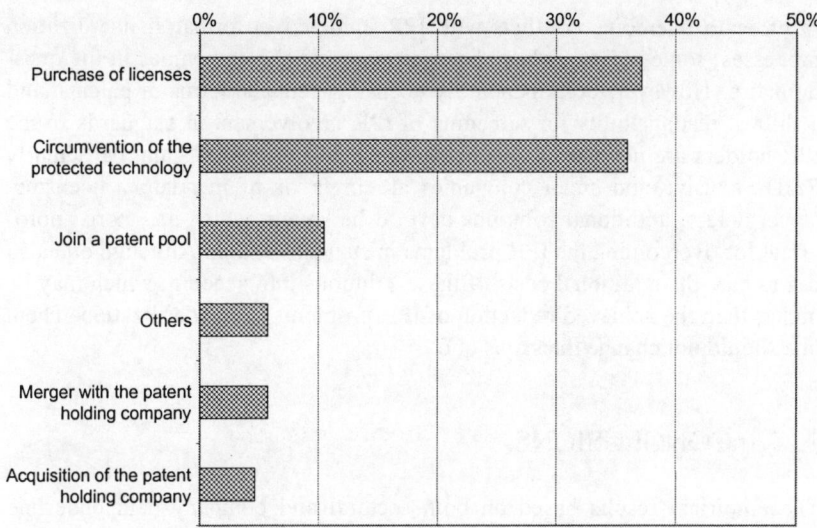

Figure 12.6   Solutions for problems with IPR in standardization processes (share of companies in per cent; multiple answers possible adjusted to 100 per cent)

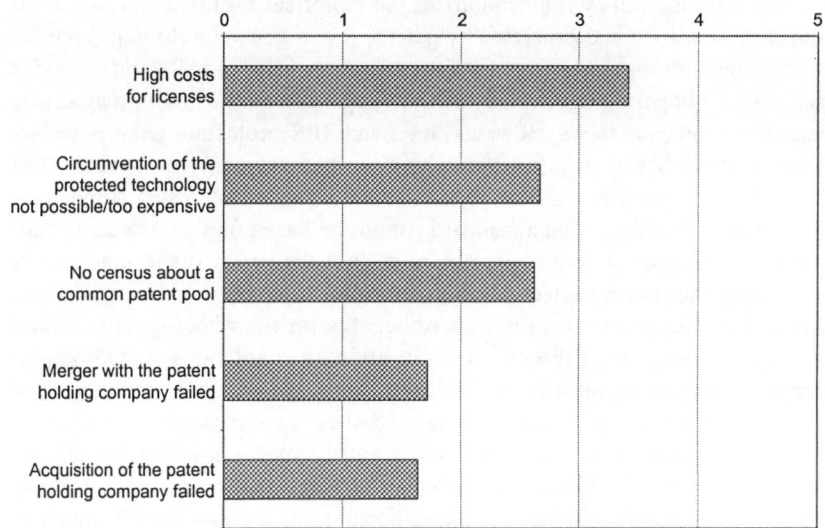

Figure 12.7   Reasons for not overcoming the problems with IPR in standardization processes (very low = 1 to very high = 5)

In order to overcome conflicts with IPR involved in formal standardization processes, some measures have been proposed to the companies in the questionnaire. However, both mandatory licensing, reduced terms of patents, and a shift of responsibility for screening of IPR involvement in standards to the IPR-holders are not assessed as being adequate solutions. Again, particularly R&D-intensive and small companies are in favor of mandatory licensing. Nevertheless, additional solutions have to be sought which are more appropriate for overcoming the IPR problems in standardization. Although one also has to take the additional costs of these solutions into account, which may be higher than the achieved reduction of IPR problems in standardization. Then, one should not change the status quo.

12.7   CONCLUSIONS

The empirical results based on both sectoral and company data underline again the dichotomy between individual rationality and social rationality. Whereas for the individual inventor in a sector it is advantageous to patent his innovation and maybe to participate in the standardization process, the whole sector is suffering from too much protection because the chance of successful standardization processes will decrease and damage the growth of the sector as a whole.

What are the policy implications of the empirical results? First of all, innovations codified and protected by patents are in general a driving force for standardization, because a successful integration in a formal standard and a satisfactory licensing agreement with the huge potential of users may create significant revenue flows. However, too much IPR protection reduces the set of possible technical specifications which can be integrated into a standard, if the patent holder is not willing to license the rights to interested companies. This means also that often a standard cannot be based on the optimal technical solution. Since it is not appropriate to limit the incentives to innovate by restricting the patent protection, the concerned companies or the whole sector, or even the government in case of benefits for the whole society, should acquire the necessary licenses.[27] First of all, market solutions should be preferred. In the above mentioned Qualcomm case, the European mobile phone company Ericsson and Qualcomm have reached a consensus by giving away licenses to each other for their patents. Furthermore, Ericsson bought the R&D department of Qualcomm. Another strategy is the way Intel goes by regularly asking the companies taking licenses to its open specifications to agree to offer royalty-free licenses to other participants for any patents that would block the specified technology.[28]

Other initiatives favor the obligated licensing, especially in the case of interface technologies (Beck 1995) with no fees at all or at least a reasonable amount in fees. However, this regulative procedure is violating the incentives for inventors and is therefore economically not efficient and, as the survey showed, not supported by the companies. In network industries, where old versions of software or hardware are substituted by updates, Kleinemeyer (1998, pp. 209ff.) suggests that the patent protection for the old version should expire when the newest version is introduced into the market, because competitors could use the old versions in order to develop superior compatible solutions, since in network industries the installed base can create a monopoly for the supplier with the IPR for the relevant technology.

In contrast to the problems caused by too much patents protection, solutions should be developed for companies which do not possess IPR protection for R&D results, to give them incentives to release their knowledge for the standardization process. In these cases, the innovative companies should receive compensation for their knowledge inputs in order to give them incentives to release their innovations.

NOTES

[1] Chapter 12 refers partly to Blind (2003) and Section 12.6 on Blind and Thumm (2002).
[2] See the discussion of Ordover (1991) about possible solutions of this dilemma by awarding strong IPR, but easing licensing agreement. Matutes et al. (1996) discuss the role of the patent regime for the incentive of inventors to disclose fundamental inventions and find that a broader patent scope is superior to a longer patent duration.
[3] Cf. Scherer and Ross (1990, p. 621 ff.).
[4] In this line of thought, Shavell and Ypersele (2001) prove that a reward system, under which innovators are paid for innovations directly by the government and the innovations pass immediately to the public domain, may be superior to granting intellectual property rights.
[5] Property rights can contribute towards the distribution of new technologies, because for example new technological knowledge is codified in patents and thus represents at least a source of information for competitors and potential users in order to create alternative or improved technical solutions.
[6] In contrast, Antonelli (1994) defines standards as non-pure private goods because they are excludable to some extent, because outsiders of the standardization process have greater distances to their products and processes.
[7] Crampes and Wolkowicz (1995) show in a duopolistic model the risks of inefficient market equilibria which justify the intervention of SDOs.
[8] Compare the discussion on the data aspects in Section 7.4 .
[9] Cf. Kleinemeyer (1998, pp. 193 ff.).
[10] See Kaufer (1989, p. 12).
[11] Cf. Harhoff (1997, pp. 348 ff.). Levin et al. (1987) find similiar results for the United States.
[12] Source: Mannheimer Innovation Survey 1993, own calculations. The scale reaches from 0 (= 0 per cent) to 1.0 (= 100 per cent) and represents the share of companies for which the respective strategy is important or very important in a 5-point Likert-scale.
[13] Valid norms are also valuable aids in defining the 'state-of-the-art' in technology and can serve to judge the technical progress represented by an invention (cf. on this topic and the following Thiard and Pfau 1989, pp. 16 ff.).
[14] In addition, it is important to know what the significance of technical standards could be for

the inventor in other ways. However, the implications of standards for innovations are not discussed here.

[15] Cf. Prins and Schiessl (1993).

[16] See also the discussion in Section 10.2.

[17] Cf. Antonelli (1994, p. 207).

[18] See Salop and Scheffman (1983, p.267).

[19] See Farrell and Saloner (1985) and Katz and Shapiro (1985).

[20] Compare Farrell (1989, pp. 43 ff.).

[21] The other informal protection strategies are disregarded, because they are not suitable to prevent standardization. However, the insufficient participation in standardization processes because of using the secrecy strategy can lead to smaller numbers of standardization processes and eventually to a lower standard output.

[22] See Section 11.4 for details.

[23] In order to avoid heteroscedasticity the weights are estimated in a preliminary regression with equal weights and then applied in weighted least squares in the second round.

[24] Cf. Blind and Thumm (2002).

[25] A trade secret is an item of information – commonly a customer list, business plan, or manufacturing process – that has commercial value and that the firm possessing the information wants to conceal from its competitors in order to prevent them from duplicating it. It is not a property in the traditional sense.

[26] This observation is also confirmed by the overview of Shapiro (2001). However, the role of patent pools in standard setting is an emerging field of research and their impacts on standardization are rather ambiguous.

[27] In such a case a reward system, under which the government pays the innovators directly for their innovations and transfers the respective intellectual property rights to the standardization development organisations, may be a solution, especially if the social value of the innovation exceeds the expected monopoly earnings of the innovator (compare Shavell and Ypersele 2001).

[28] Compare Shapiro and Varian (1999, p. 25).

13. Empirical Test at the Firm Level[1]

13.1   INTRODUCTION

The identification of the main driving forces for standardization has been performed first at the sectoral level, because standards are, in contrast to innovations and patents, not a company-specific, but a sector-specific output indicator. However, in a second stage of analysis, it is of interest to elucidate what kind of company is actively engaged in standardization processes. These insights are able to confirm or question the results of the sectoral analysis. But even if contradictory results may appear due to different rationales, those outcomes will elucidate the role of standardization for the single company and for the whole sector.

A microeconomic analysis will be performed in order to elucidate companies' driving factors for participating in standardization processes in their respective environments. In the previous section we have already found that R&D and patent intensity and export activity are the major driving forces for standards output of SDOs at a sectoral level. These results will be verified in the following section on the basis of the company-specific data using a Probit model.

The remainder of this chapter is set out as follows. Section 13.2 summarizes the limited empirical work about the determinants of standardization activities. Section 13.3 presents the data set and gives some descriptive statistics of the German data. Section 13.4 considers the model specifications in order to explain the participation in standardization processes. In a second step, Section 13.5 repeats the approach above with a data set of European companies, which also contains information about patent behavior. The chapter concludes with a summary of the results and a comparison with the sector-level results.

13.2   SURVEY ABOUT EXISTING STUDIES ON FIRM BEHAVIOR

There is only little empirical work done on standardization activities at firm level in contrast to the vast literature on R&D and innovation. The majority of empirical studies considering the determinants of standardization are based on product classes (Link 1983; Lecraw 1984) or standardization processes (Weiss and Sirbu 1990). Puzzling results came to light concerning the relationship between R&D intensity and standardization. Whereas Link (1983) finds a positive impact of the R&D intensity on the probability that a voluntary standards process will be initiated, Lecraw (1984) discovers a negative influence of the R&D intensity on standard usage, due to very short product cycles making standards quickly obsolete and strong incentives to differentiate the own R&D-intensive product against those of the competitors. Lecraw (1984) admits that his analysis 'cannot be used to determine the motivation for the higher use of standards' in different industries. The same is true for his analysis of the impact of standards on price differences between Canada and the US. Meeus et al. (2002) analyzed the standardization activities of companies in the Netherlands applying a sociological approach. They find also only a rather weak relationship between the R&D intensity and the likelihood to participate in standardization.

13.3   THE DATA AND DESCRIPTIVE STATISTICS

The empirical analysis presented here aims firstly to assess the importance of different determinants of standardization activities, in particular the role of R&D and export behavior. The data set used in this chapter is a microeconomic data set of 417 German firms which answered a questionnaire developed by the Technical University of Dresden and the Fraunhofer Institute for Systems and Innovation Research, and sent out to over 2000 firms in 1998 (Blum et al. 2000). Firms in ten manufacturing sectors were randomly chosen (see Table 13.1).

As already discussed, there are a number of different explanations for participation in standardization. The definitions of the main company-specific quantitative variables in the data set are given below:

Export intensity:	$ExpInt_{ij}$	$= X_{ij} / TT_{ij}$;
Import intensity:	$ImpInt_{ij}$	$= Im_{ij} / TT_{ij}$;
Competition intensity:[2]	$CompInt_{ij}$	$=$ Number of competitors (low $= 1$; medium $= 2$, high $= 3$);
R&D intensity:	$R\&DInt_{ij}$	$= 100 \cdot R\&D_{ij} / TT_{ij}$;
Size:	$Size_{ij}$	$= \log TT_{ij}$;
Labor coefficient:	$LabCo_{ij}$	$= 1\,000 \cdot$ Number of employees$_{ij} / TT_{ij}$;

where X stands for exports in DM (German Marks), Im for imports in DM,

TT for total turnover in DM, *R&D* for expenditure for R&D in DM. The subscript i is for the firm and j for the sector.

Table 13.1   Sample of German companies differentiated by manufacturing industries

Branches	Total	Standardizing	Exporting	R&D-active
Aeronautics and aerospace	7	5	4	3
Mechanical engineering	57	28	27	33
Chemical industry without pharmaceuticals	55	24	21	31
Electrical engineering	48	27	19	31
Manufacture of rubber and plastic goods	28	13	15	14
Construction	48	19	6	25
Vehicle construction	35	17	12	20
Manufacture of metal goods	37	20	20	24
Radio, TV and telecommunications technology	34	17	13	20
Manufacture of pharmaceuticals	17	3	6	12
Others	51	32	20	34
Total	417	205	163	247

Some descriptive statistics for the variables based on 417 observations are presented below in Table 13.2 for the two groups of standardizers and non-standardizers. The most obvious differences between the two classes are the variations in the export intensity, the R&D intensity and the labor coefficient. Whereas the higher export intensity of standardizers was expected due to the theoretical considerations, the lower R&D intensity of companies joining the standardization process supports the train of thought which assumes a substitutive relationship between the own R&D effort and the participation in standardization processes and the reluctance of R&D-intensive companies to disclose their knowledge in standardization processes. In addition, the labor coefficient of standardizing companies is significantly higher compared to the non-standardizers, which suggests that less productive companies join standardization processes with the same intention as the companies with a low R&D intensity. Finally, more than one third of those companies joining standardization processes possess a standardization department, which is more than double the share of the non-standardizers.

Table 13.2   Descriptive statistics: means (standard deviations)

	Standardizers ($n = 205$)	Non-standardizers ($n = 212$)
Export intensity	0.15 (0.25)	0.13 (0.23)
Import intensity	0.02 (0.06)	0.02 (0.05)
Competition intensity	2.59 (0.48)	2.62 (0.46)
R&D intensity	2.71 (3.67)	3.43 (5.85)
Log size	19.31 (2.10)	18.14 (1.46)
Labor coefficient	0.14 (0.60)	0.04 (0.11)
Standardization department	0.38 (0.49)	0.14 (0.35)

13.4   EXPLANATORY FACTORS FOR PARTICIPATION IN THE STANDARDIZATION PROCESS AT COMPANY LEVEL FOR GERMAN COMPANIES

The influence of various variables on the decision for or against active participation in the standardization process can be determined by means of a Probit model.[3] In general, the response of all 417 companies in Germany are taken into account in the model. By contrast to the simple representation of descriptive statistics, in a multivariate analysis – like the regression model – other company characteristics and attitudes are also simultaneously considered. Thus spurious correlations, that is apparent influences of a variable on the standardization decision, can be discovered which originate from the fact that the actual causal variable has not been considered. The observation of the significance of export intensity for the standardization decision corresponds to the comparison between two companies which only differ in their export activities, but are otherwise completely alike. The influence of the other factors can therefore be separated.

Due to the fact that standardization takes place at different regional levels, we differentiate four models. In a general model, the companies are divided between the ones which are not involved in standardization at all, while the other group participates in national, European or international standardization processes. In the other three models, the companies are differentiated on the basis of their engagement or non-engagement at the three regional levels. The share of active companies is highest at the European level with 49 per cent, and lowest at the international level with 32 per cent. At the national level, 45 per cent of the companies have been involved in standardization processes.[4]

The decision of company i from branch j to participate actively in standardization is a function of different factors. On the one hand, this decision is co-determined by company-specific characteristics, such as size, labor coefficient, and the existence of a standardization department. On the other hand, general assessments about the benefits and costs of standardization have an influence on the decision whether to join or not to join in standardization processes.[5] First, standardization is to be seen in context with the research and development activities of a company, so that on the one hand the R&D intensity and on the other hand the anticipated advantages or disadvantages for own R&D ($R\&DAdv^{exp}$) resulting from the participation in the standardization process can be seen as an explanation.[6] Second, the influence of standards on foreign trade originates from the strategic decisions in particular of those companies which are involved in export. Therefore, the actual export activities and their squares, the former expected to have a positive and the latter to have a negative impact, as well as the experienced advantages or disadvantages ($ExpAdv^{exp}$, $ExpDisAdv^{exp}$) resulting from participation in standardization processes – especially when national standards are adopted as European and international standards – are explanatory factors for participation in standardization.[7] Finally, the general attitude towards the benefits of standards for the economic development of one's own company ($EcDC^{exp}$) and own branch ($EcDB^{exp}$) is a decisive factor for collaboration in the standardization process.[8] All these answers have been recoded into dummy variables supporting or not supporting the respective attitude. The descriptive statistics are presented in Table 13.3.

Table 13.3 Descriptive statistics on attitudes on the impacts of standardization

	Mean	Standard deviation
$R\&DAdv^{exp}$	0.45	0.50
$R\&DDisadv^{exp}$	0.34	0.47
$ExpAdv^{exp}$	0.23	0.42
$ExpDisAdv^{exp}$	0.07	0.26
$EcDC^{exp}$	0.09	0.29
$EcDB^{exp}$	0.05	0.20

In addition, sector-specific characteristics $Sector_j$, expressed as dummy variables with mechanical engineering as the base, and the framework conditions for the total economy (= constant α) explain the company-specific standardization decisions. Since the R&D intensity explains the export intensity according to the results of a Hausman test, two equations are estimated separately on the one hand with the R&D intensity, and on the other hand with the export intensity.

$$Std_{ij} = a_0 + a_1 R\&DInt_{ij} + a_2 CompInt_{ij} + a_3 \log Size_{ij} +$$
$$a_4 LabCo_{ij} + a_5 ImpInt_{ij} + a_6 StdDep_{ij} + a_7 R\&DAdv_{ij}^{exp} +$$
$$a_8 R\&DDisadv_{ij}^{exp} + a_9 ExpAdv_{ij}^{exp} + a_{10} ExpDisadv_{ij}^{exp} +$$
$$a_{11} EcDB_{ij}^{exp} + a_{12} EcDC_{ij}^{exp} + a_{13} Sector_j + \varepsilon_{ij} \qquad (13.1)$$

$$Std_{ij} = a_0 + a_{1a} ExpInt_{ij} + a_{1b} ExpInt_{ij}^{2} + a_2 CompInt_{ij} +$$
$$a_3 \log Size_{ij} + a_4 LabCo_{ij} + a_5 ImpInt_{ij} + a_6 StdDep_{ij} +$$
$$a_7 R\&DAdv_{ij}^{exp} + a_8 R\&DDisadv_{ij}^{exp} + a_9 ExpAdv_{ij}^{exp} +$$
$$a_{10} ExpDisadv_{ij}^{exp} + a_{11} EcDB_{ij}^{exp} + a_{12} EcDC_{ij}^{exp} +$$
$$a_{13} Sector_j + \varepsilon_{ij} \qquad (13.2)$$

The results of the Probit estimate are presented in Table 13.4. The eight models are able to explain a significant share of the companies' likelihood to join standardization processes. The most decisive factor for participation in the standardization process is the company size. The larger a company is, the greater the likelihood that it will participate actively in the standardization process. The endowment with personnel and financial resources, which is underlined by the significant positive coefficient of a standardization department in the models explaining an activity at the European or international level, is crucial for joining standardization processes, which is similar to the size-dependent innovation activities of companies.

On the other hand, the R&D intensity is a significantly negative explanatory factor for the general model.[9] This means that companies with low R&D activities are more likely to participate actively in the standardization process. As already argued, the explanation for this can be that the participation in the standardization process compensates for the own low R&D activities, which is supported by the empirical results of Love and Roper (1999). Meeus et al. (2002) also find no significant relationship between the R&D intensity and the likelihood of joining standardization activities for a sample of Dutch companies.

Furthermore, the labor coefficient is a positive explanatory factor for the collaboration in the standardization process in the company-based assessment. Obviously, less productive companies are more inclined to join standardization work. The causes have to be sought for in the same reasons as for companies with little or no own R&D. Eventually, standardization is obviously perceived as a transfer process, not only in respect to innovations and R&D results in the narrower sense, but also for best practices concerning production processes in general.

In the general model, both the export and the import intensities are not significant. This does not support the theory that the involvement in standardization processes is one instrument in companies' export strategies and a marketing tool in the international procurement of raw materials and intermediate goods. Meeus et al. (2002) also find no positive relationship between export intensities and the level of participation in standardization.

Finally, the competition intensity is not significant in explaining participation in standardization processes. The latter results make it obvious that standardization is not used differently in high- or low-competitive environments. The inclusion of sector dummies also makes clear that companies' participation in standardization is indifferent to most sectors. However, companies engaged in vehicle construction and in the pharmaceutical industry are significantly less active in standardization.

The qualitative and opinion-driven explanatory factors, which refer to the cost advantages for R&D, have the expected influences.[10] Thus the companies which expect cost savings for own R&D by participating in standardization, or which fear cost disadvantages from non-participation, are more likely to engage actively in standardization processes. The same applies for the companies which realize high cost advantages from the adoption of national standards in European and international standards. It is surprising, however, that companies which rate standards as beneficial for the economic development of their own branch are more, although not significantly, inclined to standardize, while this does not apply to companies which rate standards as beneficial for the economic development of their own company. This confirms the role of standards as public goods for the whole branch, which are not particularly beneficial for the single company.

If one concentrates on the differences between the results of the general model and the three other models differentiated by the level where standardization takes place, the following remarkable results have to be discussed. Whereas in the comprehensive model, the R&D intensity has a significant negative coefficient, this significance disappears in the three regionally differentiated models. This difference may be an indication that currently companies with a lower R&D intensity are actively involved in standardization, whereas in the past this adverse selection was not so drastic. However, the size effect remains stable in all three models: the larger the companies, the higher the likelihood of participating in standardization processes. This effect does not disappear even at the national level. The export intensity and its square are only significant in the national and European model. This might be puzzling at first glance. The following two explanations may be helpful to explain this phenomenon. On the one hand, participation at the national level is the entry ticket for influencing European standardization, which is relevant for shaping export markets. On the other hand, especially at the international level, many multinational enterprises are involved in standardization, which have affiliates with production capacities all around the world. Therefore, these companies do not claim high export rates, which make this explanatory variable indeterminate.

Table 13.4   Regression results of Probit estimations to explain participation in standardization

Explained variable: Collaboration in the standardization process	Coefficient	z-value	Coefficient	z-value
General				
R&D intensity	-0.027^a	-1.83	–	–
Export intensity	–	–	0.815	0.82
(Export intensity)^2	–	–	-1.030	-0.79
Competition intensity	-0.058	-0.37	-0.044	-0.20
Company size (log.)	0.272^c	5.41	0.279^c	5.41
Labor coefficient	0.659^c	3.28	0.725^c	3.25
Import intensity	1.442	1.20	0.720	0.52
Standardization dept.	0.293	1.56	0.287	1.52
Influence on R&D costs (participation)	0.445^c	2.60	0.474^c	2.80
Influence on R&D costs (non-participation)	-0.721^b	-2.53	-0.703	-2.40
Cost advantages from adoption of national standards in European or international standards	0.876^c	2.83	0.852^c	2.77
Cost disadvantages from adoption of national standards in European or international standards	-0.059	-0.19	-0.097	-0.32
Influence on the economic development of the own branch	0.188	1.26	0.193	1.29
Influence on the economic development of the own company	-0.082	-0.50	-0.095	-0.58
Aero- and astronautics	0.217	0.43	0.313	0.63
Chemical industry without pharmaceuticals	-0.421	-1.48	-0.444	-1.56
Electrical engineering	0.034	0.13	0.002	0.01
Manufacture of rubber and synthetic goods	-0.096	-0.31	-0.076	-0.25
Construction, building	-0.344	-1.18	-0.279	-0.96
Vehicle construction	-0.834^b	-2.32	-0.859^b	-2.42
Manufacture of metal goods	0.016	0.06	0.0271	0.10
Radio, television and communications engineering	-0.160	0.55	-0.198	-0.67

Manufacture of pharmaceutical goods	-0.875^b	-2.18	-0.917^b	-2.23
Other	0.114	0.41	0.074	0.26
Constant	-5.005^c	-5.32	-5271^c	-5.35
Log likelihood	-228.651		-229.728	
Pseudo R^2	0.209		0.205	
Number of observations	417		417	
National				
R&D intensity	-0.011	-0.76	$-$	$-$
Export intensity	$-$	$-$	2.559^b	2.59
(Export intensity)^2	$-$	$-$	-2.845^b	-2.11
Competition intensity	0.104	0.87	0.163	1.03
Company size (log.)	0.258^c	5.60	0.277^c	5.90
Labor coefficient	0.563^c	3.37	0.673^c	3,54
Import intensity	1.081	0.89	-0.577	-0.41
Standardization dept.	0.297^a	1.67	0.295	1.63
Influence on R&D costs (participation)	0.004	0.02	0.052	0.32
Influence on R&D costs (non-participation)	-0.672^b	-2.40	-0.677^b	-2.39
Cost advantages from adoption of national standards in European or international standards	-0.053	-0.21	-0.145	-0.58
Cost disadvantages from adoption of national standards in European or international standards	0.051	0.16	0.016	0.05
Influence on the economic development of the own branch	-0.040	-0.27	-0.030	-0.20
Influence on the economic development of the own company	0.203	1.25	0.191	1.17
Aero- and astronautics	-0.075	-0.14	0.018	0.04
Chemical industry without pharmaceuticals	-0.650^b	-2.24	-0.638^b	-2.20
Electrical engineering	0.151	0.59	0.163	0.62
Manufacture of rubber and synthetic goods	0.193	0.64	0.175	0.59
Construction, building	-0.559^b	-1.98	-0.456	-1.56
Vehicle construction	-0.473	-1.48	0.466	-1.47
Manufacture of metal goods	-0.264	-0.93	-0.283	-0.98
Radio, television and communications engineering	0.266	0.88	0.312	0.97
Manufacture of pharmaceutical goods	-1.468^c	-2.70	-1.450^b	-2.59
Other	0.314	1.16	0.330	1.22
Constant	-5.181^c	-5.88	-5.869^c	-6.53
Log likelihood	-236.8840		-233.329	
Pseudo R^2	0.174		0.187	
Number of observations	417		417	
European				
R&D intensity	-0.011	-0.74	$-$	$-$
Export intensity	$-$	$-$	2.840^c	2.87

Explained variable: Collaboration in the standardization process	Coefficient	z-value	Coefficient	z-value
European				
(Export intensity)^2	–	–	–3.412	–2.51
Competition intensity	0.037	0.24	0.089	0.58
Company size (log.)	0.228[c]	5.05	0.248[c]	5.39
Labor coefficient	0.575[c]	3.19	0.707[c]	3.18
Import intensity	1.626	1.34	0.049	0.04
Standardization dept.	0.294[a]	1.65	0.302[a]	1.66
Influence on R&D costs (participation)	0.111	0.66	0.153	0.94
Influence on R&D costs (non-participation)	–0.645[b]	–2.36	–0.660[b]	–2.38
Cost advantages from adoption of national standards in European or international standards	0.090	0.36	–0.002	–0.01
Cost disadvantages from adoption of national standards in European or international standards	–0.063	–0.20	–0.112	–0.34
Influence on the economic development of the own branch	–0.020	–0.14	–0.003	–0.02
Influence on the economic development of the own company	0.141	0.89	0.123	0.77
Aero- and astronautics	0.077	0.16	0.244	0.50
Chemical industry without pharmaceuticals	–0.556[a]	–1.96	–0.546	–1.94
Electrical engineering	0.187	0.73	0.188	0.71
Manufacture of rubber and synthetic goods	0.036	0.12	0.004	0.02
Construction, building	–0.494[a]	–1.74	–0.397	–1.36
Vehicle construction	–0.405	–1.30	–0.396	–1.27
Manufacture of metal goods	–0.406	–1.44	–0.431	–1.50
Radio, television and communications engineering	0.095	0.32	0.120	0.38
Manufacture of pharmaceutical goods	–1.569[c]	–2.90	–1.554[c]	–2.77
Other	0.282	1.05	0.288	1.07
Constant	–4.354[c]	–5.03	–5.030[c]	–5.67
Log likelihood	–241.436		–237.397	
Pseudo R^2	0.164		0.178	
Number of observations	417		417	
International				
R&D intensity	–0.009	–0.59	–	–
Export intensity	–	–	1.122	1.13
(Export intensity)^2	–	–	–0.985	–0.73
Competition intensity	–0.35	–0.22	0.007	0.04
Company size (log.)	0.190[c]	3.89	0.195	3.94
Labor coefficient	0.568[c]	3.11	0.621[c]	3.24
Import intensity	1.289	1.09	0.255	0.18

Standardization dept.	0.379[b]	2.07	0.371[b]	2.01
Influence on R&D costs (participation)	−0.055	−0.33	−0.021	−0.13
Influence on R&D costs (non-participation)	−0.706[b]	−2.52	−0.691[b]	−2.44
Cost advantages from adoption of national standards in European or international standards	0.168	0.71	0.120	0.50
Cost disadvantages from adoption of national standards in European or international standards	−0.85	−0.27	−0.100	−0.31
Influence on the economic development of the own branch	0.149	0.97	0.144	0.94
Influence on the economic development of the own company	0.222	1.34	0.224	1.36
Aero- and astronautics	0.538	1.13	0.525	1.08
Chemical industry without pharmaceuticals	−0.737[b]	−2.38	−0.727[b]	−2.36
Electrical engineering	0.372	1.43	0.388	1.46
Manufacture of rubber and synthetic goods	−0.013	−0.04	−0.019	−0.07
Construction, building	−0.681[b]	−2.19	−0.617[a]	−1.95
Vehicle construction	−0.544[a]	−1.65	−0.523	−1.61
Manufacture of metal goods	−0.225	−0.77	−0.236	−0.80
Radio, television and communications engineering	0.039	0.12	0.071	0.22
Manufacture of pharmaceutical goods	−0.379	−0.93	−0.369	−0.89
Other	0.119	0.43	0.125	0.45
Constant	−4.051	−4.51	−4.371[c]	−4.77
Log likelihood	−218.706		−217.788	
Pseudo R^2	0.165		0.168	
Number of observations	417		417	

Notes: [a] = level of significance < 0.10; [b] = < 0.05; [c] = < 0.01.

The most important new insights of the national and European model, which has already been suggested beforehand in the general model, are the significant positive coefficient of the linear export intensity and the significant negative coefficient of its square. If one interprets the export intensity as an indicator of competitiveness, the likelihood of companies to join standardization processes increases with the degree of their competitiveness until a certain level. Above this high value, the degree of competitiveness has a negative impact on the probability of participating in standardization. This means that very competitive companies stay away from standardization, because the expected costs in the form of possible unintended spill-over to imitating competitors exceed the expected benefits of influencing technical specifications according to own preferences.[11] The reduced likelihood of very competitive companies participating in standardization also confirms the negative relationship between R&D intensity and involvement in standardization.

Finally, it is interesting to have a comprehensive look at the significant sector dummies. In the comprehensive model, companies engaged in vehicle construction and in the pharmaceutical industry are significantly less active in standardization; in the differentiated models this is also true for companies from the chemical sector. This corresponds to the sector-based analysis in Chapter 10, which showed that especially in patent-intensive sectors standardization is more difficult and less successful due to probable conflicts with patent holders. Furthermore, it is evident that the construction industry is less active concerning standardization at the European but especially at the international level, because it is still a domestically dominated sector. By contrast, the companies from electrical engineering which provide the infrastructure for the worldwide telecommunication networks are above average active – however insignificantly – in international standardization bodies.

On the whole, the microeconometric investigation makes it clear that large companies are more likely to be active in standardization than small ones. The export intensity of a company is a crucial factor for the participation in the national standardization process, which is in most sectors the prerequisite for exerting influence on European and international standardization. However, the companies at the leading edge do not have to rely on influencing standardization processes in order to be competitive on a worldwide scale. These are two fundamental findings which must be taken into consideration in formulating the concluding recommendations. Further, the perception to what extent the participation in the standardization process influences own R&D costs and whether the adoption of national standards in European and international standards brings advantages for one's own company is an important starting point for future strategic orientation of the standards organizations.

13.5   THE DATA AND DESCRIPTIVE STATISTICS FOR EUROPEAN COMPANIES

Besides the data set of German companies, the answers of a sample of European companies are available and allow the validation some of the German results and to answer the additional question of the role of IPR, especially patents, on the propensity to join standardization processes.

In addition to the brief description of the sample already given in Section 12.6, the sample can be characterized as follows. The sample provides a representative coverage of the European Union, including in addition some respondents from non-EU countries (Iceland, Russia, Norway, Liechtenstein and Switzerland). The highest numbers of returned questionnaires came from the United Kingdom and Germany, followed by France, Italy, and Spain. Within the following data evaluation the country distribution was one

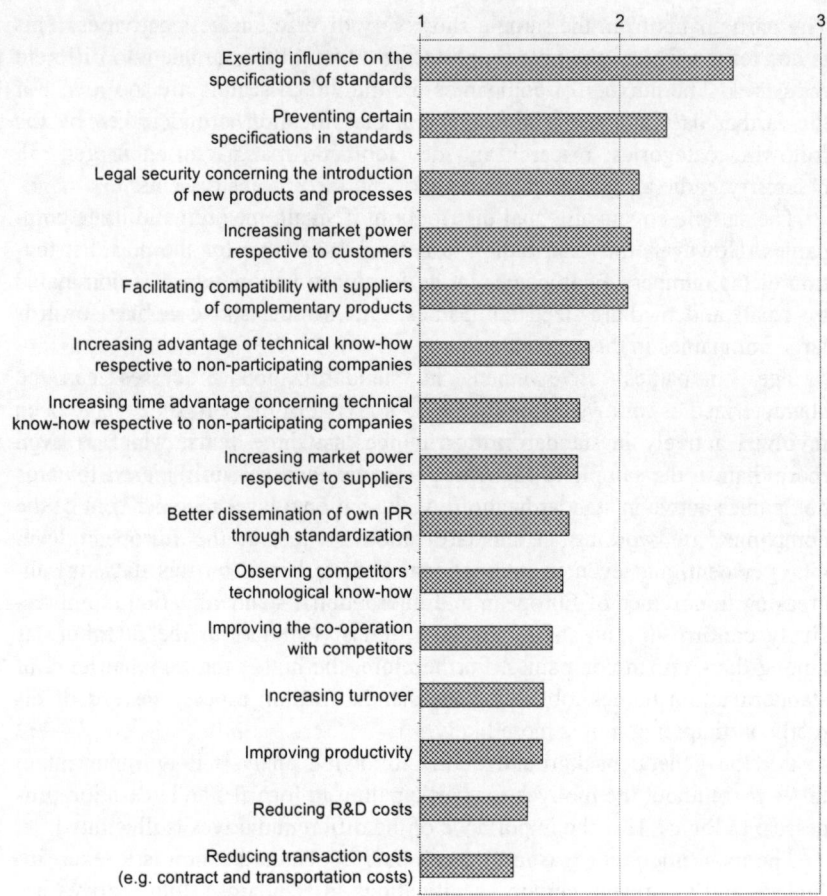

Figure 13.1   Motives for participation in standardization
(low = 1; high = 3)

criterion to analyze the sample. The low number of companies for each country makes it difficult to compare countries individually. Country groups are constructed. The underlying understanding is that we assume certain countries to show similar habits, based on similar business cultures. This is why we put Germany, the United Kingdom, Austria, Switzerland and Liechtenstein into one group; France, Italy and Spain into another group; the Benelux countries together and Scandinavian countries with Russia into one group. However, the results of the regression analysis confirmed that the location does not contribute to the explanation of participation in standardization processes. Therefore, the regional dimension will be not discussed further.

The participants from the sample show very diverse business activities. This is one reason why it was not straightforward to split the sample into different industries. The number of companies for individual sectors are too low. For the further data analysis a more general classification is undertaken by the following categories: research and development, manufacturing in general, chemistry, radio and electrical machinery and medical instruments.

The sample covers an equal distribution of small, medium and large companies. However, this distribution is not representative for the real distribution of the number of companies in the European Union, which is dominated by small and medium-sized companies. Our sample shows a bias towards large companies in this respect.

The companies' involvement in standardization processes can be characterized as follows. More than 57.7 per cent of the companies have been involved actively in standardization in the last three years, which is even more than in the sample of German companies and certainly biased towards companies active in standardization. At the national level 45.6 per cent of the companies are working in standardization bodies, at the European level 43.6 per cent, and even in international bodies. Based on this data, the increasing importance of European and international standardization is impressively confirmed. The shares are also almost identical to the distributions among the German companies. Furthermore, the higher the regional level of standardization bodies, observing the standardization process instead of directly participating in it is more likely.

For the general background of the following analysis it is important to know more about the motives for participating in formal standardization processes. In Figure 13.1 the importance of the different motives is illustrated.

The most important reason to participate in standardization is to exert influence and to prevent certain specifications in standards. Both motives are decisive for companies in the radio, television and electrical engineering industries. Furthermore, legal security is also an important motive, especially for the large companies. By participation in standardization processes, small companies also try to facilitate compatibility with suppliers of complementary products. Furthermore, companies with a high patent intensity (share of patents per R&D employee) evaluate all motives more highly compared to companies with no or only few patents per R&D employee. Most motives for participating in informal standardization processes are weaker, although among these the appropriation and observation of technological know-how are significantly higher compared to formal standardization. The motives for participation which assume a close relationship to R&D are rather weak. Both the improvement of the dissemination of own IPR and the reduction of R&D costs on the other side reach values below average.

13.6 EXPLANATORY FACTORS FOR PARTICIPATION IN THE STANDARDIZATION PROCESS AT COMPANY LEVEL FOR EUROPEAN COMPANIES

Besides the driving forces for companies to participate in standardization processes discussed in Section 10.3, the data set of the answers of European companies allows the inclusion of a further variable which is closely related to the R&D activities of companies, that is the output of the innovation process: patents. This enables us to relate these results to the sector-based results on the impact of patents on the output of standards.

Concerning R&D intensity, we have identified two different trains of thought in Section 10.3. Since standardization is a continuation of the development phase of internal R&D, companies which are actively involved in R&D should also be more likely to participate in standardization processes in order to continue their previous activities and to reach marketable products or process technologies compatible with those of other companies, especially in the case of the existence of network externalities. This reasoning suggests a complementary relationship between R&D activities and joining standardization processes. However, the involvement in standardization processes is accompanied by the danger that the other participants could use the own disclosed and unprotected technological knowledge for their purposes. Therefore, R&D-intensive companies in particular may be more reluctant to join standardization processes if their knowledge is insufficiently protected or extremely valuable not only for the company but also for other companies. Although a variety of protection instruments exists, it is difficult to measure the implementation of these instruments. One formal means of legal protection is the application for patents. Since a patent application is connected with significant costs, like fees and costs for legal advice, the expected economic value must be higher than actual expenses. Consequently, a patent application will be made only if the value of the technological know-how reaches a certain threshold. In addition, patent protection will only be looked for if the know-how to be protected may also be of value for competitors and is not only of relevance for the company itself. Therefore, the number of patent applications does not only indicate the intensity of the use of knowledge protection instruments, but also the dimension of the expected economic value of the company's know-how.

Coming back to the above reluctance of R&D-intensive companies to join standardization processes, because their know-how is both not sufficiently protected and of very high value for other companies, the following hypotheses can be derived. The higher the patent intensity the higher is the likelihood, especially of R&D-intensive companies, to join standardization processes, because their know-how which may be disclosed during the standardization process is sufficiently protected. On the other hand, it may be argued that a high patent intensity represents a very valuable stock of knowledge which the company possesses. In addition, patent protection may be not sufficient,[12] since besides the disclosure of the company's knowledge in the patent document during the discussions among the participants of the standardization process additional information may leak out to the potential competitors. Therefore, a high patent intensity may also lead to a lower likelihood of joining standardization processes. Very often large companies possessing big market shares try to establish proprietary de facto industry standards based on broad and sophisticated patent portfolios, which allow them to defend their leading position for a longer period and to reappropriate their investments spent for establishing the standard. Such a strategy automatically excludes the support of a formal standardization process.[13] Therefore, the counter-hypothesis assumes a negative relationship between the patent intensity of a company and its likelihood of joining standardization processes.

The above negative relationship between R&D respective patent intensity and participation in standardization is supported by an additional justification based on the logic of companies with low or no R&D efforts. These companies in particular may compensate for this weakness by entering standardization clubs of R&D-intensive firms and in profiting from the technology transfer there. This view is supported by the analysis of Love and Roper (1999) about the substitute relationship between own R&D and technology transfer. In general, the companies' R&D, and also patent intensity may have a negative impact in total on the likelihood of joining standardization processes.

In addition to the discussion about the role of the export intensity in Section 10.3, some modifications have to be discussed, since we have a sample of companies which is very R&D intensive, large in size and very export intensive (see Table 13.5) We have seen that due to institutional paradigms of most national standardization development bodies, the participation in standardization processes at national level is a prerequisite for having influence on the standardization at European or international level. Therefore, exporting companies which try to influence supranational standards in order to secure market shares on foreign markets are more likely to participate in standardization both at national, but also at European and international level. However, if the company has very high export quotas this indicates that it holds a very strong position on the world market, which may not be compatible with making its own product specifications align with a standard.

The definitions of the main company-specific quantitative variables in the data set are given below:

Size: $Size_{ij}$ = $\log Employees_{ij}$;

Export intensity: Exp_{ij} = $100 \cdot X_{ij} / TT_{ij}$;
R&D intensity: $R\&D_{ij}$ = $100 \cdot R\&D_{ij} / TT_{ij}$;
Patent intensity: Pat_{ij} = $100 \cdot Patents_{ij} / Employees_{ij}$;

where X stands for exports in €, TT for total turnover in €, $R\&D$ for expenditure in €. The subscript i is for the firm and j for the sector.

Some descriptive statistics for the variables are presented below in Table 13.5 for the two separate classifications of standardizers and non-standardizers. In general, the export intensity and the R&D intensity of both sub-groups are very high. Furthermore, there are several obvious differences between the two classes. The lower export intensity of standardizers was not expected due to the theoretical considerations. The reason for this sample feature can probably be found in the explanation above, which postulates that companies seeking incontestable competitive advantage stay away from formal standardization processes in order to obtain significant market shares abroad based on the uniqueness of their products.

The explanation for the lower R&D intensity of companies joining the standardization process again supports the train of thought which assumes a substitutive relationship between the own R&D effort and the participation in standardization processes and the reluctance of R&D-intensive companies to disclose their knowledge in standardization processes. Finally, the patent intensity of the companies joining the standardization process is just half of the value of the non-standardizers. This makes it obvious that companies with a strong patent portfolio are nevertheless reluctant to join standardization processes and probably try to be successful in the market without using standardized components or designs of final products.

Table 13.5   Descriptive statistics: means (standard deviations)

	Standardizers (n = 86)	Non-standardizers (n = 63)
R&D intensity	17.30 (21.88)	21.16 (24.89)
Size (log [employees])	2.66 (1.26)	2.32 (0.99)
Export intensity	45.77 (26.70)	56.77 (30.44)
Patent intensity	4.83 (9.20)	9.61 (18.14)

In order to identify the distinct variables which explain the active participation in standardization in general, at national, European, and international level, eight Probit models are tested. Besides the R&D intensity, the company

Table 13.6 Regression results of Probit estimations to explain participation in standardization

Explained variable: collaboration in the standardization process	General				National			
	Coefficient	z-value	Coefficient	z-value	Coefficient	z-value	Coefficient	z-value
R&D intensity	-0.002	-0.47	0.001	0.10	–	–	–	–
Patent intensity	–	–	-1.968[b]	-2.07	–	–	-3.404[c]	-3.21
Company size (log.)	0.269[b]	2.29	0.268[b]	2.37	0.328[c]	2.80	0.306[b]	2.58
Export intensity	-0.013[c]	-2.64	-0.016[c]	-3.05	-0.011[b]	-2.36	-0.014[c]	-2.84
Machinery	0.018	0.06	0.045	0.15	-0.089	-0.30	-0.174	-0.59
Chemistry	-0.147	-0.39	-0.007	-0.02	-0.572	-1.45	-0.476	-1.23
Electrical Engineering	0.039	0.11	0.102	0.28	-0.186	-0.52	-0.200	-0.54
Constant	0.238	0.66	0.438	1.40	-0.276	-0.75	0.153	0.46
Log likelihood	-94.485		-0.91550		-94.151		-89.337	
Pseudo R^2	0.069		0.098		0.083		0.130	
Number of observations	149		149		149		149	

Explained variable: collaboration in the standardization process	European				International			
	Coefficient	z-value	Coefficient	z-value	Coefficient	z-value	Coefficient	z-value
R&D intensity	-0.002	-0.33	–	–	0.004	0.70	–	–
Patent intensity	–	–	-1.047	-1.21	–	–	-2.569^c	-2.61
Company size (log.)	0.221^b	1.99	0.22^b	2.09	0.345^c	2.89	0.300^b	2.46
Export intensity	-0.008^a	-1.85	-0.010^b	-2.12	-0.007	-1.56	-0.009^a	-1.82
Machinery	0.015	0.05	0.042	0.15	0.054	0.18	-0.066	-0.22
Chemistry	-0.467	-1.20	-0.386	-1.01	-0.217	-0.54	-0.170	-0.42
Electrical Engineering	-0.105	-0.30	-0.066	-0.19	0.162	0.46	0.108	0.30
Constant	-0.211	-0.59	-0.128	-0.42	-0.959^c	-2.65	-0.509	-1.56
Log likelihood	-96.794		-95.995		-0.91436		-89.007	
Pseudo R^2	0.052		0.060		0.063		0.088	
Number of observations	149		149		149		149	

Notes: ^{a} = level of significance < 0.10; ^{b} = level of significance < 0.05; ^{c} = level of significance < 0.01.

size and the export intensity, the patent intensity is included as an explanatory variable in a second equation in order to avoid endogeneity problems among the explanatory variables, because the patent intensity depends on the R&D intensity.[14] A sector dummy with the rest of sector as base is included to control for sectoral influences. The regression equations are defined as follows:[15]

$$Std_{ij} = a_0 + a_1 R\&DInt_{ij} + a_3 logSize_{ij} + a_4 ExpInt_{ij} + a_5 Sector_j + \varepsilon_{ij} \tag{13.3}$$

$$Std_{ij} = a_0 + a_2 PatInt_{ij} + a_3 logSize_{ij} + a_4 ExpInt_{ij} + a_5 Sector_j + \varepsilon_{ij} \tag{13.4}$$

The results of the Probit estimate are presented in Table 13.6. The eight models are able to explain a significant share of the companies' likelihood to join standardization processes. As in the analysis based on the sample of German companies, the most decisive factor for participation in the standardization process is the company size.[16] The larger a company is, the greater the likelihood that it will participate actively in the standardization process. The endowment with personnel and financial resources is crucial for joining standardization processes. In addition, the benefits of an active involvement in standardization are higher for larger entities.

The R&D intensity of the companies has no significant impact on their likelihood to join standardization processes. This result also confirms the observation for the German companies.[17] In contrast to the German model, we have included a further company-specific innovation variable, the patent intensity. With the exception of the European model, the patent intensity has a strong negative impact on the probability to participate actively in standardization. Consequently, the hypothesis that sufficient patent protection decreases the danger of unintended knowledge spill-over to competitors and therefore increases the likelihood of companies to join standardization processes must be rejected. On the contrary, the counter-hypothesis experiences a strong affirmation. The patent intensity is obviously an indicator for the value of the companies' stock of knowledge and their competitive strength, which enables them to abstain from standardization and to be successful without influencing specifications of formal standards or even to establish proprietary de facto standards.

The significant negative impact of the patent intensity, as an indicator of competitiveness, finds its confirmation by the negative coefficient of the export intensity, which represents the strength of the company on foreign markets. Obviously, very competitive companies with high export shares do not have to rely on influencing the specifications of formal standards, but are able either to cope with the specifications set by others, or for their company-specific product design to prevail with customers. These idiosyncrasies make them successful in foreign markets. This result corresponds to the findings based on the sample of German companies which makes it evident that companies with an average success in foreign markets are more likely to join standardization processes, whereas companies with very high export shares do not have to rely on influencing standardization as a measure to support their other marketing activities abroad. For their success it is even favorable to abstain from standardization in order to save the idiosyncrasies of their products compared to those of their competitors.

13.7 CONCLUSIONS

This chapter has presented new insights about the explanatory factors for the company-related decision to join formal standardization processes or to abstain from them. We have used two different empirical bases, a sample of German companies and a smaller sample of European companies. Although the questionnaire-based surveys have not been identical, a comparison of major explanatory variables is possible.

First, the participation in standardization increases significantly with the size of the company. The explanation for this phenomenon can be found in cost-related reasons. To run a standardization department or to employ staff responsible for standardization causes fixed costs. This kind of cost can more easily be carried by larger companies. Furthermore, the benefits are also more favorable in larger companies, since standards have a public good feature which makes them more valuable in large entities. In summary, the cost–benefit ratio of an active participation is much better for large companies compared to small companies. This is impressively reflected by the results both of the descriptive statistics and the Probit analyses.

The second major observation is the negative relationship between the competitive strength of companies and their likelihood of joining standardization processes. Competitiveness is measured by several indicators. First, we find a negative impact of the R&D intensity in both samples, which is not always significant. Second, the labor productivity in the German sample is also negatively related to the likelihood of an involvement in standardization work. Third, for the sample of European companies, we find that the patent intensity, an output indicator of the innovation process, is strongly negatively correlated with being active in standardization. Finally, the likelihood of companies to join standardization increases with their export activities at least in the range of low to medium export intensities. However, companies at the leading edge denoted by very high export intensities abstain from standardization.

These major results have to be taken into account when the impacts of standards are discussed and confirmed by empirical analysis, since we have obviously very biased participation behavior in standardization processes, which may influence both the amount and the quality of the standards produced in an ambiguous manner.

13.8   COMPARISON WITH RESULTS AT THE SECTORAL LEVEL

This section is devoted to relating the results regarding the explanatory variables of participating in standardization processes on the micro-level presented in Chapter 13, to the insights on the determining factors of sectoral output of standards, since the understanding about the motivation of companies to join standardization may also enrich the sector-specific factors explaining the annual output of standards. We will concentrate in the following analysis on three crucial dimensions for explaining standardization: R&D, export and company size.

On the microeconomic level, we find, at least in the general model, a significant negative coefficient of companies' individual R&D intensity concerning the explanation of the probability of joining standardization processes. At the sectoral level, it turns out that the R&D expenditure of the sector correlates positively with its standards output, since the higher the R&D expenditures in a sector, the higher the likelihood of technical innovations which make existing standards obsolete and call for the publication of a revised document or which cause demand for a completely new field of standardization, such as in biotechnology. These kinds of technological developments are not only driven by the R&D activities of single companies, but they rely often on essential preceding scientific or technological insights built up by universities or other public or semi-public research institutes. In the Probit models it is controlled for the sectors of the companies. The sectoral dummies reflect these general framework conditions. Therefore, there is no contradiction between the company and sector results regarding R&D. In a comprehensive multivariate approach which includes the input indicator R&D expenditure and the output indicator of the innovation process patent applications, the R&D variable is negatively connected with the standard output. The patents have a positive coefficient, since the better the generated know-how is protected, the smaller is the danger of uncontrolled knowledge spill-over. Therefore, sectors with a high R&D activity but little protection efforts hesitate to transfer know-how into standardization processes. However, too much patent protection causes major obstacles for the standardization process, since too many of the possible technical specifications for a standard are then controlled by a patent-holder who may prohibit a successful standardization process for strategic reasons. The micro-based approach confirms this tendency, since companies from the patent-intensive pharmaceutical and chemical sectors are less likely to join standardization activities.

Since export activities depend very often on the acceptance of domestic preferences and specifications in the target countries for imports, the involvement in standardization at the European or international level, which in general presupposes an engagement at the national level, represents a strategy to find a common understanding of safety and quality preferences, but also concerning compatibility questions. At the sectoral level, the bivariate analysis shows a significant positive correlation coefficient between the export intensity and the amount of standards produced. In the multivariate regression, the export intensity loses its significance. The same is true for the results of the micro-based analysis. Companies active in standardization have significantly higher export intensity than the non-standardizers. In multivariate approaches, this discriminating feature loses its significance due to its positive correlation with company size. However, the multivariate results contain hints which confirm the important role of export for standardization. First, the export intensity of the companies explains positively, at least in the national model, the likelihood of joining standardization processes. Second, companies from the construction sector, which can be regarded as a branch providing mostly non-tradables and therefore being less active in export, are significantly less involved in international standardization activities.

Finally, the individual company size and the average company size in the different sectors as determining variables of the micro-level and the sector results have to be compared. In the sectoral models, the average company size has no explanatory power for the produced output of standards. On the micro-level, company size is decisive for participation in standardization since smaller companies in general cannot afford to join standardization activities. However, the different indications do not represent a contradiction, since we have controlled for the sector in the Probit model. On the contrary, the tendency of larger companies to be active in standardization fits very well with the positive impact of enterprise concentration in the sector on the sectoral standard output. Since the larger companies join in standardization activities, the smaller the number of companies involved and the easier the negotiation process for producing a new standard is.

Summarizing the comparison between the results of the analyses in preceding sections and the findings of driving forces of standardization at the sectoral level, two main conclusions can be derived. First, there exist differences between the determinants for the likelihood of participation in standardization processes at the company level and the driving forces for the output of standards at the sectoral level. Second, these differences do not necessarily represent contradictions, since the annual output of standards is also mainly influenced by the direction and the speed of technological change not only caused by the activities of the companies, but also by the research results of publicly funded research institutes. The relevant pool of technological knowledge is not only nourished by new insights produced in domestic research institutes, but also in institutions abroad performing basic research. This external factor with its global dimension cannot be analyzed on a company level, but only on a sectoral or even macroeconomic level. Finally, the individual rationality of a single company determines its decision concerning joining standardization processes. However, the interplay of different companies following various standardization strategies may lead to different driving forces at the aggregate sectoral level. It is an open question whether the interaction of rather similar companies in standardization processes may lead to more standards compared to a set of rather heterogeneous companies, whose characteristics may complement each other. Therefore, it was necessary to analyze the significant characteristics of companies being active in standardization taking into account the sectoral framework as a dummy and the sectoral output of standards depending on sectoral characteristics.

NOTES

[1] Chapter 13 refers to Blind (2001a) and Blind (2002d).

[2] This variable relies on the self assessment of the companies.

[3] In the Probit model based on a filter question in the questionnaire it is simply assumed that the variable in question – in this case the decision about cooperation in a standardization committee – can only assume two values: no or 0 and yes or 1.

[4] In the general model, the dependent variable of active participation in standardization is based on the recent behavior. The three regional models are based on the answers of participation in a more general sense (including the observation of standardization processes) in the last three years. This may lead to discrepancies between the results of the general model and the three regional models.

[5] A similar approach is applied by Meeus et al. (2002).

[6] The answer to the question of the impact of (non-)participation on the own R&D costs reaches on a 5-point scale from very negative (−2) to very positive (+2).

[7] The answer to the question of the cost advantages or disadvantages for the own company due to the conversion of national standards into European or international standards reaches from none over temporary to lasting on a 3-point scale.

[8] These answers vary from total rejection to total support on a 5-point scale.

[9] In order to integrate the sectoral degree of innovation, in one other model the patent intensities of the sectors were included as dummies. However, no significant sector impact was found.

[10] Meeus et al. (2002) find that 'cognitive' factors explain better the involvement in standardization and that both R&D and export intensity explain these 'cognitive' factors significantly.

[11] In the international model, the signs of the coefficients of the export variable are confirmed, but lose their significance. By contrast, the existence of a standardization department is

obviously more important for the participation in standardization compared to the national or European level.

[12] In most studies, secrecy is the most important protection strategy, whereas patenting is only of secondary importance (Cohen et al. 2000; Arundel 2001).

[13] However, there are examples where companies tried to influence the formal standardization process into specifications which seemed to be very good in the technological sense or not very promising concerning market acceptance.

[14] However, the correlations between these two variables is rather low and a model including both variables does not lead to different results. The relationship between R&D and patenting behavior is discussed in Janz et al. (2001) and Blind and Thumm (2003).

[15] In contrast to the model applied for the German companies, the export variable is included only in its linear form, since the high mean of the export intensity reflects the general strong competitiveness of the sample. This makes the incorporation of the square of the export intensity superfluous. Furthermore, statistical tests elucidated that there is no significant positive relationship between R&D or patent intensity and export intensity. Therefore, we included them both in the regression equations (13.3) and (13.4).

[16] Significant sector-related effects cannot be observed.

[17] For Dutch companies Meeues et al. (2002) also do not find a significant relationship between R&D intensity and the involvement in standardization processes.

14. The Role of Standards in the Service Sector[1]

14.1 INTRODUCTION

A prominent feature of the current structural change in the economy and society is the growing significance of the service sector in terms of market volumes and labor market, reflected in the growing share of services in the gross national product and employment. The services sector is the largest single employer in all EU member states: in 1996, for example, 60 per cent of the workforce in the European Union were employed in this sector. Its share in the gross domestic product within the EU amounted to 62 per cent in 1996, while that of industry and agriculture was only 35 per cent and 3 per cent, respectively. Active participation in this structural change presents a great challenge. Services are going to be a crucial factor in international competition for markets and locations. Commissions for services will not be restricted by national frontiers. Globalization and regionalization describe a process in which global presence is linked with the delivery of services individually tailored to meet local needs.

In line with this trend, standardization is also extending its range beyond its traditional, technical fields to include the service sector. Where such activities do not originate as European or international standardization projects, national standards may, in appropriate areas (for example where services are offered across national borders), later be submitted to CEN or ISO as the basis for a draft proposal. In the past, standardization was mainly directed at products and production processes. Services, however, form an emerging field for standardization.

After becoming the most important sector in the industrialized countries, the service sector is meanwhile attracting the focus of academic and empirical research. The research activities in services have started only since the 1990s, after the service industry became the dominant sector in the whole economy. However, the focus was and still is directed to innovation activities and patterns in the different service sectors. Standardization in services is a rather new field of empirical research mainly because of the very few and only recently started standardization activities related to services. Whereas for the manufacturing sector analyses are possible about the impact of the existing stocks of standards on growth and foreign trade, the few service standards do not allow such analyses, but only more case-oriented studies.

Besides the small empirical base, in the service literature standardization has also received little attention, because of the technocratic paradigm associated with standards (Gustavson 2000). In addition, the problems related to formalization of services, characterized by their high degree of individuality, represented in the recent tendency towards customization of services in order to meet clients' preferences better, causes a reluctance to standardize. However, the tendency in services towards decreasing labor intensity, decreasing direct consumer interaction and decreasing customization leads to a higher need for standards (Schmenner 1992), because the competitive pressure forces service companies to increase their productivity in the service production respectively to industrialize their services (Normann 1991). The first empirical evidence about the distribution between mass-produced, partially customized and individualized services can be found in Tether et al. (2001), which supports the existence of a large share of standardized services already.

De Vries (1999) gives a short overview of the more descriptive work about service standards. Since some service sectors have a longer tradition of standardization, like the financial sector, first descriptions (Darsie 1990) and analyses have been performed (Karapetrovic and Willborn 2001). Another well established service sector is the transport market. Holler et al. (1997) present an overview of the issues related to standardization in the European transport market in the presence of the trends towards deregulation and disintegration of network and service provision. Furthermore, Prebezac (1997) discusses the role of standards for the quality of air transport services. The large body of literature on standards in telecommunication is until now mainly related to technical standards and not primarily on the service core itself. Since the technical related literature does not provide new insights concerning service standards in the narrower sense, it is not included in this literature review.

Across service sectors, standards are mostly analyzed in the context of quality. Berry et al. (1992) define standards as customer expectations stated in a way that is meaningful to employees in their guideline for improving service quality. The literature about the most widely accepted quality standard, the ISO 9000 series, was mainly focused on manufacturing sectors (Davis 1997; Karapetrovic et al. 1997; Resetarits 1997; Withers et al. 1997). Meanwhile, there are also studies about ISO 9000 in the service sector, like Docking and Dowen (1999), Karapetrovic and Willborn (2001) for financial services, Chu and Wang (2001) for public sector services in Taiwan, and earlier Gilpin and Kalafatis (1995) for the leisure industry and Johannsen (1995) in professional information services. Besides ISO 9000, service quality management itself is a research field of growing interest (Page and Spreng, 2002). Obviously, the quality aspect dominates the analyses of service standards so far.

Since not many service standards exist as yet and consequently no data about service companies joining standardization processes, the focus of this chapter is not on standardization in services in general,[2] but particularly on the diffusion of the quality standards ISO 9000ff series in innovative service companies.[3] After describing the genesis of ISO 9000ff and its contents in Section 14.2, theoretical hypotheses are postulated concerning driving forces for the introduction of ISO 9000ff. In Section 14.4, the data source is described and first descriptive statistics are outlinded, before in Section 14.5 results based on Probit estimations are presented. The chapter concludes with a short summary and perspectives for future research.

14.2   THE HISTORY AND CONTENT OF ISO 9000FF

The International Organization for Standardization (ISO) is the specialized international agency for standardization at present, uniting the national standards bodies of 91 countries. ISO is made up of approximately 180 Technical Committees. Each Technical Committee is responsible for one of many areas of specialization, ranging from asbestos to zinc. The purpose of ISO is to promote the development of standardization and related world activities to facilitate the international change of goods and services, and to develop co-operation in intellectual, scientific, technological, and economic activity. The results of ISO technical work are published in international standards.

ISO Technical Committee 176 was formed in 1979 to harmonize the increasing international activity in quality management and quality assurance standards.[4] Whereas subcommittee 1 was established to determine common terminology, subcommittee 2 was set up to develop quality systems standards – the result being the ISO 9000ff series, published in 1987 and revised in 1994.

The ISO 9000ff series is a set of five individual, but related, international standards on quality management and assurance.[5] They are generic, not specific to any particular products. Manufacturing and service industries alike can use them. These standards were developed to effectively document the quality system elements which are necessary in order to maintain an efficient quality system in companies. The ISO 9000ff series standards do not themselves specify the technology to be used for implementing quality system elements.

ISO 9000ff provides the user with guidelines for selection and use of ISO 9001, 9002, 9003, and 9004. ISO 9001, 9002, and 9003 are quality system models for external quality assurance. These three models are actually successive subsets of each other. ISO 9001 is the most comprehensive, covering design, manufacturing, installation, and servicing systems. ISO 9002 covers production and installation, and ISO 9003 covers only final product inspection and testing. These three models were developed for use in contractual situations such as those between a customer and a supplier. ISO 9004 provides guidelines for internal use by a producer developing his own quality system to meet business needs and take advantage of opportunities.

The choice of which model to implement depends on the scope of the operation. For example, if the company designs its own product or service, it should consider ISO 9001. If it only manufactures (working off someone else's design), it may wish to think about ISO 9002. Finally, if the firm neither designs nor manufactures, it may wish to consider ISO 9003.

In order to introduce ISO 9000ff, it is necessary to have an independent third party conduct an on-site audit of the company's operations against the requirements of the appropriate standard. Upon successful completion of this audit, the company will receive a registration certificate that identifies its quality system as being in compliance with ISO 9001, 9002, or 9003. The company will be listed in a register maintained by the accredited third-party registration organization. It is possible to publish the registration and use the third-party registrar's certification mark and the accreditation body's mark on the advertising, letterheads, and other publicity materials, but not on the products. The accredited third-party registrar will perform periodic inspections to assure that the quality system is being maintained. Many registrars also require a full re-audit after a specified time. If the company fails to maintain the quality system, the registrar will suspend or cancel the registration.

There are several benefits for companies in implementing this series. For example, it will guide management to build quality into their products and services and avoid costly after-the-fact inspections, warranty costs, and re-working. In addition, it may also be able to reduce the number of audits customers perform on the operations. Increasingly, customers are accepting supplier quality system registration from an accredited third-party assessment based on these standards. In order to set up a system of driving forces, theoretical hypotheses concerning the introduction of ISO 9000ff are deliberated in the following section.

14.3 THEORETICAL HYPOTHESES ABOUT THE ROLE OF QUALITY STANDARDS FOR SERVICE COMPANIES

In order to focus on the role of ISO 9000ff as a quality standard in service companies, an iterative procedure will be performed. First, the general role of quality standards[6] is outlined, and second, the special standardization issues in providing services are elaborated. Based on these preliminary thoughts, theoretical hypotheses concerning driving and hindering forces for the introduction of ISO 9000ff can be derived in the last step.

14.3.1 The Role of Quality Standards in General

First, benefits of the ISO 9000ff have already been sketched. Nevertheless, to be able to determine service sectors or even service companies in which a need for ISO 9000ff might be expected, general theory on service marketing and service quality is combined with general knowledge about quality standards.[7]

In most markets, we observe information asymmetries between the consumers and the suppliers of goods and services. Secondly, inside the company, we observe information deficits of the company owner or management concerning the performance of the employees. A remedy for both problems, internal and external, can be found in quality standards.

In the first stage, the more common external information problem will be discussed. Information asymmetries between producers and consumers of goods and services are abstracted from the standard economic textbook case of homogeneous goods with completely transparent product characteristics. However, in most cases this assumption does not hold. In particular, the attributes of services are difficult to grasp in advance, as we will elaborate in more detail later. Besides the information screening activities of the demand side, the supply side may use signals to reduce the information asymmetries. Signals can be voluntary warranties about the characteristics or performance of the products. When product attributes are difficult to observe prior to purchase, consumers may plausibly use the quality of products produced by the firm in the past as an indicator of present or future quality. Furthermore, the company can commit itself to a regular investigation of its production process by an independent and officially authorized third party which awards the tested company or products with an official certificate.

Due to these effective mechanisms for reducing information asymmetries, even in a competitive market, high quality goods can be sold over cost because of the cost of building up a reputation. As the time lag between successive sales of the product and hence the period after which consumers detect the real product quality approaches zero, the price–quality schedule comes close to the one under perfect information.[8] The information problems are most acute when products are infrequently bought, long lags in the detection of quality exist, reputations are slowly updated, and quality attributes are difficult to detect.

The more severe these issues are in a market or for a product or service, the more monopolistic product differentiations, obligatory public minimum quality standards or supply-side strategies of quality commitments can be observed.[9] Quality standards are an instrument for signaling the quality of products and services. Depending on the market conditions, a service company is confronted with signaling as an effective strategy to keep or expand its market share. Therefore, the likelihood or the probability of introducing ISO 9000ff depends on the sector of the service industry and its characteristics. Additionally, if the majority of the competitors is establishing ISO 9000ff as a quality signal, then the company will be forced to do the same in order to keep up with the others.

Besides this external effect of quality standards because of information asymmetries between suppliers and customers, information problems also exist inside an organization like a company.[10] Here, the principal–agent theory has to be cited. With quality standards, the management or the proprietors of a company own an instrument to control the performance of their employees better.[11] The more difficult it is to observe the quality of a product or a service delivered by the employees, the higher is the pressure to introduce a system of quality standards which makes sure that in each phase of the production or service-providing process a certain minimum quality is guaranteed. Because of the intangible nature of services and the information asymmetries thus caused between management and service provider, the need to introduce quality standards for each stage of the service production is especially high. Standards also have the function of ensuring compatibility between the different components of a product. In the case that a product or a service goes through different production stages, quality standards guarantee, especially in the case of services, a process with little friction between the involved employees or organizational units, whereas for products compatibility standards are additionally more suitable and easier to control.

We leave this general discussion about the role of quality standards in decreasing information asymmetries both externally to the customers and internally between management and employees, and direct the focus of the next section to standardization of services.

14.3.2   Standardization of Services

In order to build hypotheses about the suitability of ISO 9000ff as quality standards for service companies, some general features of services have to be outlined. According to Sirilli and Evangelista (1998) the following attributes are specific for services:

- A close interaction between production and consumption – so-called co-terminality. This makes it difficult to distinguish between processes and products, because the service product is often a set of procedures. Furthermore, due to the non-physical existence of services, the relationship is blurred between what is produced and the means of production.
- A high information content and intangible nature of the service output. The high information content of services has led research to focus on information technologies, especially in innovation in services (Barras 1986). The intangible nature of most services inspires thoughts about strategies which customize service outputs and adapt them to the needs of the users.
- The key role of human capital in the provision of services. Service production depends heavily on the knowledge and skills of the people involved in the process of production and innovation. Despite the fact that many services depend on information and communication or on transport networks, human resources are decisive for the success of the production and innovation of services.
- A critical role played by organizational factors in firms' performance. Product innovations in services are related to the improvement of the relation between the user needs and the services supplied.

Services and the service process are characterized by their high degree of individuality, represented in the recent tendency towards customization of services in order to meet clients' preferences better. Another conflicting goal is to increase the productivity in the service production by standardization, or industrialization of services respectively. In the following, different aspects or stages of standardization in the service process are discussed, in order to derive final hypotheses concerning the use of ISO 9000ff in service companies:[12]

- Standardization of the Performance Capacity: Human and physical capital are necessary for the production of services. Therefore, standardization can take place in the selection of the technical equipment and the employed service providers by checking the compliance with technical standards and formal qualifications of the employees. Concerning human capital, that is the employees, congruence between their formal qualifications and their actual performance in providing services is not guaranteed.

- Standardization of the Service-providing Process: The standardization of the process of providing services seems to be more effective. Here, quality management by ISO 9000ff is adequate. However, too strong reglementation can be negative for the motivation of the employees.
- Standardization of the Service Output: Thirdly, standardization of the results of the service process is possible. Here, the supplier may define quality criteria which have to be met or which the customer is expecting. The former approach is easier to realize because of the uncertainty concerning the degree of satisfaction of the customer.
- Standardization of External Factors and of Situative Circumstances: Fourthly, the external factor can be standardized by standardization of the expectations of the customers. The more markets can be segmented, the more homogeneous they become. Because of the high credence qualities[13] of services the communication with the suppliers is very important. If certain quality features are promised, probably only those customers who expect just these features will react to this. However, by segmenting the market into too small niches, the market potential of these niches is low and the danger that they cannibalize each other is high, if fencing is not effective. Finally, the situative factors like the location and the time of the service provision can be standardized.

Although some of the factors mentioned in the process of service provision may be standardized just with products complying with adequate technical standards or staff having the formal qualifications, the immaterial and intangible nature of services makes an integrated approach containing a set of quality standards still sensible. The less the above-mentioned formal technical and qualification standards concerning the input factors play a role, the more an integrated system of quality standards covering the whole service process makes sense.

14.3.3 Driving and Hindering Forces for the Introduction of ISO 9000ff

The general discussion of quality standards and standardization of services allows us now to derive hypotheses concerning the use of ISO 9000ff in service companies. The different aspects of standardization of services lead to the following hypotheses:

ISO 9000ff is less important in services which are more likely to hire highly qualified people and use technology-intensive equipment. Furthermore, too much regimentation by standards is a disincentive in markets where creativity is required. To the contrary, service providers who need only low-qualified employees would require standards for the performance potential of the human resources engaged.

When services are characterized by missing or hardly recognizable feedback loops between the service providers and the customers, ISO 9000ff may help to standardize the service results at least from the perspective of the supply side.

Finally, the expectations of the customers in niche markets are more homogeneous, which makes the introduction of ISO 9000 less necessary. Because of the heterogeneity of services, the relationship between service provider and customer is different in the subsectors. Official quality certificates become more important when the distance between the two is greater than in cases where the service is produced in a close cooperation between the two. Therefore, sectors offering business services will show a higher propensity to introduce ISO 9000ff than sectors offering consumer services with high frequencies of personal contacts.[14]

Nowadays, services are based very often on an intensive use of technologies. These technologies also affect the quality of the service and therefore the well-being of the customers. However, technologies can be distinguished by the level of risk for health and safety they may cause. Therefore, quality standards are more likely for technologies with a risk potential for the customers or the environment in general, because they are a sign for the safety reputation of the service provider and are able to raise the confidence of the consumer and the supervising governmental agencies in the technology and the service.

Another discriminatory factor is the company size. In small service companies, the customers are more likely to be served by the same employees. Therefore, they collect experience about the service supplied by the staff and the need to standardize is lower compared to search goods provided by continuously changing staff in large firms.

Additionally, for small and medium-sized companies the costs of ISO 9000ff registration, from training the staff to final registration, have been reported to be from $50 000 to over $250 000.[15] These costs will reduce the inclination of small and medium-sized companies to introduce ISO 9000ff compared with large companies.

However, both large and small companies with international business perceive the ISO 9000ff series as a route to open markets with an improved competitiveness. ISO 9000ff is an international standard generated by the international standardization organization and agreed on by the over 100 member states. Because of the higher information asymmetries between the home country and foreign countries, companies with ISO 9000ff have a quality signal which is accepted all over the world, which makes it easier for them to export services abroad. On the other hand, foreign service companies with ISO 9000ff will have an easier access to the domestic German market. The domestic service companies also have to introduce ISO 9000ff to keep up with these foreign competitors. Therefore, companies exporting services or expecting increased international competition are supposed to introduce ISO 9000ff sooner or more frequently.

Finally, in the age of globalization and internationalization, many multi-national enterprises have emerged which are active in different national markets. In order to present a common quality standard to their customers and to unify their internal service quality control as well, the introduction of ISO 9000ff in the affiliates all over the world makes sense.

In service markets with a high price competition, companies are more likely to introduce ISO 9000ff. In markets with small price margins and nearly homogeneous services, ISO 9000ff serves two functions. First, it is an additional quality signal, which may make the difference to other competitors without a further attribute. Second, ISO 9000ff may help to support a cost-efficient production of the services, which may increase the company's profit margin above the sector average. In high quality segments of the service market, ISO 9000ff does not play any significant role, because ISO 9000ff represents a minimum quality standard which does not have any meaning for customers who demand high quality services.

Where flexibility and in-time production is demanded the service companies are more likely to introduce ISO 9000ff. Here, the positive internal effects of ISO 9000ff have a positive influence on the above demanded product characteristics, although ISO 9000ff requires sustainable high quality work from the employees, which consequently puts a certain pressure on their performance. This perspective may cause rigidities in the organizations against ISO 9000ff and other innovations.

Concluding the explanations of the theoretical hypotheses concerning factors for the introduction of ISO 9000ff, it has to be stated that from the theoretical point of view, positive and negative forces exist. In our empirical analysis we try to verify the different theoretical hypotheses based on a survey of companies in the service industries.

14.4   THE DATA AND DESCRIPTIVE STATISTICS

Surveys of companies in the service sector on topics connected with techno-logical and organizational change used to be very uncommon. The Oslo Manual (OECD 1997c), conceived as a handbook for innovation surveys in the industrial sector, mentions that there are problems involved in simply transferring the definitions and concepts related to the service sector. In order to identify the specific features of the service sector, and to test provisional definitions and concepts, the Mannheim Innovation Panel (MIP) on services was preceded by a detailed pilot phase (Licht et al. 1997). The database of the VVC (German Confederation of Associations for Credit Reform) was used to build up the gross random sample. The files contained in the database enable us to obtain a random sample of legally autonomous companies, weighted by size and industry.

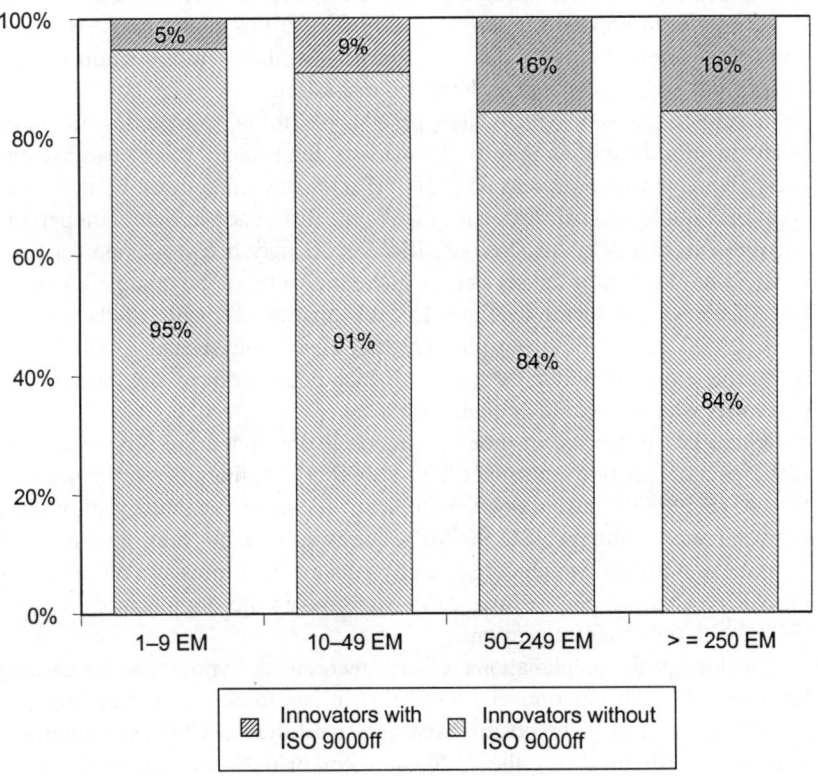

Figure 14.1   ISO 9000ff and the size of the companies in number of employees (EM)

After a pretest in July 1995, questionnaires were sent out to the 11 600 companies randomly selected in early October 1995. After several reminders, 2896 companies finally returned the questionnaire.

For our analysis of driving forces for the introduction of ISO 9000ff, we use the subsample of over 2100 innovators,[16] because we only have data about their introduction of ISO 9000ff in the past three years.[17] Furthermore, we leave out the banking and insurance services in the Probit estimations, since in these sectors there are either zero companies or only a very small number of innovators who introduced ISO 9000ff.

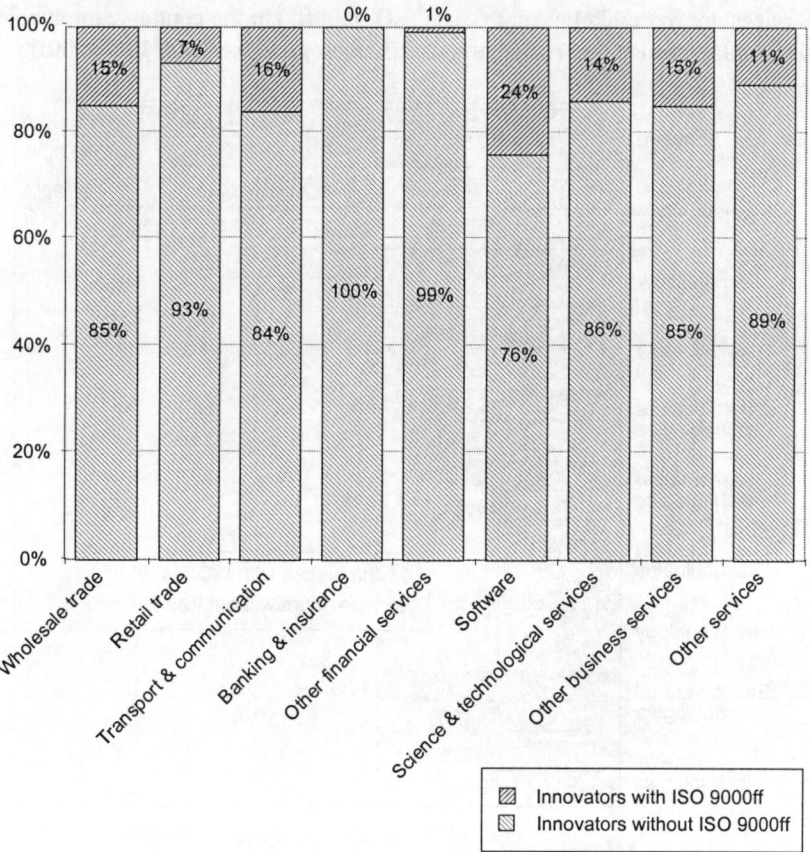

Figure 14.2   ISO 9000ff in different service sectors

In order to give first insights about the role of ISO 9000ff in different size groups, service industries and technology user groups, the following figures are presented. Figure 14.1 underlines that the propensity to introduce ISO 9000ff is higher in companies with more than 50 employees. A supply side explanation is the high costs of introducing ISO 9000ff in a company. From the demand side, it can be argued that in small companies the connection between suppliers of the services and their customers is closer. The consequence is a relationship which rests more on informal confidence which makes quality standards obsolete.

The average propensity of innovators to introduce ISO 9000ff is around 12 per cent. However, especially companies in the wholesale trade, the transportation and communication sector, the software business and other business services are more likely to introduce ISO 9000ff. On the contrary, the whole financial sector is more reluctant to certify their processes after ISO 9000ff.

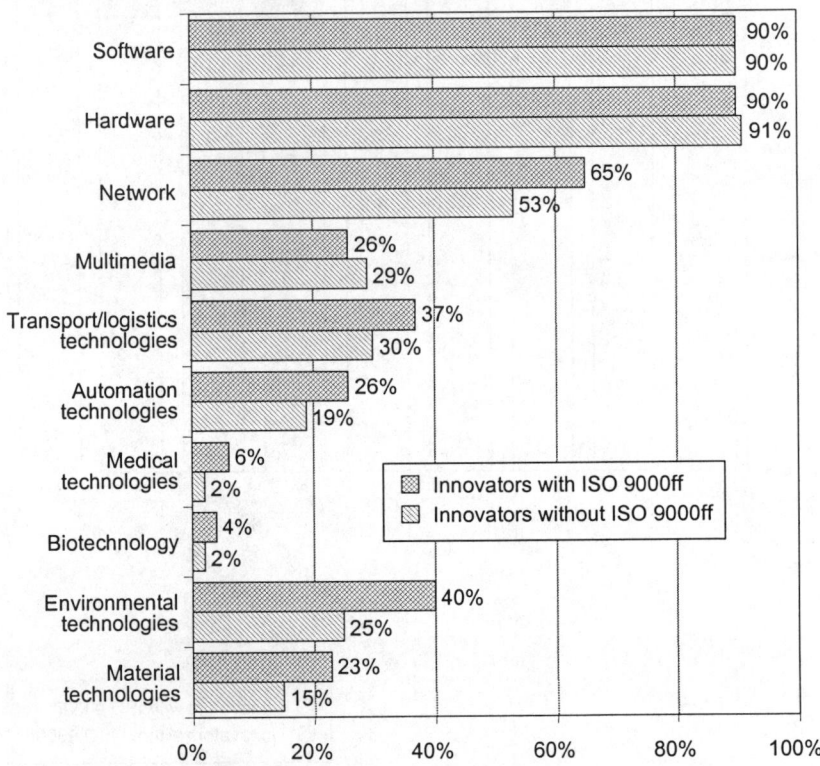

Figure 14.3   Importance of different technologies for innovators with/ without ISO 9000ff

In Figure 14.3 it becomes obvious that the use of some technologies has an impact on the introduction of ISO 9000ff. Whereas hard- and software use is not a distinguishing factor for the introduction of ISO 9000ff, companies which use very sophisticated communication network technologies are more likely to have ISO 9000ff. The same is true for traffic and transport technologies. These results support the above analysis by sectors. Furthermore, the use of measurement and medicine technology makes it more likely for companies to introduce ISO 9000ff. The same is true for environmental technologies. In all three cases, the careful production and delivery of the service is important because in many cases the safety of the customers and other people can be endangered. However, the use of multimedia increases the likelihood of not introducing the standard, which may be explained by the little safety and risk impacts of this technology.

First preliminary conclusions can be drawn from the descriptive statistics. The size effect is obvious, which is caused by both the high introduction costs of ISO 9000ff and the greater information asymmetries between service provider and customer in larger companies. The last consideration also leads to the differences between the analyzed service sectors. Finally, health and safety-relevant technologies are also encouraging the introduction of the quality standard.

14.5 A COMPREHENSIVE PROBIT MODEL OF DRIVING FACTORS FOR THE INTRODUCTION OF ISO 9000FF

After the simple descriptive statistics, a comprehensive Probit model is tested. Probit specifications are designed to analyze qualitative data reflecting the choice between two alternatives. The endogenous variable is either one for companies introducing ISO 9000ff and zero for not considering ISO 9000ff.[18] The Probit specification models the probability of the event (= introduction of ISO 9000ff) as depending on a linear combination of the observed exogenous variables. This information is taken from the answers to the questionnaire on several innovation-related questions.

First, dummy variables for the industry, the size, and the technology used should explain the probability of introducing ISO 9000ff. Second, the external sources of knowledge are additional explanatory factors. When the service company is embedded in a tight network of knowledge flows from customers (manufacturing and services) and suppliers and integrated in a whole group of enterprises, then the likelihood of introducing ISO 9000ff should be significantly higher. Therefore, in order to meet the needs and preferences both of the customers and the suppliers, the service companies tend to implement quality standards. Third, the impact of some desired effects of innovations as driving forces for ISO 9000ff, such as flexibility, and improved productivity both of customers and employees, is integrated. Innovations are often hindered by several factors. Information is used about the importance of the role of internal resistance against innovations, which may be caused by anticipated performance pressure through the introduction of a quality-ensuring system. Secondly, another disincentive for innovations is the danger of easy imitation by competitors. In these cases, a quality certificate can help to distinguish the own product or service from those of the competitors. Finally, some features for the success of the services, like low prices and high quality, are used as exogenous variables for explaining the probability of introducing ISO 9000ff.

Table 14.1 Probit estimation to explain the implementation of ISO 9000

	A comprehensive model	Coefficient	Significant
Industry	Wholesale trade	0.9385	0.000
	Retail trade (base)	–	–
	Transportation/communication	1.0943	0.000
	Software	1.5370	0.000
	Science and technical services	1.1379	0.000
	Other business services	0.9747	0.000
	Other services	0.7429	0.000
Size group	1–19 employees	−0.5313	0.002
	20–49 employees	−0.2601	0.032
	50–249 employees (base)	–	–
	250 and more employees	0.2046	0.101
Technology	Software	0.0810	0.651
	Hardware	−0.1791	0.316
	Networks	0.2926	0.007
	Multimedia	−0.2851	0.014
	Transport/logistic technology	−0.0700	0.573
	Automation technology	−0.0769	0.541
	Medical technology	0.6714	0.005
	Biotechnology	−0.3613	0.184
	Environmental technology	0.2762	0.020
	Material technology	0.0186	0.894
External sources of knowledge	Customers (manufacturing)	0.0432	0.722
	Customers (services)	0.2185	0.038
	Suppliers	0.1152	0.285
	Competitors	−0.1241	0.246
	Enterprises within the group	0.2181	0.026
	Marketing or consultancy	−0.1775	0.143

Effects of innovations	Improved internal flexibility	0.1295	0.292
	Improved user-friendliness	0.0632	0.538
	Improved reliability of the services	−0.1071	0.367
	Improved time availability of the services	−0.0451	0.679
	Improved speed of the production/delivery	0.0686	0.570
	Improved capability to meet safety requirements (technical regulations, data protection)	0.2676	0.008
	Improved ecological, medical, ergonomic features	−0.1458	0.303
	Improved performance of the customer	−0.1083	0.338
	Improved productivity of the customer	0.0806	0.493
	Improved productivity of the employees	−0.1921	0.106
Factors hampering innovation	Organizational rigidities	−0.3133	0.037
	Too easy to copy innovation	0.2513	0.011
Features for the success	Low prices	0.0304	0.751
	High quality	−0.3413	0.035
	Flexibility	0.2969	0.079
	Schedule effectiveness	0.3811	0.030
	Image	0.9077	0.427
	Novelty of the product	0.1687	0.097
Foreign competition	in the home market	0.2898	0.004
Sale of services	in foreign markets	0.0912	0.398

Notes: Prob > chi2 = 0.000; Pseudo R2 = 0.1820; Constant = −2.7640 (0.000); Number of obs = 1436.

Source: ZEW/FhG–ISI: Mannheim Innovation Panel – Service Sector.

The results of the Probit estimation are presented in Table 14.1. With the retail trade and the size group of 50 to 249 employees as the basis, all other industries have a significantly higher propensity to introduce ISO 9000ff, with the highest probability being software companies. Furthermore, small and medium-sized companies are less likely to introduce the standard. Concerning the use of technologies, only communication networks, medical and environmental technologies cause a higher inclination for ISO 9000ff. These technologies are supposed to have a higher and immediate impact for health and safety. In particular, communication networks are confronted with the risk for safety, which cannot be observed immediately or at all. In contrast, multimedia use reduces the chance for ISO 9000ff, because obviously no strong risks endanger the users. The other technologies do not have any impact on the use of the quality standard.

A significant influence from the competitors as an external source of knowledge does not exist, therefore the introduction of ISO 9000ff seems not to be an imitation strategy. Furthermore, close connections to customers in the producing sector and to suppliers are not significant. In the case of being in close contact with customers from the service sector then the introduction of ISO 9000ff is more likely. The assurance and the signaling of a certain quality level for customers of the service sector is obviously caused by a tight integration in the knowledge flows of the service sector. However, if the company is advised by marketing agencies or consultants, then the likelihood for ISO 9000ff decreases slightly. The only explanation for this phenomenon could be that these companies have already introduced ISO 9000ff because of the advice of the consultants.

Most of the consequences of innovations are not significant. However, the reliability of the services and the ability to meet safety requirements generate significant results, because ISO 9000ff makes it easier for the companies to reach these goals and it serves as well as a seal of approval in order to increase the confidence of the customers. Additionally, quality standards are surprisingly intended to improve the productivity of the customer more than the productivity of the own employees. This result underlines that ISO 9000ff serves more the image of the company among the customers and regulatory bodies than the productivity of the service provision.

However, most of the innovation obstacles are not significant explanatory variables for the introduction of ISO 9000ff. Nevertheless, two significant results become visible and can be explained on the grounds of the theoretical hypotheses. First, when organizational rigidities usually prohibit innovations, then the likelihood of introducing ISO 9000ff decreases too, because of this internal resistance. Second, the greater the problem of innovations being imitated by competitors, the higher the probability of introducing ISO 9000ff. This makes it evident that in a market with homogeneous service products and narrow price margins, the certification according to ISO 9000ff represents a strategy to save the market share by having a quality seal which segments the market in service providers with and without quality certifications and therefore even allows a little price mark-up.

Concerning market characteristics, due to the productivity-enhancing effects of a quality standard system, flexibility towards customers' preferences and schedule effectiveness are positive determinants for ISO 9000ff, whereas the need for high quality services negatively influences the introduction of the ISO 9000ff as a kind of minimum quality standard.

The empirical result underlines the sector-, company size- and technology-specific results of the above analysis by the relatively high pseudo R^2 of over 18 per cent. Respecting the external sources of knowledge the customers in the service sector are important for the introduction of ISO 9000ff, which underlines its role as a quality seal. In groups of enterprises ISO 9000ff assures the compatibility of internal business processes between the different affiliates. Furthermore, ISO 9000ff is obviously supporting the capability of meeting safety requirements from legal regulations. Still, organizational rigidities are a negative factor for introducing ISO 9000ff. In line with the theoretical hypotheses, it can be postulated that ISO 9000ff has the image of a minimum standard, but it is not a seal for high quality services, because service companies active in these market segments are unlikely to introduce the standard. However, in markets with homogeneous and easily imitated services, ISO 9000ff serves as an additional quality sign in order to protect market shares. Moreover, the quality standard ISO 9000ff seems to increase the flexibility and the schedule effectiveness of the service companies.

14.6   CONCLUSION

The first analysis of the role of the quality standard ISO 9000ff in German innovative service companies, based on general hypotheses concerning the role of quality standards, produced elucidating results. Besides sector- and size-specific differences, the use of 'risky' technologies positively influences the probability of introducing ISO 9000ff. This quality standard has another twofold impact: first, as expected, it is a quality seal for the customers of the service company, especially in markets with homogeneous products and average qualities. Second, the introduction of ISO 9000ff has impacts on the internal processes of the service companies. In contrast to a conventional product standard, it supports the management in being flexible especially towards the preferences of the customers and in reaching project deadlines. However, the introduction may also increase the pressure on the employees, who are therefore evidently more reluctant about its introduction.

Based on these interesting first results, the analysis of further driving forces for the introduction of ISO 9000ff is recommended. However, the survey was not designed explicitly to analyze the introduction of ISO 9000ff and because of this we are able to analyze the driving forces of the introduction of ISO 9000ff only in innovative service companies. Comparisons with non-innovative companies and firms of the producing sector are not possible. Therefore, specific surveys based on questionnaires on this topic and interviews with relevant companies would support the empirical research in this field.

NOTES

[1] Chapter 14 refers to Blind (2002b) and is partly reprinted from *Technological Forecasting and Social Change*, Blind, K. and C. Hipp (2003), 'Driving Forces for the Introduction of Quality Standards in Innovative Service Companies: an Empirical Analysis for Germany', pp. 653–69, with permission from Elsevier.

[2] Cf. Muehlbauer (2001) for the actual standardization projects in service industries and Tether et al. (2001) for standardization and specialization in services.

[3] Cf. Hipp et al. (2000) on general incidences and effects of innovation in services.

[4] See Tamm Hallstrom (1996) for an explanation of the evolution of ISO 9000 as a management standard.

[5] Cf. Resetarits (1997, pp. 97 ff.), for the description of ISO 9000ff.

[6] The analysis of the impact of the other management standards Total Quality Management (TQM) and Just-In-Time (JIT) on ISO 9000 implementation in 500 US firms by Withers et al. (1997) revealed a positive interdependence. However, the complementary or substitutive relationships between different quality management systems will not be further discussed in this chapter.

[7] For specific literature about standardization in service sectors, see CEN (1996), Hartlieb and Behrens (1996), ISO (2001), and DIN (2002).

[8] For the formal proof see Shapiro (1983, p. 669) and the following (p. 673).

[9] Cf. Hauser (1979, p. 756 f.) and Jones and Hudson (1996) on the impact of standardization on the costs of assessing product quality.

[10] For a discussion of the role of ISO 9000 within a firm in order to assure quality, see Davis (1997).

[11] Another principal–agent problem exists between the shareholders and the management of a company. Concerning the positive reaction of US firms' stock price to the announcement of ISO 9000 registration see Docking and Dowen (1999).

[12] Cf. Pepels (1999, pp. 703 f.).

[13] Cf. Nelson (1970).

[14] Cf. de Vries (1997a, p. 318).

[15] Cf. Reserarits (1997, p. 100).

[16] Innovators are identified by a filter question in the survey which allows a self-assessment of the companies concerning the introduction of innovative products, processes or organizational structures.

[17] A distinction between ISO 9001, 9002, 9003, and 9004 is not possible due to the data restrictions.

[18] However, we are not able to determine whether or not ISO 9000ff was already introduced. Therefore, we assume that the probability of introducing ISO 9000ff remains constant for the remainder of the companies.

PART D

The Economic Impacts of Standards:
Empirical Evidence

15. Introduction to Part D

As we have seen in Part A, there are numerous impact dimensions of standards depending on the type of standards. Furthermore, even the differentiation by the types of standards does not lead to unequivocal conclusions regarding the impact dimensions, since they are mostly connected with both positive and negative effects. Whereas in Part C, the empirical approach tried to build up a comprehensive set of explanatory variables for participating in standardization and producing standards, the empirical analysis of Part D follows another logic. Since it is both theoretically and empirically[1] impossible to analyze all the effects of standards in a comprehensive and holistic way, we will concentrate on a selective set of impact dimensions, which will be analyzed in depth.

A crucial selection criterion for the restriction of the analysis of impact dimensions is the adequacy of the empirical treatment of the theoretical effects of standards. We concentrate on those impact dimensions which allow us to test the hypotheses derived from theory or set up ad hoc as precisely as possible. Effects of standards, which are challenging and interesting from a theoretical perspective, are neglected, if there is no sound data-base available which allows the quantification of the impacts and therefore the testing of the respective hypotheses. Therefore, impact dimensions with sets of hypotheses only testable in a qualitative manner based on a case-by-case approach are not further analyzed.[2] This means that consequences of standards described in very rigorous theoretical models with very strict assumptions cannot be followed.

Furthermore, the focus of the effects of standards is on the role of standards in the dynamics of technologies and innovations and on the impact of standardization and innovation on international competitiveness and growth. The relationship between standardization and market structures is not in the kernel of the analysis, especially since adequate quantitative data to test this rather complex relationship are not easily available. And standardization and standards are only one aspect among various others having influence on the market structure.[3]

Finally, the intangible impact dimensions of standards, such as on health, safety and environment, are also left out because of the serious measurement problems which deserve a single methodological based analysis, a challenge that cannot be tackled in this study. On the other hand, the theoretical hypotheses regarding these impact dimensions are rather straightforward and positive, which make this shortcoming not so severe.

First of all, standardization is a crucial phase in the innovation process, therefore we concentrate in Chapter 16 on the interrelationship between innovation in the narrower sense of technological change and standardization. As we have already seen in Part C about the driving forces of standardization, the dynamics in technology have a decisive influence on standardization activities. The cross-section analysis revealed that significantly more standards are produced in R&D- and patent-intensive sectors. However, a simple cross-sectional approach is not able to disentangle dynamic relationships between innovation and standardization. Therefore, after presenting some descriptive cross-sectional data we apply a time-series approach, which analyzes more in depth the influence of technological change and innovation on the production of standards documents. Since we rely mostly on patents as indicators for technological innovations, we are able to match very precisely the technological dynamics with the standardization activities and to run time-series regressions. This allows us also to deal adequately with the effects of the stock of standards on technological change and to analyze an important impact dimension of technical standards.

The second major impact dimension, which is empirically analyzed in Chapter 17, concentrates on foreign trade. Here, we are confronted with different theoretical frameworks, from which rather contradicting hypotheses regarding the impacts of standards are derived. Therefore, it is likely that the empirical analysis cannot reveal significant results, if the ambiguous effects outweigh each other. According to the analysis of the impact of standards on innovation, the results of some cross-sectional analyses are presented, before longitudinal data is investigated in order to identify the impacts of standards. Besides this time aspect, we start the time-series analysis on a high aggregated level, before we differentiate the approach by trading partners and sectors of trade.

Finally, we have a look in Chapter 18 at the macroeconomic impact of standards on growth relying on the roots of the endogenous growth theories, which acknowledged for the first time that besides the traditional production factors, capital and labor, the existing base of technological knowledge is a key determinant for sustainable growth. Since we examine growth rates, cross-sectional analyses are not appropriate and we just focus on time-series analysis. A further differentiation by countries or sectors is not feasible because of the restricted availability of long time series necessary for the empirical estimation of comprehensive growth models.

NOTES

1   For an analysis of the impacts of standardization on the whole economy, many quantitatively non-ascertainable aspects must be considered, which often make only qualitative statements possible. The difficulty of quantifying the 'societal benefits', for example by the impossibility of including the preferences of future generations, has always presented a real problem for economic cost and benefit analyses. For standardization, cf. Weise (1994, pp. 118–21).

2   Hawkins et al. (1995) provide an overview of further impact dimensions of standards in the area of information technology and environmental issues.

3   Gruber (2000) analyses the role of product standards in the market for semiconductors and finds that product standards lead to market concentration. His analysis is based on firm data of the most important semiconductor producing companies.

16. The Relationship Between Technical Change and Standardization: Virtuous or Vicious Circle?[1]

16.1 INTRODUCTION

In this first chapter about impact dimensions of standards, two questions should be investigated and answered: first, whether standardization activities follow the direction and speed of technical change and secondly, whether standards make technical change generally possible, or rather prevent it. The technical change, however, has besides the economic also social, ecological, political, historical and cultural dimensions, so that a purely economic approach focussing on the innovation process can only present a segment and cannot claim to be complete.[2]

It must again be pointed out here that the focus of this overview is directed above all on formal standards/technical rules, which have been set up by standards development organizations and are available to the general public. The differentiation between formal standards and industry standards which have evolved in (private enterprise) market processes will be made explicitly clear.

In Chapter 4, we have already presented the theoretical literature on the interrelationship of standardization and technical change. The three main issues analyzed concentrated on the general problem of transition from old to new technologies in network industries based on standards, the timing of this transition and the impact of the different types of standardization on technical change. The objective of this chapter is to provide quantitative empirical evidence about the relationship between technical change and standardization. The reminder of this chapter is structured as follows. First, we try to arrange the process of standardization with the innovation process covering both the invention of new technologies and products until their broad diffusion within the market. This step makes it obvious that there are close links and manifold feedback loops between the different phases of the innovation process and standardization. This complex relationship is also taken into account by the introduction of an institutional innovation within the traditional standardization process, the so-called development-related standardization, which will also be introduced in Section 16.2. The following Section 16.3 gives an overview of the small body of empirical literature on the issue, before Section 16.4 presents the set-up of the empirical analysis and its results.

16.2   TIMING OF STANDARDIZATION IN THE INNOVATION PROCESS

The question of timing in the innovation process will be pursued in this section in two parts. In the first part, standardization should be categorized in the innovation process and the ideal-typical anchoring of the standardization process in the innovation process presented. In the second part, the concept of development-related standardization (Entwicklungsbegleitende Normung) is introduced which arose from the necessity of changing the standardization practice to be closer in time to the innovation process, especially regarding interface compatibility and market introduction.

16.2.1   Fitting Standardization into the Innovation Process

Researchers, developers, construction engineers and marketing experts utilize the standardization documents as important sources of information about the state of the art in technology,[3] regarding restrictions from standard-related regulations and as an instrument for generating ideas for example for strategic innovations.[4] Only with a completed innovation, to which the beginning of the diffusion process must be counted, does the standardization process intervene as suggestions for a standardization of the new technological findings are submitted.[5] Baskin et al. (1998) differentiate in this connection between anticipatory, participatory and responsive standards, which can be equated with early/premature (shortly after the market introduction), respectively accompanying (including the users) and respectively belated (after product diffusion). They admit that the corresponding structures do not exist for participatory standards, but could be provided through the Internet. Figure 16.1 depicts – in a very simplified fashion – the chronological connection between standards, standardization process and innovation process.

It becomes obvious hereby that the standardization process and the innovation process are linearly related, that is one process follows the other and both are independent of each other initially. The only exception is the active collaboration in standardization committees which can deliver an input for the current innovation process. From this, Malisius and Weidner (1998) derive the future growing significance of enterprise-own standardization departments which assess existing external standards as a source of information for the innovation process but also actively introduce company-specific standards and represent the interests of the company in the standardization process. Standardization which runs parallel and accompanies the innovation process, however, is not foreseen in this system.

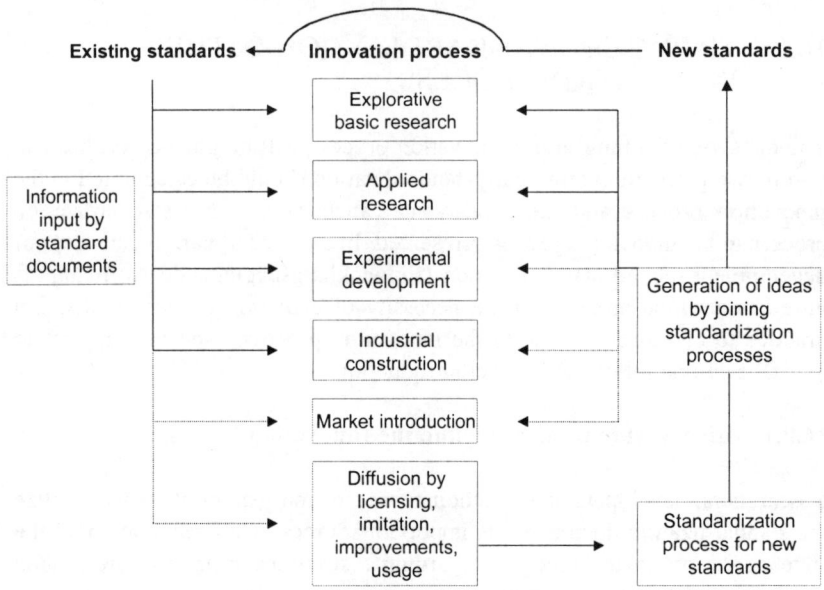

Figure 16.1   Fitting standardization into the innovation process

16.2.2   The Development-related Standardization (Entwicklungsbegleitende Normung = EBN)

In order to understand the origins of the concept of development-related standardization (EBN), it makes sense first of all to describe the inadequacies of the standardization practice in SDOs up till now (Hartlieb 1993). As already mentioned in Section 4.2, standardization at the wrong time can lead to economic inefficiencies. If it takes place prematurely, then there is a danger of a transition to a sub-optimal technology which will then possibly be further developed and the actually superior technology abandoned. If the standardization process takes place too late, the costs of the transition to the standard solution can be too high for many customers and suppliers, and an obsolete technology would continue being used for an unnecessarily long period. In both cases, standardization influences technical change negatively.

These theoretical findings can be observed in the reality of standardization. With the ever-increasing shortening of product life cycles, a concentration on systems and their elements and the increasing networking of R&D and productions on an international level due to globalization, standardization is faced with requirements which – in its original form, as sketched in Section 8.2 – it can no longer adequately fulfill.[6] At the beginning of the 1980s, the standardization procedure took five years because of a standards tailback which could not be worked off too long a time to guarantee the optimal transition to a new technology. The national SDOs were accused in many cases of being too slow, too bureaucratic and too little application-related.[7] Through the safety and health policy as well as technical insurance implications attached to the standards, not only the network externalities but also consumer policy and legal rules are the decisive obstacles to the successful market introduction and diffusion of new, not standardized, products and processes.[8]

The concept of development-related standardization (EBN) evolved in the mid 1980s as a reply to the above-mentioned challenges.[9] In principle, three tasks should be performed: it should check the need for standardization in running R&D and innovation projects and provide information about the standards for suppliers and (potential) customers, examine existing standards for practicality and flaws and perform active industry policy (promoting innovation through EBN in the industrial branches involved).[10]

EBN takes place in the course of standardizing activities (for example DIN expert reports, expert reports, pre-standards, drafts and from 1997 per PAS), that is standardizing documents, which can be drawn up within circa three months to two years and must not necessarily be transformed into a conclusive standard. These standard-similar documents are drawn up in a curtailed process, parallel to the R&D or innovation process, with the co-operation of researchers and developers from companies and government research institutes. Standardizing activities are coordinated within the framework of the EBN, financial and personnel advice and support provided, and above all committees set up at an international level, which stimulate the uniform, joint normative activities. Under these conditions EBN is also regarded as a suitable vehicle for technology transfer between industry and research. The EBN process takes place in three phases. Figure 16.2 should make this clear.

EBN thus differs from traditional standardization, not only because of the point in time when it takes place, but also the place of origin. Traditional standards evolve above all from the market processes and as quasi de facto standards are officially confirmed by the standardization institutes, whereby they are awarded with an official character. The main focus of this type of standardization lies on the proximity to application. The normative documents which emerge in the EBN process are hardly based on market processes and have a strong technology bias, they determine rather what the optimal technology-wise solution is.[11] The optimal solution economy-wise is thus pushed into the background. However, in Kristiansen's sense (1998), EBN is economically efficient because it removes the pressure from an 'innovation race' which is undesirable for the economy as a whole.

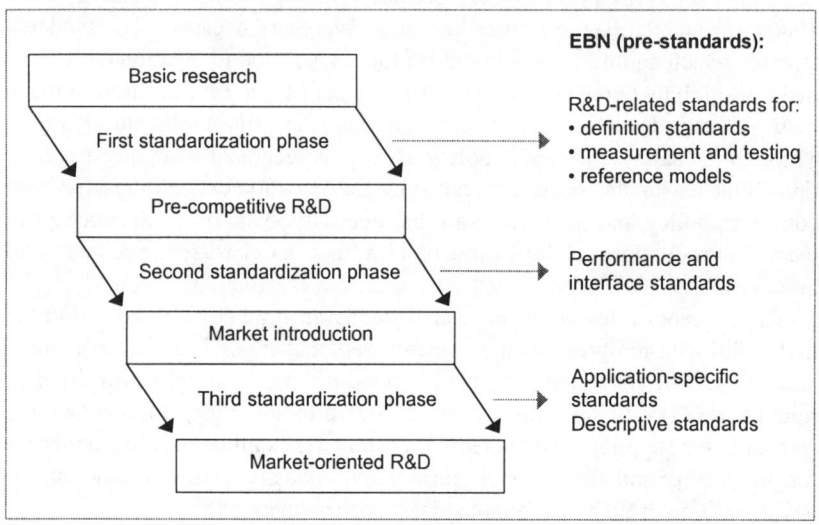

Figure 16.2   Layer model of development-related standardization[12]

The international standardization organizations also recognized the basic need to recognize the future demands on standardization in time and to adapt to them.[13] For this reason the two working groups ABTT (ISO/IEC Presidents' Advisory Board on Technological Trends) and LRPG (ISO/IEC ad hoc Group on Long Range Planning) were established in 1987.[14] The ABTT came to the conclusion that the fundamental changes in the technical innovation processes, especially their accelerated course, mean new demands on standardization. Therefore standardization should become more flexible with the help of provisional standards, faster by means of simplified procedures and more transparent via the cooperation of private and public R&D institutions. The LRPG tried by means of an international survey in twelve fields with a high scientific and economic significance to find out where the activities of international standardization should be concentrated. The areas of highest priority proved to be information technology, biotechnology, health care and environmental protection. It was recommended on principle to discuss and coordinate the survey results with running and planned standardization projects. The recommendation of ABTT and LRPG were then put into practice in the international and national standardization institutions, among other means by extending the EBN.

16.3   OVERVIEW OF EMPIRICAL INVESTIGATIONS

Empirical investigations on the influence of standardization on technical change can often not be detached from the influence of standardization on other economic variables. The complexity of the economic relationships allows this just as little as the resulting multi-functionality of many indicators regarding the presentation of the relations between economic issues such as growth, foreign trade, technical change and so on. The problem of this complexity was referred to already in Chapter 4, discussing the various impact dimensions of formal standards. In the following section some empirical approaches on the general economic impacts of standards should firstly be summarized before special investigations and results on the influence of standardization and standards on technical change are presented.

Many existing empirical studies on the economic effects of standardization refer to the effects of standardization on company costs and yields. From an economist's point of view, usually the particular entrepreneurial costs and benefits are added, supplemented by the additional costs of the standardization institutes and the social benefit of environmental and health effects added on. The latter category impedes a purely quantitative measurement of the effects of standardization on the economy; the social benefit frequently cannot be estimated qualitatively. Approaches will be presented in this section which attempt to evaluate the benefits and costs of standardization, but to tackle the challenge to quantify the social benefits would go beyond this study.

There are a number of studies which investigate the 'benefits and costs' of standardization, or its economic efficiency. Among others, these are studies by Pokorny (1974), Stübler et al. (1984), Christoph (1980) as well as Händel (1980). The authors concentrate in their studies on the comparison of the entrepreneurial costs arising from participation in standardization work or the creation of relevant standards and the benefit via saving. In a study by Coursey and Link (1998), the work of a standardization institute or rather its fundamental indicators were assessed in a case study. The efficiency of standardization is determined by the quotient of benefit and costs. The most important cost categories are named as procurement cost (participation in committee work, membership fees, costs for standardization documents) and application costs (search for relevant standards, data retrieval, application R&D, production), as benefit–cost savings from the application of the standards. The authors, however, are agreed that even on a company level the benefits of standardization are difficult to quantify. Through the positive external effects of standardization, the benefit is also not unambiguously classifiable to the costs. The most significant result of this survey can be resumed as follows: the economic viability of standards lies far above 100 per cent, the authors name numbers from 300 to 700 per cent; the setting up and application of standards is accordingly a worthwhile investment.[15] A criticism of these studies is that by dint of the exclusively company viewpoint, important effects of standardization for the economy as a whole are not taken into consideration and therefore the claim to quantify the benefits and costs of standardization for industrial sectors and the total economy is not justly dealt with.

Empirical studies on the interrelationship between standardization and technical change are thin on the ground, more or less non-existent. Allen and Sriram (2000) present four historical cases, which underline the ambivalent role of standards in innovation. The lack of broad empirical evidence besides anecdotal cases is also reflected in the very limited theoretical basis on this special subject area. In various studies – already mentioned above – the attempt is made to set out the benefit and costs of standardization from a company standpoint with overall economic additions, however a criterion 'influence on the technical change' is not to be found in these studies.

The studies of Meyer[16] tend in the direction of the focal point of this study, and are concerned with the aspect of product variety and measure and evaluate the success of variety-reducing standards (rationalizing standards). But in these studies the cost aspects are also more in the foreground, and the reduction of types/varieties is seen as a chance for internal savings. Great potential for savings is seen in a study of the Institute for Management Practice in Winterthur, and also in the interface standardization of computer-aided construction.[17] In another survey by Greenstein (1997), the economic inefficiencies of standardization caused by the effects of *lock-in* to the whole economy are estimated. In a study by Harhoff and Moch (1997) price indices are used to examine what value network externalities of an established product (*code compatibility* of software) have in comparison to new products. This study reaches the conclusion that the network effect causes a significant price difference and thus goods with network effects are preferred to new, technologically improved products. On the same subject, another study by Gandal (1995) examines to what extent compatibility standards influence the price of software. Gandal establishes that because of indirect network externalities, compatibility exercises a positive influence on the price.[18] Finally, Weiss and Sirbu (1990) examine the decisions to standardize in voluntary standardization committees to determine to what extent the size of the participating enterprises, their market influence, individual support measures, as well as the installed base and the technological quality of the product in question leads to certain technological decisions. Weiss and Sirbu reach the conclusion that the size of the supporting enterprise is more significant than the influence in the market; however, the influence in the market plays a great role for the demand and for this reason market-oriented decision-making criteria take center stage. The technological quality is not decisive, neither is the installed base.

A study which explicitly emphasizes the influence on technical change as an assessment criterion for standardization could not be found. This deficit is explainable by two arguments: on the one hand, the quantification of technical change is a task which only very few empirical scientists want to undertake. There are indicators referring to R&D outputs such as publications in scientific journals and patent applications in technology fields which have matured only in the last years, for example according to Schwitalla (1993) or Grupp (1997) and are coming into more widespread use. Input indicators such as R&D personnel and expenditure are long known, and can be consulted in supplementation of the output indicators. On the other hand, standardization is difficult to quantify and to operationalize for empirical analyses.[19]

16.4   EMPIRICAL INDICATOR-BASED ANALYSIS OF THE RELATIONSHIP BETWEEN STANDARDIZATION AND TECHNICAL CHANGE

The overview of the theoretical and empirical work on standardization in the context of technical change has shown – as in many other economic questions – that technical standards can trigger off not only positive but also negative effects for overall economic development. This can depend, on the one hand, on the specific form and application of the standards, such as the timing of standardization and interface compatibilities, and therefore can be corrected. On the other hand, many positive standardization effects possess directly negative implications, such as for example the variety-reducing effect of cost-cutting rationalizing standards. Which effects predominate overall can theoretically only be determined depending on certain framework parameters, such as cost digression effects or the preference for heterogeneous products.

A comprehensive and integrated empirical examination of all effects of technical standards on technical change is not possible because of ambivalent theoretical causalities and insufficient empirical data. For this reason, only a selection of important hypotheses on standardization in the context of technical change which are compatible with the data available, are presented on an aggregated level.

Figure 16.3 gives an overview of science, technology, innovation and standards as diffusion indicators. Focusing on innovation two approaches are available to trace innovation activities on the micro-level of the firm and the macro-level of sector and whole economies (among others Grupp 1997). On the one hand, one can look on the input into innovation activities, like R&D expenditure. One the other hand, patent applications are a good measure for assessing the output of R&D activities, although they do not indicate anything about the economic value (Scherer 1998).

Regarding standardization this twofold approach could also be applied. However, there are no data available about the real input into standardization processes, since the time of engineers spent in joining standardization processes cannot even roughly be assessed. Therefore, we use the output of SDOs, the number of standards documents published.

The standardization of a product, a component, an interface or a process is to be understood in the entire system of the innovation process at least as a provisional final act. In an ideal case, standardization is understood as the promoter of the diffusion of new products and production processes. At this point, at the latest, it crystallizes whether a product or a technology will find general application, that is will become an economically relevant market success. Then the standardization mechanism takes effect, which leads to further product improvements through standardized quality, interfaces to economically compatible products or to mass production made possible through variety-reducing standards. Only in a few cases does standardization take place during the development process (EBN); this form of standardization will not be dealt with here.[20]

In order to examine the correlations between technical change and standardization or standards statistically, the most suitable indicators must be defined for each case. Figure 16.3 provides an overview of the possible options. To measure the extent of standardization, recourse is made to the publication of standardization documents, because in the processed standards database PERINORM a data bank is available which reflects the output of the standardization process very well, both in the technical and time dimension. There is no adequate qualitative and quantitative information available on the input into the standardization process. The decision to use the patent applications as an indicator of technical change is one of several options and requires a certain explanation. For with the expenditure on research and development, with respect to the personnel engaged therein, so-called resource indicators exist. The decisive disadvantage of these indicators on the operative level is, however, that they can only be disaggregated to a very limited extent and thus adapted to the international classification of standards (ICS). In this manner sector-specific studies are subject to inaccuracies which can lead to significant distortions of the results.[21] In addition, it must be taken into account that the economic effect of the technical change is determined rather by the output of innovations and different R&D productivities lead to distortions.

As an output indicator for research and development, scientific publications are also available. Here the output of basic research is meant above all, which can only indirectly be categorized according to the economic sector-oriented ICS classification. Besides, the activities in basic research are usually too far removed in the time dimension from the standardization process, so that the probability of disruptive influences and distortions is very high and thus the likelihood of significant correlations rather low. Standards must be seen, however, as diffusion-promoting yield indicators of technology development.

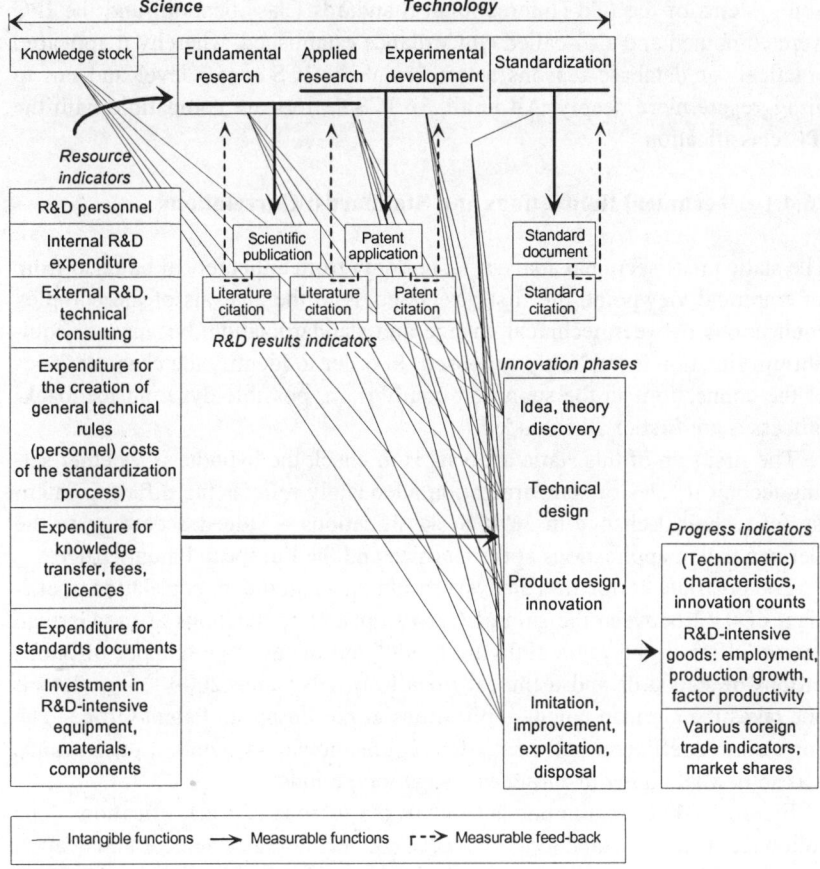

Figure 16.3   Systematic of innovation indicators[22]

Unlike in the United Kingdom[23] no comprehensive database containing all relevant technical innovations exists for Germany and other European countries, so the patent applications[24] remain as indicators for the technical change over a longer period of time. In particular the possibility to be able to search for the patent applications from Germany[25] at the German Patent Office disaggregated according to the IPC (International Patent Classification) and to form long time series from 1972 until 1995 for Germany are important common factors, along with the standardization documents as indicators for standardization.

These common features allow us to define the interfaces between standardization and technical change relatively accurately, not only in a factual but also in a time respect. For this purpose, as a first step the different classification systems of the ICS (International Standards Classification) and the IPC were combined and a so-called concordance established, whereby it appeared practical for database reasons to remain at the ICS sector level and not to disaggregate more deeply. All in all, 36 ICS sectors are compatible with the IPC classification.[26]

16.4.1   Technical Innovations and Standards: Correlations

The static cross-sectional analysis is, not only from a theoretical but also from an empirical viewpoint, the first direct access to the analysis of the complex connections between technical change and standardization, because an equilibrium situation is implicitly supposed. In order to identify the characteristics of the connections in the supposed equilibrium, possible dynamic feedback-processes are first of all excluded.

The first step of this static analysis is to check the hypothesis whether setting technical rules or standardization adequately reflects the differing intensity of technical change in 36 ICS classifications – judged according to the German patent applications at the German and the European Patent Office.[27]

The bivariate correlation analysis results in a significant correlation coefficient of 0.42 between the sum of German patent applications at the German Patent Office in the years 1997 until 1999 and the average number of publications of standards and technical rules from 1995 until 2000.[28] It is 0.48 if one takes the German patent applications at the European Patent Office. The correlation coefficient between patent applications and standard publications was somewhat higher in earlier investigation periods.

Figure 16.4 shows the percentages of the various ICS classifications. The following features stand out: the network technologies telecommunication and electrical engineering have the highest numbers of standards which are not complemented by respective patent shares; the cross-sectional group of environmental, safety and quality standards is as expected also significant, while the corresponding patent applications are to be found relatively speaking in midfield. Conversely, chemical process technology and vehicle

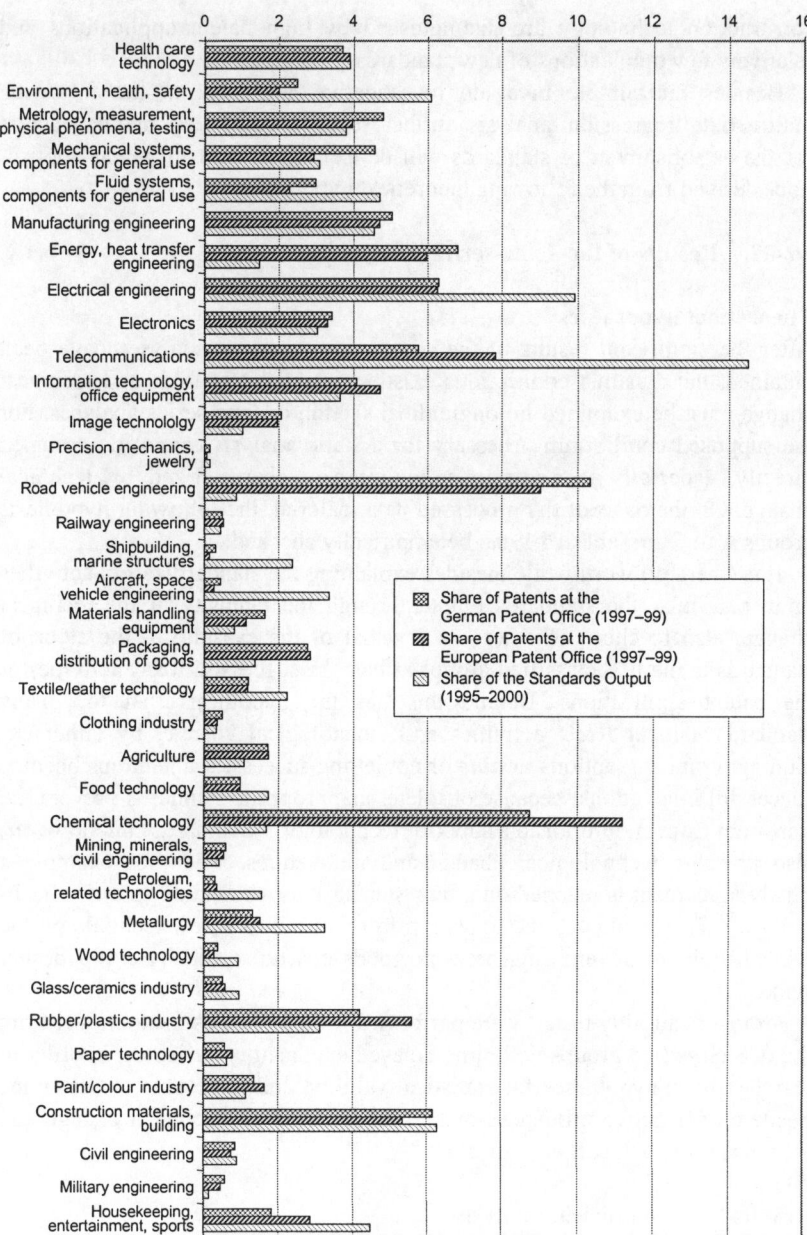

Figure 16.4   Share of German patent applications at the German and European Patent Office (1997–99) and of standards output (1995–2000)

construction technology are distinguished by high patent applications and relatively few publications of new standards.

Besides this simple bivariate investigative approach, in the following multivariate regression analysis further sector-specific explanatory factors for the establishment of standards will be examined, next to patent applications derived from the following theoretical hypotheses.

16.4.2   Results of the Time-series Analysis

Theoretical hypotheses
After the significant results of the static cross-sectional analyses have been obtained, the dynamic connections existing between standards and technical change must be examined in longitudinal section or time-series analyses. For the supposed equilibrium necessary for a static analysis cannot be assumed directly, especially in a dynamic environment characterized by technical change. On the basis of the processed data material, the following hypothesis groups A to C in Table 16.1 can be empirically checked.

It is (thereby) worthwhile, besides explaining the standardization activities or in particular the standards output through the intensity of the technical change, also to check whether – a reversal of the causality – the stock of standards in the ICS classified groups affects the successful R&D activities or the patent applications. Behind this lies the hypothesis that too many standards hamper R&D activities and technological change, by either excluding promising options *ex ante* or not letting successful inventions become successful innovations, because obsolete, inappropriate standards prevent the transition from an inferior to a superior technology.[29] Standards can, however, also promote technological change and innovations, when for example a timely agreement is reached on a new standard corresponding to a new technology.[30] This facilitates the transition from an old to a new technology; the typical problem of renewing network goods can be solved by compatibility standards.

Granger causality tests[31] were performed for a pooled estimate model from the ICS classified groups with time-delayed explanatory variables, in order to test the above hypotheses for statistical validity, and to be able to determine positive or negative influences of the standards on the technical change and vice versa for each ICS classification.[32]

Results of the empirical analysis
The period which forms the basis of the analysis encompasses the years between 1972 and 1995, on account of the restrictions in the patent search possibilities. In the following investigations for an analysis related to economic sectors, the time series will be restricted to 1973–93, if the expenditures on

R&D are brought in as an additional explanation, in particular for the development of the patent applications. The main question to answer is whether and to what extent technical change leads also to the generation of new standards and whether conversely positive or negative impulses emanate from the respective stock of standards in the subject areas onto technological change and innovation activities in the form of patent applications.

Table 16.1   Hypotheses about the relationship between technical change and standardization

	Influence of technical change on standard output and stocks
A0:	Accelerated technical change in the form of higher patent applications leads subsequently to a significant increase in new standardization documents.
A0a:	Accelerated technical change in the form of higher patent applications leads subsequently to a significant increase in the stock of standards.
A1:	Accelerated technical change in the form of higher patent applications does not lead subsequently to a significant increase in new standardization documents.
A1a:	Accelerated technical change in the form of higher patent applications does not lead subsequently to a significant increase in the stock of standards.
	Positive influence of standard output and stocks on the technical change
B0:	An increase of the stock of standards leads subsequently to a significant acceleration of the technical change in the form of patent applications.
B0a:	An increase of the standard output leads subsequently to a significant acceleration of the technical change in the form of patent applications.
B1:	An increase of the stock of standards does not lead subsequently to a significant acceleration of the technical change in the form of patent applications.
B1a:	An increase of the standard output does not lead subsequently to a significant acceleration of the technical change in the form of patent applications.
	Negative influence of the standard output and stocks on the technical change
C0:	An increase of the stock of standards leads subsequently to a significant slowing down of the technical change in the form of patent applications.
C0a:	An increase of the standard output leads subsequently to a significant slowing down of the technical change in the form of patent applications.
C1:	An increase of the stock of standards does not lead subsequently to a significant slowing down of the technical change in the form of patent applications.
C1a:	An increase of the standard output does not lead subsequently to a significant slowing down of the technical change in the form of patent applications.

Significant correlation coefficients do not necessarily prove the existence of causalities. Besides the calculation of cross-correlations, by means of the Granger causality tests[33] it can also be tested whether alterations of one variable are the reason for alterations of another variable.[34] The basic idea behind the Granger causality test is simple: if Y is caused by X, then alterations of X should follow alterations in Y. In order to be really able to say that X is the cause of Y, two conditions must be fulfilled. First, in a regression of Y compared to past values of Y, the addition of preceding X values as explanatory variables should significantly increase the explanatory power of the estimation. Secondly, Y should not contribute to the explanation of X, for in case X contributes to the explanation of Y and vice versa, it is probable that either a feedback process exists or third variables influence not only X but also Y.[35]

Table 16.2   List of sectors

- Metrology and control instruments; optical and photographic appliances; clocks/watches
- Metal products
- Mechanical engineering
- Electrical engineering
- Electronics
- Data processing and office machinery
- Vehicle construction
- Shipbuilding
- Aeronautics and space craft construction
- Textile industry
- Food industry
- Chemical industry (without pharmaceuticals)
- Oil processing
- Metal production
- Wood industry
- Glass industry
- Rubber and plastics industry
- Paper industry
- Household goods (jewelry, sports equipment, music instruments).

In order to be able to judge whether these two conditions are met, the zero hypothesis is tested, that one variable does not contribute to the explanation of the other. For example, for the test of the zero hypothesis, that the patent applications do not influence the standards output, in a first step the standards output is regressed in t compared with past values of the standard output in $t - 1$ to $t - m$. In a second step the past values of the patent applications are used in $t - 1$ to $t - m$. A simple statistical F-test (Wald test) can then be used to check whether the past values of the patent applications contribute significantly to the explanatory power of the regression. If this is the case, then the zero hypothesis can be rejected and it can be judged from the data that the patent applications have a real influence on the standards output. Accordingly, the converse hypothesis is tested whether the standards output influences the patent applications. The same procedure is also chosen for the connection between patent applications and stocks of standards.

Regarding the validity of the results, it must be noted that the investigation period from 1972 to 1995 contains only 24 observation points and thus the necessary normal distribution assumption cannot even be roughly assumed. As data for patent applications and standards publications/stocks exist for the period from 1972 to 1995 in 36 ICS classified groups, it is possible to pool the cross-section and time-series data.[36] Assuming that the 120ICS classified group-specific connections between technical change and standardization are constant in the period under observation, more efficient estimation parameters will result, as 24 data points for 36 ICS classified group and thus a maximum of 864 observation points are the basis for the analysis.

Because of the heterogeneity of the various ICS classified groups which became evident during the cross-correlation analysis, logically models are estimated in which sector-specific influences (fixed effects) are admitted regarding the connection between technical change and standardization, which also cover the differences in level of the autoregressive processes based on the analogy to the theory of technological accumulation.[37] Hereby the hypothesis was checked in a first model (Model 16.1) for all ICS classified groups i whether the standards outputs sto_{it} are influenced by chronologically preceding alterations in the speed of technical change.[38] As represented in Equation 16.1, the change in patent applications pat of the two preceding years is also referred to for the determination of the standards outputs sto besides an autoregressive process up to the third order:[39]

$$sto_{it} = C_i + \alpha_1 \cdot sto_{i,t-1} + \alpha_2 \cdot sto_{i,t-2} + \alpha_3 \cdot sto_{i,t-3} + \\ \beta_1 \cdot pat_{i,t-1} + \beta_2 \cdot pat_{i,t-2} + \gamma_i \cdot TREND_i + u_{it} \quad (16.1)$$

The calculation model has an adjusted R^2 of 0.88, and its F-statistics reveal a value of 144 for a level of significance under 0.01 (see Table 16.3). In order to test the significance of the influence of the technical change on the standards output, a Wald test for the coefficient value of zero was carried out for the coefficients of the patent applications of the two previous years. This was rejected with a value of the F-statistics of 3.09 for a significance level under 0.05. The sum of both coefficients is additionally positive, so that a significantly positive influence of the technical change on the annual standards output can be assumed.

Table 16.3   Results of the regression analyses explaining standardization by technical change/innovation

	Model 16.1	Model 16.2		Model 16.3	Model 16.4
Log [sto (–1)]	0.23^c [6.14]	0.14^b [2.26]	Log [stst (–1)]	1.00^c [26.90]	1.00^c [18.11]
Log [sto (–2)]	0.13^c [3.66]	0.02 [0.33]	Log [stst (–2)]	–0.14^c [–2.61]	–0.19^b [–2.42]
Log [sto (–3)]	0.06 [1.60]	0.12^b [2.05]	Log [stst (–3)]	0.05 [0.96]	0.06 [0.84]
–	–	0.09 [1.46]	Log [stst (–4)]	0.00 [0.09]	0.07 [1.06]
–	–	–0.11^a [–1.93]	Log [stst (–5)]	–0.10^c [–3.34]	–0.16^c [–3.40]
Log [pat (–1)]	0.28^b [2.37]	0.29 [1.45]	Log [pat (–1)]	–0.00 [–0.02]	–0.03 [–1.42]
Log [pat (–2)]	–0.23^a [–1.95]	–0.41^b [–2.11]	Log [pat (–2)]	–0.04^c [–2.69]	–0.05^b [–2.12]
–	–	–	Log [pat (–3)]	0.02^a [1.72]	–0.01 [–0.59]
–	–	–	Log [pat (–4)]	–0.01 [–1.10]	0.01 [0.34]
–	–	–	Log [pat (–5)]	0.04^c [2.82]	0.02 [1.29]
–	–	–	Log [pat (–5)]		
Log [R&D (–1)]	–	0.12^b [2.44]	Log [R&D (–1)]	–	0.02^c [4.20]
n = 756		n = 684	n = 304		n = 304
R²adj	0.88	0.89	R²adj	0.998	0.998
F-statistic	144.11	97.91	F-statistic	9093.93	7259.39
DW statistics	2.02	2.07	DW statistics	1.88	2.18

Notes: ^a = level of significance < 0.10; ^b = level of significance < 0.05; ^c = level of significance < 0.01.

As not only the development of patent applications, but also R&D expenditures are an important indicator for technical change, it is appropriate to explain the annual standards output, in addition to the R&D output indicator patent applications, also by the R&D input indicator R&D expenditure. According to the ICS classification, however, there are no R&D data available, so that we must fall back on the R&D expenditure classified according to ISIC 2nd rev. of the OECD database ANBERD for Germany (old federal *Laender*) from 1973 to 1993.[40] This means that a re-distribution/grouping of the patent and standard time series to the following 19 classes of the international economic sector classification was performed.[41]

The pool estimate on the basis of these 19 economic sectors was carried out for the following calculation model (Model 16.2):

$$
sto_{it} = C_i + \alpha_1 \cdot sto_{i,\,t-1} + \alpha_2 \cdot sto_{i,\,t-2} + \alpha_3 \cdot sto_{i,\,t-3} +
$$
$$
\alpha_4 \cdot sto_{i,\,t-4} + \alpha_5 \cdot sto_{i,\,t-5} + \beta_1 \cdot pat_{i,\,t-1} + \beta_2 \cdot pat_{i,\,t-2} +
$$
$$
\rho \cdot R\&D_{i,\,t-1} + \gamma_i \cdot TREND_i + u_{it} \qquad (16.2)
$$

This model shows an adjusted R^2 of 0.89 and an F-value of 97 (see Table 16.3). The Wald test for significance of the three indicator variables for the technical change underline once again the positive influence of the technical change on the current standardization activities with a highly significant F-statistic of 3.40. If the input into and thus also the output from R&D processes increases, then standardization activities also increase on principle.

With the same procedure it was also examined whether the changes in the standards stocks $stst_{it}$ were significantly influenced by technical change. The following model calculation (Model 16.3) appeared to be the one with the highest F-value on the basis of the given data available.

$$
stst_{it} = C_i + \alpha_1 \cdot stst_{i,\,t-1} + \alpha_2 \cdot stst_{i,\,t-2} + \alpha_3 \cdot stst_{i,\,t-3} +
$$
$$
\alpha_4 \cdot stst_{i,\,t-4} + \alpha_5 \cdot stst_{i,\,t-5} + \beta_1 \cdot pat_{i,\,t-1} + \beta_2 \cdot pat_{i,\,t-2} +
$$
$$
\beta_3 \cdot pat_{i,\,t-3} + \beta_4 \cdot pat_{i,\,t-4} + \beta_5 \cdot pat_{i,\,t-5} + \gamma_i \cdot TREND_i + u_{it} \qquad (16.3)
$$

This model has a very high adjusted coefficient of determination of 0.998, and its F-statistic shows a value of 9.094 by a significance level under 0.01 (see Table 16.3). The Wald test for the significance of the five patent application variables underlines also a positive influence of the technical change for the stock of standards. By comparison with the standards output, the stock of standards is influenced to a similar extent by the alterations in speed of the technical change. The same analysis on the basis of the economic sectors (Model 16.4) again provides significant results, where however only the R&D expenditures are a significantly positive explanatory factor for the development of the stock of standards:

$$stst_{it} = C_i + \alpha_1 \cdot stst_{i,\,t-1} + \alpha_2 \cdot stst_{i,\,t-2} + \alpha_3 \cdot stst_{i,\,t-3} +$$
$$\alpha_4 \cdot stst_{i,\,t-4} + \alpha_5 \cdot stst_{i,\,t-5} + \beta_1 \cdot pat_{i,\,t-1} + \beta_2 \cdot pat_{i,\,t-2} +$$
$$\beta_3 \cdot pat_{i,\,t-3} + \beta_4 \cdot pat_{i,\,t-4} + \beta_5 \cdot pat_{i,\,t-5} +$$
$$\rho \cdot R\&D_{i,\,t-1} + \gamma_i \cdot TREND_i + u_{it} \qquad\qquad (16.4)$$

Finally, the important question about the benefits of standardization for the whole economy must be answered, whether conversely the standards and technical rules also exert significant influence on the technical change, and if so, which. The reservation must be made in advance, however, that standards are only one possible determining factor in a series of many other determinants, such as R&D expenditures for example.[42] In a first model, therefore, not only the influence of the standards output *sto* but also of the R&D expenditures *R&D* on the patent applications *pat_{it}* will be estimated, based on the 19 economic branches. The following model (Model 16.5) provided the best estimated results:[43]

$$pat_{it} = C_i + \alpha_1 \cdot pat_{i,\,t-1} + \alpha_2 \cdot pat_{i,\,t-2} + \alpha_3 \cdot pat_{i,\,t-3} +$$
$$\alpha_4 \cdot pat_{i,\,t-4} + \alpha_5 \cdot pat_{i,\,t-5} + \beta_1 \cdot sto_{i,\,t-1} + \beta_2 \cdot sto_{i,\,t-2} +$$
$$\beta_3 \cdot sto_{i,\,t-3} + \beta_4 \cdot sto_{i,\,t-4} + \beta_5 \cdot sto_{i,\,t-5} +$$
$$\rho \cdot R\&D_{i,\,t-1} + \gamma_i \cdot TREND_i + u_{it} \qquad\qquad (16.5)$$

The adjusted coefficient of determination of this equation is 0.994, the F-test shows a highly significant value of 1.510. The Wald test for the influence of the six standard output variables is significant (F-value of 3.2), so that here too a positive influence cannot be rejected (see Table 16.4).

The second model (Model 16.6) utilizes instead of the standards output the stocks of standards *stst_{it}* to explain the development of the patent applications *pat_{it}*. It is tested on the basis of the ICS classified groups, assuming that the stocks of standards represent an indicator for the existing R&D potential:

$$pat_{it} = C_i + \alpha_1 \cdot pat_{i,\,t-1} + \alpha_2 \cdot pat_{i,\,t-2} + \alpha_3 \cdot pat_{i,\,t-3} +$$
$$\alpha_4 \cdot pat_{i,\,t-4} + \alpha_5 \cdot pat_{i,\,t-5} + \beta_1 \cdot stst_{i,\,t-1} +$$
$$\beta_2 \cdot stst_{i,\,t-2} + \gamma_i \cdot TREND_i + u_{it} \qquad\qquad (16.6)$$

This model has an adjusted coefficient of determination of 0.988 and a significant value of 1.275 for the F-test. The Wald test to prove the influence of the two standards stocks variables is significant with an F-value of 2.52, at least at the 10 per cent level of significance. This result also proves that the stock of standards cannot be rejected as a positive factor for the development of patent applications (see Table 16.4).

A further model (Model 16.7) utilizes, in addition to the stock of standards *stst*, R&D expenditures to explain the development of the patent applications

pat_{it}. It is assumed hereby for all economic sectors observed, that the dynamic interactions are identical between the input indicator R&D expenditures and the output indicator patent applications and the static differences are covered by the constant C ('fixed effects').

$$pat_{it} = C_i + \alpha_1 \cdot pat_{i,t-1} + \alpha_2 \cdot pat_{i,t-2} + \alpha_3 \cdot pat_{i,t-3} +$$
$$\alpha_4 \cdot pat_{i,t-4} + \alpha_5 \cdot pat_{i,t-5} + \beta_1 \cdot stst_{i,t-1} +$$
$$\rho \cdot R\&D_{i,t-1} + \gamma_i \cdot TREND_i + u_{it} \qquad (16.7)$$

This model has an adjusted R^2 of 0.994 and a significant value of 1685 in the F-test. The Wald test for the influence of the standards stocks variable is however no longer significant on the 0.10 error level with an F-value of 2.35 (see Table 16.4). Without the R&D expenditures, the stock of standards in this model is however highly significant with an F-value of 6.33 in the Wald test. Furthermore, the Wald test produces highly significant results for the joint significance of standards stocks and R&D expenditures. This indicates that the stock of standards and R&D expenditures represent a similar explanatory factor. This can be explained by the fact that R&D expenditures finally flow in part into the standardization process and the standards stocks thus represent the aggregation of a part of the R&D expenditures in the past.

In order to determine the sector-specific effects of the stocks of standards on the innovation potential, on the basis of the 36 ICS classified groups and the 19 economic branches, sector-specific and no joint effects regarding the standards stocks are assumed for the following equation:[44]

$$pat_{it} = C_i + \beta_{1i} \cdot stst_{i,t-1} + \beta_{2i} \cdot stst_{i,t-2} + \beta_{3i} \cdot stst_{i,t-3} +$$
$$\beta_{4i} \cdot stst_{i,t-4} + \beta_{5i} \cdot stst_{i,t-5} + \gamma_i \cdot TREND_i + u_{it} \qquad (16.8)$$

The equation shows an adjusted R^2 of 0.99 for an assessment of 315 in the F-statistic. For the following ICS classified groups, significant coefficients for the stocks of standards could be determined by means of the Wald test.

With the exception of the paper industry, precision engineering and railway technology, the stocks of standards in shipbuilding, aeronautics and aerospace, clothing industry, agriculture, mining, oil technology, metallurgy, wood processing, glass and ceramics industry, rubber and plastics industry, construction and military technology on the whole exert a positive influence on the development of innovations and patent applications respectively. For most of the ICS classified groups, however, no statements can be made.[45]

As these results were determined on the basis of a highly aggregated, sectoral analysis approach, they require microeconomic support through the assessment of company data in order to arrive at sound results and conclusions. In the next section, the assessment of companies about the quality of the stocks of standards and their impact on their R&D activities are presented.

Table 16.4   Results of the regression analyses assessing the influence of
standards on technical change respective innovation

	Model 16.5		Model 16.6		Model 16.7
Log [pat (−1)]	0.26^c [4.15]	Log [pat (−1)]	0.42^c [10.42]	Log [pat (−1)]	0.30^c [4.82]
Log [pat (−2)]	0.01 [0.21]	Log [pat (−2)]	0.22^c [5.00]	Log [pat (−2)]	0.10 [1.58]
Log [pat (−3)]	−0.01 [−0.08]	Log [pat (−3)]	0.02 [0.71]	Log [pat (−3)]	−0.004 [−0.06]
Log [pat (−4)]	0.02 [0.25]	Log [pat (−4)]	−0.03 [−0.62]	Log [pat (−4)]	−0.05 [−0.79]
Log [pat (−5)]	−0.24^c [−4.03]	Log [pat (−5)]	−0.07^a [−1.80]	Log [pat (−5)]	−0.15^b [−2.44]
Log [sto (−1)]	0.03 [1.55]	Log [stst (−1)]	0.23^b [2.01]	Log [stst (−1)]	0.09 [1.53]
Log [sto (−2)]	0.01 [0.29]	Log [stst (−2)]	−0.17^a [−1.66]	Log [R&D (−1)]	0.03^a [1.73]
Log [sto (−3)]	0.01 [0.71]	−	−	−	−
Log [sto (−4)]	0.02 [0.97]	−	−	−	−
Log [sto (−5)]	0.00 [0.09]	−	−	−	−
Log [sto (−6)]	−0.07^c [−3.81]	−	−	−	−
Log [R&D (−1)]	0.15^b [2.59]	−	−	−	−
n = 285	−	n = 684	−	n = 304	−
R^{2}adj	0.99	R^{2}adj	0.99	R^{2}adj	0.99
F-statistic	1510.45	F-statistic	1274.98	F-statistic	1684.54
DW statistics	2.00	DW statistics	2.04	DW statistics	1.98

Notes: ^{a} = level of significance < 0.10; ^{b} = level of significance < 0.05; ^{c} = level of significance < 0.01.

Table 16.5   ICS groups with significant coefficients of the stocks of standards

ICS Classified Group	F-test	Significance level	Sum of coefficients $(\beta_1 + \beta_2 + \beta_3 + \beta_4 + \beta_5)$
Precision engineering, jewelry	9.3945	0.0000	−
Railway technology	23.2306	0.0000	−
Shipbuilding, marine technology	9.1432	0.0000	+
Aeronautics and aerospace technology	4.0128	0.0014	+
Clothing industry	4.0814	0.0012	+
Agriculture	2.3273	0.0418	+
Mining, mineral resources	3.7312	0.0025	+
Oil and related technologies	4.7028	0.0003	+
Metallurgy	2.0030	0.0769	+
Wood processing	4.4703	0.0005	+
Glass, ceramics industry	9.9824	0.0000	+
Rubber and plastics industry	4.3959	0.0006	+
Paper industry	3.3460	0.0056	−
Construction industry and materials	3.2460	0.0068	+
Military technology	28.4187	0.0000	+

16.5   COMPARISON AND INTERPRETATION OF SECTORAL AND MICRO-BASED RESULTS[46]

16.5.1   Selection of Data and Methodological Procedure

The comparison between the sectoral and the microeconomic results is based on the responses of the companies questioned in Germany (see Section 13.3). Further, the comparison concentrates on the effects of formal standards which are exclusively examined in the sectoral or technology-based analysis. For only by this focus can a comparison be justified between the elaborated sectoral results and the results of the company survey.

As a rule, the depiction is limited to descriptive statistics differentiated according to eleven branches (see Table 13.1 in Section 13.3). Besides the descriptive statistics, it will be tested to what extent the participation in the standardization process is able to have an influence on research and development activities in Section 16.5.3. The corresponding connections have been proved on an aggregated level, but the question remains open whether this overall economic pattern can be explained by the sum of the actions of individual economic units, the so-called micro-funding.

Analogous to the approach to determine the driving forces for standardization in Part C, both the annual output of standards and the number of patent applications published are also mainly influenced by the direction and the speed of technological change not only caused by the activities of the companies, but also by the research results of public funded research institutes. The relevant pool of technological knowledge is not only nourished by new insight produced in domestic research institutes, but also in institutions abroad performing basic research. This external factor with its global dimension cannot be analyzed on a company level, but only on a sectoral or even macro-economic level. Finally, the individual rationality of a single company determines its decision concerning joining standardization processes. However, it is an open question whether the interaction of rather similar companies in standardization processes may lead to other R&D and innovation activities compared to a set of rather heterogeneous companies, whose characteristics may complement each other.

16.5.2   Comparison of the Sectoral and Micro-based Results

From the sectoral analysis of the connections between standardization and technical change based on ICS classified groups or economic branches, it became clear on the one hand that standardization adapts adequately to the technical change, and on the other hand the thesis has been refuted that standards prevent innovations. Whether these results are reflected in the responses of the companies questioned should be examined on the basis of the answers to the following questions.

Age and Scope of the Stock of Standards

The question whether the stock of standards is too old and is therefore obsolete was answered by the majority of companies questioned either with a negative or 'partly-partly' response. As Figure 16.5 makes clear, no great differences can be ascertained between the branches.[47] Therefore, it can be assumed in principle that the formal standards are regularly revised by the SDOs and correspond to the state-of-the-art in science and technology.[48] If the responses to the company survey are compared with the life cycles of historical, that is already withdrawn standardization documents (see also Figure 16.5), then in principle a positive connection emerges which, however, does not reach a significant level.

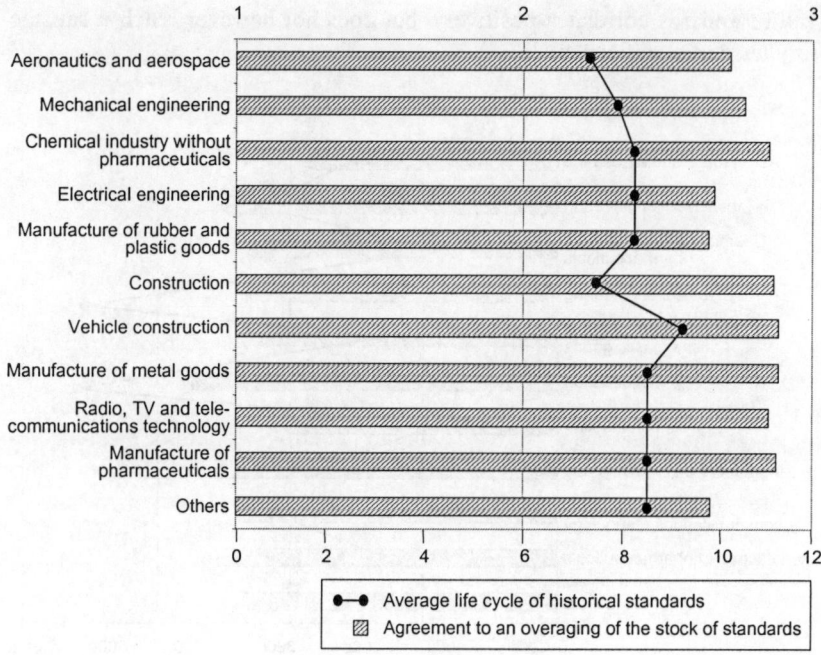

Figure 16.5   *Answers on the age of the stock of standards and average life cycle of formal historical standards*[49]

Besides the up-to-dateness of the stock of standards, its scope is of crucial significance for the innovation activities of the companies. It emerges from the time-series analysis according to ICS classified groups that with increasing standards output or stocks the patent applications and thus the innovations do not decrease but rather tend to increase. In a further question, a general assessment of the statement 'The number of formal standards is too large' was called for, without specifically targeting the subject of innovations. This must be kept in mind for the comparison with the aggregated results.

The mean of all branches lies between 'partly-partly' and 'agreement'. This means that the companies and in particular the 'non-standardizers' and companies conducting R&D are basically in favor of a reduction of the stock of standards. In addition, there are significant differences between the branches. Whereas the enterprises of the aeronautics and space construction industry and vehicle construction regard the scope of the stock of standards rather neutrally, the attitude expressed towards the number of standards in particular in building, mechanical engineering and in metal goods manufacture is very skeptical (see Figure 16.6). The connection between the companies' call for reduction in the number of standards with the actual scope of the valid standards correlates positively, but does not however reach a satisfactory level of significance.

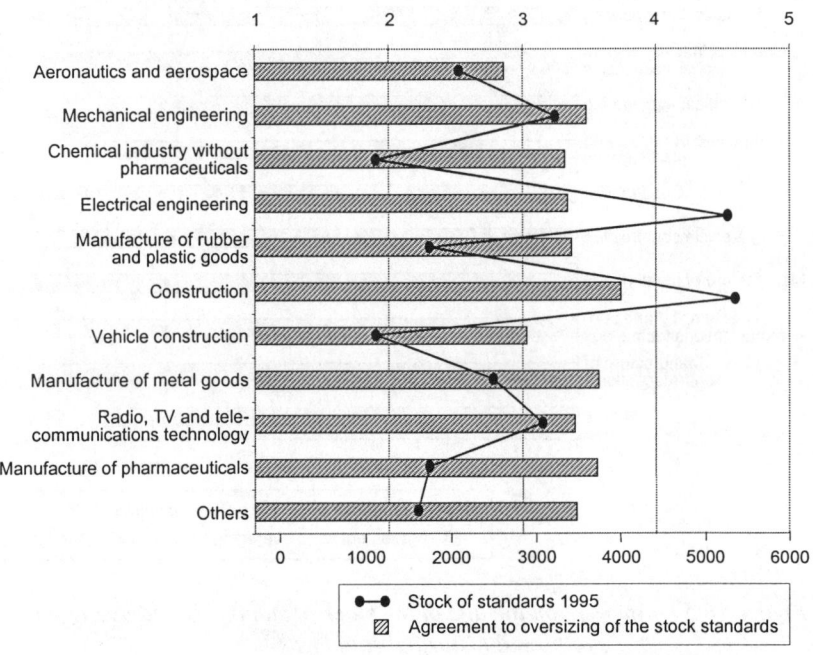

Figure 16.6   Answers on the scope of the stock of standards and actual stock of formal standards, 1995[50]

Standards as obstacles to innovation

Due to the positive correlation between standardization and innovation activities it can hardly be expected that standards present barriers to innovation. In order to be able to ascertain the meaning of formal standards as barriers to innovation compared to other innovation barriers, one question in the company survey dealt with the significance of various hampering factors. In all branches it was determined that the main barriers were seen in economic risks, the long approval procedure and the legal requirements (see Figure 16.7). On the other hand, neither the existing formal standards nor their lack presented serious barriers to innovation.[51] This also does not apply to industry standards.

Between the branches, however, a number of significant differences can be ascertained (see Figure 16.8). Especially in the construction industry, the existing stock of standards is regarded (at least partly) as an obstacle to innovation. On the other hand, formal standards present rather few barriers in radio, television and communications technology. It is further surprising that in the building/construction industry the lack of formal standards is regarded as an obstacle to innovation, unlike the other branches. If one excludes misunderstandings in replying to this question, then this result means that in the building sector not only existing standards but also lacking or non-existent standards hamper innovations and so the current stock of standards does not meet the current requirements of the branch. This result agrees with the general estimate, that the stock of standards in the building industry is too large.

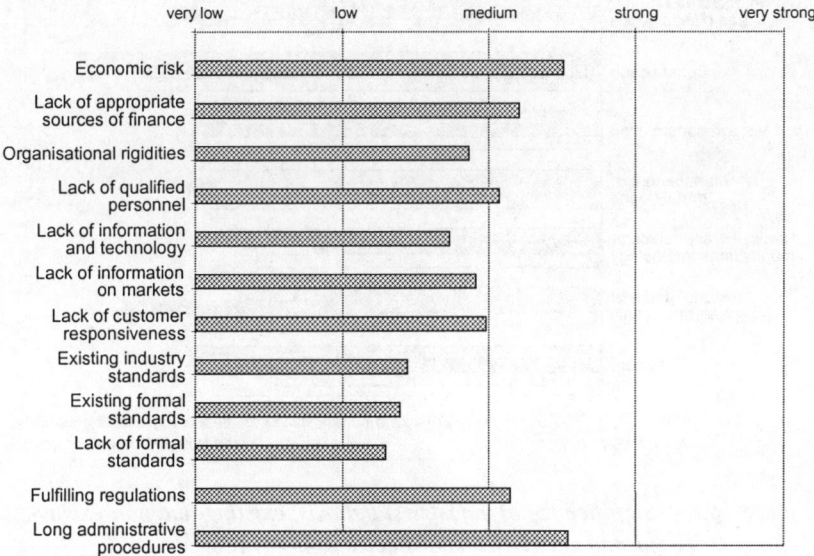

Notes: On the five-point scale: 1 = very strong; 2 = strong; 3 = neutral; 4 = slight; 5 = very slight.

Figure 16.7   Factors hampering innovation in the German manufacturing sector

The estimates of small, medium-sized and large enterprises do not differ fundamentally from each other. Only large enterprises, in comparison with the other companies, do not regard the lack of formal standards as obstacles to innovation at all. In case a certain standard does not yet exist, then a larger enterprise has better chances to initiate appropriate standardization processes. The 'non-standardizers' among the companies questioned consider not only the existing industry standards, but also the formal company standards to be greater obstacles to innovation. For this sub-group, however, the other innovation obstacles are also much more important.

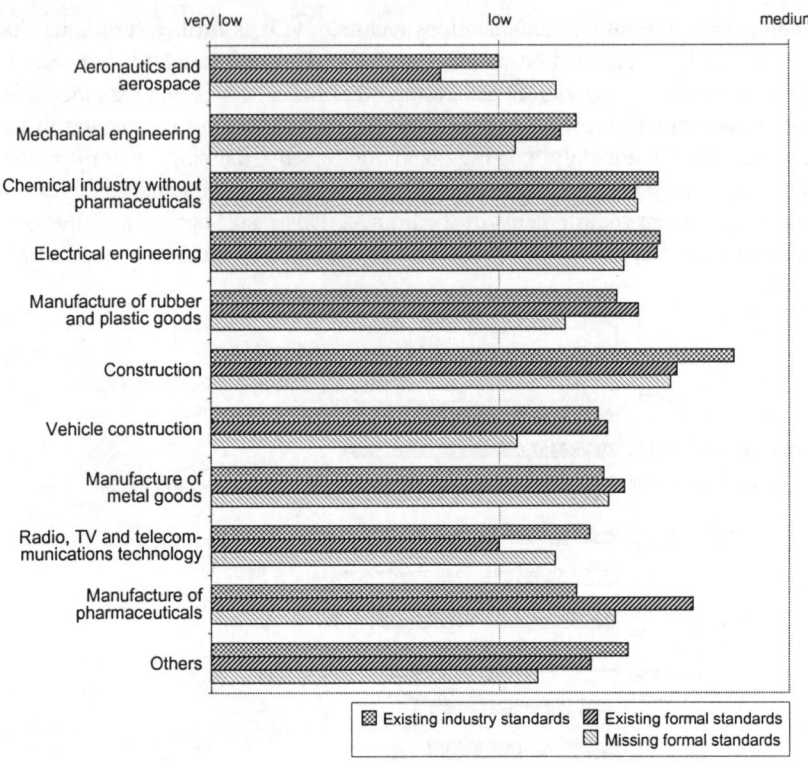

*Figure 16.8   Significance of industry standards, existing and non-existent
          formal standards as obstacles to innovation*

16.5.3   Impact of Standards and Standardization on the R&D Activities of Companies

Introduction

In order to conclude the analysis about the relationship between innovation
and standardization, a comprehensive, multivariate model to describe the
R&D activities of the surveyed companies is presented. In particular, it will
be examined to what extent (besides other factors) the active collaboration in
standardization (*Std*) can explain R&D activity or the lack of it which, once
again, is crucial for success in a competitive environment.[52] Additionally, the
actively utilized stock of standards (log*StdStock*) may be used as a substitute
for own R&D or may be an indicator for one part of the company's codified
knowledge base which is able to support its R&D efforts. Therefore, both the
collaboration in standardization and the active use of standards have ambiguous impacts on a company's decision to perform R&D. Not only the export

but also the import intensity will be included as explanatory factors for involvement in R&D, because on the one hand, the strong links to international markets increase the pressure to innovate. On the other hand, high technology can be procured in foreign markets, imported and used to complement or substitute own R&D.

Like export behavior, active involvement in R&D is also accompanied by high fixed costs. For this reason it should be easier for larger companies to amortize this pool of costs by fixed cost digression. Furthermore, human capital is a crucial pre-condition to conducting successful R&D. The average wage of the branches (*AvWag*) is therefore included as an additional variable for the quality of the labor force in the branch. After all, companies in highly competitive markets (*CompInt*) are forced to innovate in order to reach a temporary monopoly at least in niche markets, therefore these companies will tend to conduct more R&D (Scherer 1967).[53]

Table 16.6   Regression results of Probit and Tobit estimation to explain R&D activities

Explained variable: R&D behavior	Probit estimation		Tobit estimation	
	Coefficient	z-value	Coefficient	t-value
Company size (log)	−0.005	−0.13	−0.053	−0.80
Competition intensity	0.024	0.16	−0.465	−0.58
Export intensity	1.611[c]	3.67	4.850[c]	2.80
Import intensity	6.770[c]	2.94	17.958[c]	2.67
Average wage of the branch	0.000	−0.10	0.000	1.64
Active collaboration in standardization	−0.000	−0.29	−0.009	−0.25
Stock of standards (actively utilized)	−0.009	−0.37	−0.045	−0.76
Constant	0.077	0.09	−2.635	−0.56
Log likelihood	−250.249		−948.241	
Pseudo R^2	0.112		0.017	
Number of observations	417		417	

Notes: [c] = level of significance < 0.01.

Results of Probit and Tobit regressions

In order to determine the influence of the individual explanatory factors on the R&D decision, the following Probit regression model is used:[54]

$$R\&D_{ij} = \alpha_0 + \alpha_1 logSize_{ij} + \alpha_2 CompInt_{ij} + \alpha_3 ExpInt_{ij} +$$
$$\alpha_4 ImpInt_{ij} + \alpha_5 AvWag_j + \alpha_6 StdPart_{ij} + \alpha_7 logStdStock_{ij} + \varepsilon_{ij} \quad (16.9)$$

The R&D commitment of the companies questioned can be explained by the selected indicators only to a very limited degree. The export and import intensity of a firm alone significantly increase the probability of R&D activities. Companies which have strong links to international markets through their procurement and sales activities, have not only better prerequisites, but also more pressure to conduct R&D actively. Neither the collaboration in the standardization process nor the actively utilized stock of standards exhibit significant coefficients.[55] Whereas R&D may be a substitute for the participation in standardization, the latter and an actively used stock of standards are not sufficient to make own R&D redundant. Therefore, the results indicate that participation in standardization and an actively used stock of standards are both complements and substitutes for own R&D efforts. Finally, the average wage as an indicator of the quality of the employed human capital does not provide the anticipated positive explanatory factor for R&D.

In analogy to the analysis of the export activities, besides the decision to perform R&D it is interesting to see whether the R&D intensity is influenced by the proposed variables:

$$R\&DInt_{ij} = \alpha_0 + \alpha_1 logSize_{ij} + \alpha_2 CompInt_{ij} + \alpha_3 ExpInt_{ij} +$$
$$\alpha_4 ImpInt_{ij} + \alpha_5 AvWag_j + \alpha_6 StdPart_{ij} + \alpha_7 logStdStock_{ij} + \varepsilon_{ij} \quad (16.10)$$

The performed Tobit estimation to explain the R&D intensity rather than the R&D decision has an even smaller explanatory power, compared to the Probit estimation. However, the coefficients of the export and import intensities do not lose their significance. In addition, the average wage as an indicator of the quality of the employed human capital is close to being a significant variable for explaining the R&D intensity.[56] On the other hand, companies in a highly competitive environment are confronted with a negative – but not significant – impact on their R&D intensity. If we assume in the long run that companies performing R&D are also more innovative, then they may be able to change their market conditions from nearly perfect competition with a large number of competitors to monopolistic competition with a much smaller number of suppliers of similar goods. However, in the short run analysis, the degree of competition is an exogenous variable. The results suggest that lower competition leaves the companies more freedom to devote resources to R&D. These considerations confirm Schumpeter's hypothesis that companies need a certain degree of market power to perform R&D successfully (Schumpeter 1934), which could not be empirically verified by Scherer (1965) for companies in the United States.

Summary

Coming back to our initial intention in Part C to identify variables which determine the likelihood of companies to join standardization processes, we have assumed the exogeneity of the R&D intensity. The result of this last model supports this central assumption, since standardization – both the participation and the stock of standards actively used – has no influence on the R&D activities of companies. Based on the results presented in Section 13.4, depending on its R&D intensity the company decides about whether it should join a standardization project or not. And the more a company is engaged in R&D, the less it is inclined to commit resources for standardization activities. The cost–benefit ratio is obviously less favorable for those kind of companies.

Regarding the impact of participation in standardization and of the use of standards on R&D activities, we find on the company level no evidence for this assumed link. However, with the exception of the involvement in international trade no other variables attribute significantly to the explanatory power of the models.

16.5.4   Conclusion

The survey about the literature dealing with the relationship of innovation and standardization emphasizes their ambivalent relationship. On the one hand innovation and technical change challenge the existing stocks of standards. One the other hand, standards may restrain innovations activities by freezing the technical state of the art. Nevertheless, standards are also a necessary condition for innovations in network industries in order to realize the necessary positive network effects. In total, the relationship between technical change/innovation and standardization is, according to the various theoretical approaches, rather ambivalent. Therefore, approaches to analyze this relationship empirically do not promise to come up with clear and significant results.

In order to tackle the empirical problem in a more comprehensive way, different approaches are applied. First, a sectoral and indicator-based aggregated approach is performed. Based on time-series analyses using patents and standards documents a positive impact of technical change and innovation on standardization activities can be found. The other way round, increasing stocks of standards have only a weak, but still positive impact on innovation, with respect to the growth of patent applications. In total, innovation and standardization represent more a virtuous than a vicious cycle.

In order to complement the sectoral aggregated analysis, a micro-based approach using answers from a company survey is performed. The answers suggests a more critical attitude on the adequacy of the stock of standards especially in relation to innovation. However, compared to other obstacles for innovation formal standards have only a small importance.

Finally, in multivariate models participation in standardization and the use of standards within the company have an ambivalent impact on R&D activities, which means that both substitutive and complementary relationships exist in companies.

Summarizing this empirical evidence, it turns out that standards are no serious obstacles for innovation, but on the other hand it cannot be proved that they have a strong positive influence on the innovation activities and successes both at the sectoral and the company level. In total, their relationship is more positive, although changes in the standardization processes may improve the positive influence of standards on innovation.

NOTES

[1] This chapter refers in parts to Blind et al. (1999a).

[2] Reihlen (1998, p. 14), refers to the importance of considering these non-economic framework conditions in standardization work as a basic pre-condition for successful standardization.

[3] Thus a standard document is comparable to a patent document which also contains the state-of-the-art of technology and discloses the technological know-how entailed. Cf. also Thiard and Pfau (1992, pp. 16–18).

[4] Beitz (1986, p. 87), cites for this the example of the deliberate deviation from existing standards, also to achieve competitive advantages.

[5] On the course of the standardization process, see for details Chapter 8.

[6] Cf. Tassey (1995, p. 170) and Holler (1996, p. 145).

[7] Cf. Steffensen (1997b, p. 61).

[8] An example of this is laser technology which was neither permitted for use in welding seams in ships' hulls (objections on insurance grounds) nor in operations (health insurers do not pay for the more expensive operation, although the recuperation time in hospital is reduced). See on this also Kleinaltenkamp and Marra (1994, pp. 74–7).

[9] The term 'development-related standardization' was coined by the former German research and technology minister Heinz Riesenhuber (on this topic see also Riesenhuber 1989, pp. 121–23 himself).

[10] Cf. Kleinaltenkamp (1995, pp. 81–101).

[11] Cf. Kleinaltenkamp (1995, pp. 94 ff.).

[12] Cf. the presentation of Eichener and Voelzkow (1993, p. 766). The arrows point to transitional stages; it is not intended to imply a sequential model.

[13] Cf. Pfau (1991, pp. 68 ff.) on this issue.

[14] Cf. ISO and IEC (1990) for the documentation of the whole results.

[15] Pecher (1996, p. 728), even reckons with a minimum value of 1.000 per cent for the waste water releated technical rules rules relevant in sewage technology.

[16] Cf. Meyer (1995, especially pp. 141–53) and Hesser and Meyer (1993).

[17] Cf. Scharf (1999). This potential exists above all in the automotive industry, in aerospace and mechanical engineering.

[18] Gandal (1995) proves positive network effects empirically for spreadsheet programmes, Harhoff and Moch (1997) for database software, Saloner and Shepard (1995) for the adoption of automated teller machines, and Hartman and Teece (1990) for the minicomputer industry.

[19] See the explorative study of Hawkins (1996).

[20] The EBN are regarded by many experts as an important additional option in standardization

practice, as at present standardization takes place content- and time-wise too far removed from the innovation process, and is thereby delayed and not optimal technologically. On EBN see Section 16.2.2 and Steffensen (1997a, pp. 159–67).

21   In single steps of the study we shall return to the industrial expenditure for research and development, if additional insights are to be expected herefrom.

22   Modified according to Grupp (1997, p. 145).

23   For the United Kingdom a database exists of all relevant technical innovations between 1945 until 1983 (see University of Sussex, Science and Policy Research Unit 1982).

24   Cf. among others Brouwer and Kleinknecht (1996) on strengths and weaknesses.

25   All patent applications are counted with the home of the applicant in Germany and thus the domestic and not the native concept applied.

26   The concordance is attached in Appendix I.

27   Source: Blind (2002c).

28   Due to institutional conditions and the low number of released standards by comparison with the patent applications, the five-year mean was formed.

29   A further, rather formal argument is advanced by Thiard and Pfau (1991, p. 18), to interpret formal standards which represent the latest status of technology, also as a minimal criterion for the novelty of innovations when awarding patents.

30   Cf. Farrel and Saloner (1987, p. 15). Timely standardization narrows the spectrum of possible product characteristics, lowers thereby the risk of innovation projects and so increases the success of R&D expenditure (that is patent applications).

31   With the Granger causality test it can be tested whether one time series is explained by another. The term 'causal relationship' is therefore used here as usual in econometrics and does not correspond exactly to the colloquial sense.

32   In addition, by means of a Cox regression Blind et al. (1999a) examine to what extent the life span of standards depends on the technical change dominant at the same time and find that the higher the dynamic of technical change the shorter the life span of standards.

33   Cf. Pindyck and Rubinfeld (1991 pp. 216 ff.), on this.

34   Granger causality tests were also used for example to test whether causal relationships exist between the expenditures for public infrastructure structure and economic growth in the Federal Republic of Germany. Cf. Schlag (1997).

35   For applying the Granger causality test, the time series must be stationary.

36   Cf. Pindyck and Rubinfeld (1991, pp. 223 ff.) on pooling cross-sectional and time-series data.

37   According to this, the capabilities for technological innovation depend on the collective experience in dealing with technology in the past. Technological innovations are therefore the result of a process of learning and experience (cf. on this point Münt 1996, pp. 24 ff.). If one regards the standardization process as a part of the innovation process, the same connection applies for it.

38   In order to meet the stationarity condition, the natural logarithms were formed for the time series of the standards output, the standards stocks, the patent applications and the R&D expenditure and in each case a trend variable introduced. The critical values of Levin and Lin (1992, p. 45), for stationarity tests with panel data were applied.

39   This model provided the best estimates in view of the available data.

40   The existing time series in absolute DM figures were deflated/revalued with the price index of the capital goods sector (Statistisches Jahrbuch, various annual volumes).

41   The concordance is attached in Appendix III.

42   On the basis of the ICS classified groups therefore no significant effect of the annual output of standards can be determined on the patent applications.

43   The theory of technological accumulation calls on the other hand for an autoregressive process. Münt (1996, pp. 309 ff.) proves this estimation approach empirically in an international comparison for selected branches.

44   Ideally the models of Equations 16.6 and 16.7 should be estimated with sector-specific coefficients, however this approach leads to a supercritical number of coefficients to be estimated, in view of the limited length of the time series.

[45] The analysis on the basis of the 19 economic branches did not lead to any contradictions with the presented results. For this reason they are not presented separately.

[46] The following presentations refer partly to part 5 of Blind and Grupp (2000).

[47] According to the nonparametric Mann–Whitney U test, there are no significant differences. This test is applied in the following for all questions which permit interval-scaled answers. The Chi square test is applied for yes/no answers.

[48] The companies conducting R&D however estimate the stock of standards to be older than those not conducting R&D.

[49] In the following charts on questions about attitude the numbers along the axles represent the following: 1 = do not agree at all, 2 = do not agree, 3 = partly-partly, 4 = agree, 5 = agree absolutely. As there are only very few standards in pharmaceuticals, the average life cycle of the ICS group 'Medical Technology' is taken as proxy. The category 'Others' is compared to the average of all standards.

[50] On the stock of standards in pharmaceuticals and the other companies, see footnote.

[51] This result coincides basically with the findings of the Mannheim Innovation Panel, whereby in the survey on the overall economic benefit of standardization the question was split into 'Legal requirements' and 'Formal standards', which makes clear that legal requirements and long administrative and approval procedures above all are the (main) barriers to innovation (cf. Janz and Licht 1999, pp. 44 ff.). Also Berger (1998) proves the negative overall economic implications of regulation for the Federal Republic of Germany.

[52] A more adequate approach is to estimate a set of equations determining R&D activity and standardization participation simultaneously. However, the available data set does not allow this more sophisticated approach.

[53] The counter argument is the so-called Schumpeter hypothesis, which postulates that companies need a certain degree of monopolistic power in order to conduct innovative activities (Nelson and Winter 1982).

[54] In the Probit model, the variable R&D behavior takes only two values, with yes = 1 and no = 0.

[55] However, in contrast to the negative impacts of governmental regulations on macroeconomic growth (Berger 1998), involvement in standardization does not negatively affect R&D activities.

[56] In the models not presented, where the involvement in standardization processes is differentiated by the regional level, the average wage level is able to explain the R&D intensity.

17. The Impact of Technical Standards on Foreign Trade

17.1 INTRODUCTION

After the ambivalent impact of standards on innovation, we continue with a second impact dimension of standards, which again has to be analyzed closely in the context of the technological strength and the innovative capacity of a country. Whereas the traditional neo-classical theory of trade assumes common technologies across countries and therefore excludes the possibility of technology gap-motivated trade, there is a growing literature which considers differences in technology as an important driving force for trade. This literature consists of both the formulation of new theories of trade and the reformulation of traditional approaches to trade. In the latter group is the so-called neo-endowment theory of trade, which extends the traditional two-factor model of trade to include a greater number of narrowly defined input factors, including science and technology, while keeping the assumption of an identical world production function. In the former group are theories considering technology as an endogenous factor, including new trade theory and the evolutionary approach (Grossmann and Helpman 1991).

The last approach provides the background for this chapter, in which we estimate (in response to the claim of Helpman 1998, 1999) for more technologically-oriented trade theories and empirics, the importance of differences in technology in explaining the trade volumes on the basis of technology sectors. They are based on the International Classification of Standards (ICS) because, besides the incontestable positive role of technology and innovation, we are especially interested in the assessment of the trade impact of technical standards published by legitimated standardization bodies following recent approaches of Swann et al. (1996) and Maskus and Wilson (2000).[1]

While in the past the role of tariff barriers in export trade, such as customs duties, formed the focus of the political and economic discussions (GEWIPLAN 1988), the emphasis has in the meantime shifted to post-GATT and WTO agreements to center on around non-tariff trade barriers (Langhammer 1998; Baldwin 2000). Standards and technical rules are included in this category (Ritterbusch 1999). The standardization organizations were

...ged in the WTO agreements to ensure that the standards they publish should not represent unnecessary barriers to international trade. Furthermore, they should adapt their standards to the already existing international standards.

The remainder of this chapter is structured as follows. In Section 17.2 hypotheses on the mechanisms of the national innovation potential and of standards on export trade are presented; the data which formed the basis of the empirical study will be explained in Section 17.3. In Section 17.4 the effects of the innovation profile and the stocks of standards on trade will be examined, first for the German, Austrian and Swiss bilateral trade relations and then for the trade in the technology sectors.

Since a simple cross-sectoral analysis is not sufficient for the investigation of the causal relationship between standards and export performance, we also apply various time-series analyses progressing from a very general to a very differentiated level of analysis. First, we rely in our examination on the total trade of Germany with the rest of the world and compare these results with insights from a similar approach applied to the situation of Switzerland. Second, we analyze the bilateral trade flows of Germany with the United Kingdom and France, since this also allows us to take into account the stock of standards in these two countries. Finally, we have a close look at Swiss trade with measurement and testing products, which is affected by additional impacts of standards. In contrast to the approaches based on aggregated statistics, we finally have a look at the impact of the engagement in standardization on the export performance of companies at the micro level.

All these approaches enable us to derive a comprehensive and differentiated picture regarding the impact of standards on foreign trade. Consequently, the chapter concludes with the attempt to evaluate the influence of innovations and standards on foreign trade and to provide corresponding recommendations for standardization practice.

The time horizon of the time-series analyses covers the period from the beginning of the 1980s until the middle of the 1990s. As already mentioned in the introduction, the globalization of the economy in terms of the increasing number of multinational companies and growing transnational trade volumes intensifies the need for uniform international standards and questions the existence of idiosyncratic standards. Within the European context, the target of the completion of the Single Market reflects this trend by an institutional transformation. In the long run, international standards will crowed out national standards almost totally leaving just a small fraction of standards released by national SDOs, although the standardization activity itself is still distributed among the national SDOs. However, this new stable equilibrium has not yet been reached and we are still in a phase of transition. Therefore, the time-series analyses cover a period until the middle of the 1990s, until which the completion of the Single Market and the substitution of national standards by European and international standards were just at their beginnings and their impacts on trade flows were only partly effective.[2] The restriction of the time series also enables us to differentiate between the effects of national and European or international standards, a distinction which will not be feasible any more after a complete harmonization of standards specifications at least within Europe, but the results are very useful in respect of the trade policies with the United States of America.[3]

17.2 HYPOTHESES ABOUT THE IMPACT OF INNOVATIONS AND TECHNICAL STANDARDS ON FOREIGN TRADE

Since a comprehensive theoretical model covering the various and often contradicting impacts of technical standards has so far not yet been developed,[4] a set of hypotheses derived from theoretical considerations is presented closely related to the discussion of Section 4.4. Based on the categorization of Swann et al. (1996) and with the explicit addition of the dimension innovation and its impact on trade, three perspectives of the impacts of standards and technical rules on foreign trade can be differentiated.

In the context of competitive advantage, technical innovations protected by intellectual property rights like patents have unambiguous positive impacts on trade performance (Maskus and Penubarti 1998), because they restrict others from using the protected technologies or products. Formal standards are close to be general public goods and can be used by every producer worldwide. However, due to high adaptation costs and restricted absorptive capacity, other companies not involved in the standardization process itself have disadvantages to apply them. Consequently, formal standards, especially national ones, are able to promote the competitive advantage of domestic companies by facilitating economies of scale or compatibility between components of complex systems. Furthermore, the application of national, and even more international, standards can increase the perception of the quality of domestic products.

Formal standards released by national SDOs can also create competitive disadvantages both for domestic and foreign suppliers leading to trade distortions. The latter can be hindered in their export activities by artificial non-tariff barriers to trade caused by country-specific idiosyncrasies in formal standards. Domestic producers may suffer from additional formal national standards, which create an extra burden or discrepancies with international standards and no further advantages regarding costs or quality of products. Innovations of domestic suppliers have in general no negative impacts on trade. They have a trade deterring effect in the way that imports may be re-duced due to the higher domestic competitiveness, but this is based on the market mechanisms of price and quality competition.

Finally, the dimension of intra-industry trade dominates the trade flows among the highly industrialized countries. In particular internationally har-monized formal standards provide the ground for the expansion of intra-industry trade. Common compatibility or minimum quality standards facilitate the specialization in product variations and so allow a better exploitation of economies of scale. The impacts of variety-reducing standards are ambiva-lent. Additional product and process innovations generally increase the pro-duct diversity, lower production costs and therefore increase both domestic and foreign demand. Therefore, they cause more intra-industry trade (Saviotti 1996).

Table 17.1   Effects of innovations: a comparison of two theoretical approaches

Theoretical approach	Innovations indicator: patents	Economic effects	Impacts on exports	Impacts on imports
Competitive advantage	International: yes, possibly stronger	Create temporary monopoly in case of new products	+ +	–
	National: yes	Improve quality and/or reduce costs of national products	+	–
Intra-industry trade	International: yes; National: possible	Increase product diversity	+ + +	– –

Tables 17.1 and 17.2 summarize the outlined perspectives, or show the hy-potheses which will be empirically tested. In Table 17.1, the impact of patent applications, as indicators for innovation, in the theoretical approaches of competitive advantage and intra-industry trade is outlined. The differentiation between national and international patent application confirms the higher importance of applications in more than the origin country or at the European patent office. Therefore, we rely in our empirical analysis on the patent appli-cations at the European patent office.

The positive implications of international standards on foreign trade are certainly stronger than those of the idiosyncratic national standards.[5] However, the existence of the latter is preferable to a standardless situation. Besides, the substitution of idiosyncratic national standards by international ones basically promotes trade, but can lead to a reduction of product diversity, and thus slow down the incentives for intra-industry trade. Although each formal standard implicitly reduces variety more or less, the positive impacts of the other types of standards are likely to outweigh the negative intra-industry trade impact of variety reduction.

Table 17.2   Effects of various types of standards: a comparison of three theoretical approaches

Theoretical approach	Type of standards	Economical effects	Impacts on exports	Impacts on imports
Competitive advantage	International: yes national: yes, even stronger	Improve quality and/or reduce costs of national products	+	–
Trade distortion or competitive disadvantage	National product standards	Reduce openness of national markets and the marketing chances on foreign markets	–	– –
	National process standards	Increase production costs for domestic producers	–	+
Intra-industry trade	International: yes national: possible			
	Compatibility standards	Increase the openness of national markets	+ +	+
	Quality standards	Increase the openness of national markets	+ +	+
	Variety-reducing standards	Reduce product diversity	–	–

From Table 17.2 (compare Swann et al. 1996) it appears that the impact dimensions of the idiosyncratic national and the assumed international standards on foreign trade are not completely unambiguous. If one assumes that the German stocks of standards have, on the one hand, a high reputation and the role of 'international best practice', and on the other hand that standards generally fulfill the function of an indicator for innovative technological performance for the analysis, a stronger positive impact on the exports by comparison with the imports can be assumed.

17.3   THE DATA

In addition to the clarification between the role of innovations, especially protected by intellectual property rights, and formal technical standards, the following improvements with respect to the approach of Swann et al. (1996) are based on the availability of a broader[6] and more precise database considering standard data taken out of the PERINORM database,[7] especially bilateral trade data on an in-depth SITC level (OECD 1998a, 1998b) and additional data on a further well established and proved technology indicator, the number of patent applications at the European Patent Office.

The PERINORM database contains both current and historical documents, but we focus at first on a cross-country and cross-section analysis based on the data in 1995. Our approach is based on the international classification of standards (ICS) and not on an industry classification. In order to analyze the impacts of innovations and standards on economic performances, a clustering derived from the technical aspect of standardization seems more appropriate, in concordance with Helpman's conclusion, and is likely to generate more significant results.

Based on the ICS we have generated a concordance to the international classification of patents (IPC) (see Appendix IV). According to the results presented in Chapter 15, it turned out that this concordance produced positive significant correlations between standard and patent numbers. From the theoretical point of view, and already well established in the empirical literature, it is obvious that patents should be the better indicator for technological strengths compared to standards, because patent applications, especially at an international patent office like the European one, indicate the intensity and the success of a country's research activities. Furthermore, patents concede a restricted monopoly, whereas standards can theoretically be bought at a limited price by everyone, including the companies of other countries, and the specifications can be used without further payments. European standards are even obligatory for all European countries and international standards are adopted on a voluntary level all over the world. Nevertheless, standards, especially national ones, are still an important indicator for technological capacity, but patents are likely to be the better indicator.

17.4 EMPIRICAL ANALYSIS OF THE IMPACT OF TECHNOLOGY ON FOREIGN TRADE BASED ON BILATERAL FOREIGN TRADE RELATIONS[8]

17.4.1 Analysis of the Bilateral Trade Relations of Germany, Austria and Switzerland

Proceeding from research approaches outlined in the empirical literature on foreign trade and innovation (Wakelin 1997) in respect of technological performance, we shall examine to what extent the bilateral balance of trade depends on the relations of the standards stocks in the exporting or importing country, or whether the above-mentioned hypotheses derived from the theory of competitive advantage and trade distortion apply. We shall examine whether a surplus or a stronger specialization of the standards stocks according to the ICS classified groups lead to a relative balance of trade surplus, because standards not only positively influence quality and price competition, but can also create non-tariff trade barriers for importers. As a comparison, the existing relations between the patent applications at the European Patent Office and the foreign trade relations will be cited. For the technological performance the innovation potential has – as already proven – a significant influence for the industrialized countries on the specialization profile in foreign trade.

The bilateral balance of trade for 1995 in the 34 ICS classified groups, on the one hand, should be differentiated and explained by the relations of the stocks of standards in the exporting and importing country according to idiosyncratic national and adopted or adapted international standards.[9] On the other hand, two indicators based on patent statistics will be used as explanatory factors.[10]

The OLS estimation equation for bilateral trade on the basis of the 34 ICS classified groups is defined as follows:

$$TP_{pqs} = \alpha + \beta \cdot STDR_{pqs} + \varepsilon_{pqs} \qquad (17.1)$$

whereby TP is the bilateral trade balance between the export country p and the import country q, defined as the quotient between the difference of the export flows and the sum of the exports in the ICS classified group s:[11]

$$TP_{pqs} = (EX_{pqs} - EX_{qps}) / (EX_{pqs} + EX_{qps}) \qquad (17.2)$$

This index lies in the interval between $+1$ and -1, and therefore is not distorted by the differing significance of the various industrial branches. The utilization of the difference between export and import abstracts also from the significance of the intra-industry trade in the bilateral trade relations.

The standard stocks STD can be differentiated initially according to two different performance indicators $R = P$ respectively $R = POP$. The former is the 'standards performance indicator' $STDP$ between the exporting country p and the importing country q:

$$STDR_{pqs} = STDP_{pqs} = (STD_{ps} - STD_{qs}) \, / \, (STD_{ps} + STD_{qs}) \qquad (17.3)$$

The second indicator is defined as the quotient of the per capita stock of standards between the exporting country p and the importing country q:[12]

$$STDR_{pqs} = STDPOP_{pqs} = (STD_{ps} \, / \, POP_p) \, / \, (STD_{qs} \, / \, POP_q) \qquad (17.4)$$

In order to determine the various effects of the idiosyncratic national standards and the adapted international standards, besides the total standards stock STD, the stock of national standards $STDNAT$ and the stock of adapted international or European standards $STDINT$ in the ICS classification s are also differentiated.[13]

In order to integrate the theoretical and empirically proved significance of the national innovation potential for foreign trade in the analysis, the results for the following OLS regression equation are presented:

$$TP_{pqs} = \alpha + \beta \cdot INN_{pqs} + \varepsilon_{pqs} \qquad (17.5)$$

The structure of bilateral foreign trade is explained by the relative innovation potential INN in both countries surveyed. Here also two indicator concepts are applied. On the one hand, analogous to the stock of standards per capita, the patent applications per millions of inhabitants of the export and import country are juxtaposed:

$$INN_{pqs} = PAT_{pqs} = (PAT_{ps} \, / \, POP_p) \, / \, (PAT_{qs} \, / \, POP_q) \qquad (17.6)$$

In contrast to the stocks of standards, for which the worldwide population is not known, the relative patent share RPS can be calculated:

$$INN_{ps} = RPS_{ps} = 100 \cdot \tanh \ln$$
$$[(PAT_{ps} \, / \, \Sigma_s PAT_{ps}) \, / \, (\Sigma_p PAT_{ps} \, / \, \Sigma_{ps} PAT_{ps})] \qquad (17.7)$$

This index measures whether the patent share of a country p corresponds in a certain ICS classified group s to the share which this sector holds in its total worldwide population, that is all the applications at the European Patent Office, or whether this share is above or under average. The logarithm ensures a symmetrical range around the neutral point 0, the tangens hyperbolicus leads to a limitation of the range ±100. For the regression equation, the difference of the *RPSs* between export and import country (Δ*RPS*) will be utilized.

Table 17.3 presents the results of the estimation of the simple OLS regression equations (17.1) and (17.5) based on 34 observations.

The results of the analysis make it fundamentally clear that the technological specialization profile (*RPS*) is suited to explain the balance of trade according to ICS classification groups for the broad majority of the bilateral relations. This means that the national innovation potential is of decisive significance for international competitiveness. In those countries which already co-operate very closely together economically, or have similar specialization profiles and whose bilateral foreign trade flows are accordingly strongly dominated by intra-industry trade – as in the case of Germany and France – the technological specialization profile hardly contributes any more to the explanation of the trade structure. These estimated results underline that the selected categorization in ICS classification groups and the concordance of the patent classification with the SITC product groups are based on a solid statistical base.

By contrast, the stock of standards and technical rules is a significant explanatory factor for the bilateral foreign trade structure in only around one-third of the relationships examined. However, with the exception of Germany–Austria,[14] the expected positive correlation between the standards and the balance of trade surplus is confirmed. A systematic difference between the significance of national and international standards cannot be determined, so that the total stock of standards turns out to be the most appropriate indicator.

Whereas the national innovation potential, measured according to the specialization indicator *RPS*, represents the decisive explanatory factor for the bilateral balance of trade, formal standards can be used to explain bilateral foreign trade flows only up to a limited extent. This result can be explained as formal standards and technical rules contribute not only to quality improvement and price reduction, but also fulfill other functions which are not necessarily reflected in a competitive advantage, because standards can present safety standards, the observance of which can contribute to the protection of consumers, employees and the environment, and trigger off positive effects for the economy. The resulting requirements of the producers however can – as postulated in the hypothesis of the theory of the competitive disadvantage – restrict their international competitiveness and thus exercise a negative influence on the balance of trade. In order to better record the various functions of formal standards and technical rules, and their impacts on the bilateral foreign trade flows, a differentiated analysis according to the ICS classification will be carried out on the basis of 36 bilateral trade relationships. For the distribution according to the kind of standards outlined in Chapter 3 is not identical in the ICS category groups.

Table 17.3   Determining factors of the bilateral trade of Germany, Austria and Switzerland with selected countries (t-values in brackets)

Indicators	STDP	STDP POP	STD NATP	STD NAT POP	STD INTP	STD INT POP	PAT POP	ΔRPS
Germany								
Germany	–	–	–	–	–	–	–	–
Austria	−0.09	−0.11	0.07	−0.004[a]	−0.16	−0.37[a]	0.003	0.14
	[−0.39]	[−1.21]	[0.34]	[−1.70]	[−0.85]	[−1.71)	[0.07]	[1.24]
Switzerland	0.47	0.002	0.43	0.002	0.58[b]	0.41[a]	0.42[c]	0.34[c]
	[1.62]	[0.15]	[1.51]	[0.79]	[2.56]	[1.92]	[3.91]	[3.25]
Spain	−0.05	−0.004	0.16	−0.002	−0.12	0.06	0.01	0.03
	[−0.13]	[−1.20]	[0.39]	[−0.64]	[−0.58]	[0.68]	[1.64]	[0.31]
France	0.19	−0.02	0.20	−0.02	−0.05	0.05	−0.03	−0.05
	[1.07]	[−1.39]	[1.18]	[−1.21]	[−0.31]	[0.58]	[−0.71]	[−0.49]
Netherlands	0.75[b]	0.10	1.48[b]	0.001	0.41[b]	0.48	0.07[b]	0.28[c]
	[2.44]	[1.42]	[2.51]	[0.79]	[2.46]	[1.66]	[2.74]	[3.55]
Great Britain	0.49[a]	0.005	0.52	0.0004	0.14	−0.06	0.01	0.30[c]
	[2.02]	[0.91]	[1.59]	[0.22]	[0.75]	[−0.27]	[1.19]	[3.22]
USA	−0.25	0.00	−0.19	0.00	0.01	0.00	0.01	0.21[b]
	[−1.07]	[−0.62]	[−0.92]	[−0.52]	[0.02]	[−1.14]	[0.73]	[2.46]
Japan	−0.03	0.003[b]	−0.03	0.003[b]	−0.02	0.0005	0.06[c]	0.49[c]
	[−0.06]	[2.08]	[−0.07]	[2.05]	[−0.04]	[0.74]	[3.84]	[4.69]
Austria								
Germany	–	–	–	–	–	–	–	–
Austria	–	–	–	–	–	–	–	–
Switzerland	0.30[a]	0.06	0.21[a]	0.06	0.12	0.05	0.76[c]	0.38[c]
	[1.72]	[1.64]	[1.73]	[0.95]	[0.52]	[1.02]	[3.36]	[3.68]
Spain	−0.31	0.00	−0.07	−0.07	−0.21	−0.02	0.01	0.13
	[−1.38]	[−0.27]	[−0.41]	[−0.26]	[−0.88]	[−0.97]	[1.24]	[1.21]

France	0.45^b	0.003^a	0.53^c	0.003^a	0.42^b	0.02	0.11^b	0.29^c
	[2.38]	[1.88]	[3.02]	[1.90]	[2.20]	[1.06]	[2.16]	[2.95]
Netherlands	0.70^c	0.02^b	0.19	0.002^b	0.74^c	0.26^a	0.10^c	0.38^c
	[2.77]	[2.67]	[0.99]	[2.71]	[2.77]	[1.78]	[3.06]	[4.24]
Great Britain	0.21	0.0002	0.19	0.0001	0.21	0.02	0.02^b	0.41^c
	[0.81]	[0.41]	[0.95]	[0.42]	[0.87]	[0.70]	[2.45]	[3.97]
USA	0.05	0.00	0.25	0.00	0.16	-0.0001^b	0.03^b	0.36^c
	[0.24]	[1.24]	[1.24]	[1.29]	[0.31]	[−2.18]	[2.04]	[3.67]
Japan	0.13	0.003	0.25	0.00	0.73	0.0001	0.04^a	0.45^c
	[0.55]	[1.33]	[1.16]	[0.02]	[1.65]	[1.49]	[1.96]	[4.38]
Switzerland								
Germany	–	–	–	–	–	–	–	–
Austria	–	–	–	–	–	–	–	–
Switzerland	–	–	–	–	–	–	–	–
Spain	0.54^c	0.40^c	0.47^c	0.02^c	0.49^b	0.37^b	4.86^a	0.23^b
	[2.81]	[3.85]	[2.85]	[3.09]	[2.27]	[2.43]	[2.02]	[2.10]
France	0.52^a	0.03	0.41	0.002	0.46^a	0.36^a	0.12	0.25^b
	[1.97]	[0.93]	[1.52]	[1.00]	[1.79]	[1.97]	[1.47]	[2.16]
Netherlands	0.79^c	0.21^c	0.25	0.005	0.71^b	0.22^c	0.52^b	0.29^b
	[3.72]	[3.03]	[1.69]	[1.44]	[2.39]	[3.35]	[2.10]	[2.27]
Great Britain	0.13	0.10	0.06	0.003	0.55^a	0.16	0.33	0.34^c
	[0.57]	[1.03]	[0.33]	[0.22]	[1.84]	[1.30]	[1.35]	[3.29]
USA	0.002	0.10	0.24	0.01	0.41^a	13.36^a	0.47^b	0.28^b
	[0.01]	[0.16]	[1.00]	[0.24]	[2.03]	[1.89]	[2.13]	[2.36]
Japan	0.05	0.30	0.18	0.03	0.31	7.63^a	1.15^c	0.53^c
	[0.23]	[0.81]	[0.70]	[1.47]	[1.38]	[2.03]	[5.48]	[4.49]

Notes: [a] = a significance level < 0.10; [b] = a significance level < 0.05; and [c] = a significance level < 0.01.

As formal standards and technical rules account for the structure of the bilateral trade relations in only a few cases, it can also be concluded from this that standards are only to a certain extent suitable to seal off national markets systematically from foreign competition. This conclusion is underlined by the negligible differences between the influence of the indicators for the national and international stocks of standards. Further, the correction of the stocks of standards according to the respective total population produces no better explanation for the foreign trade performance, so that the importance of the absolute number of standards is thus emphasized, and so indirectly the characteristic of the stock of standards as a public good is not refuted.

17.4.2 Analysis of Foreign Trade in Technology Fields on the Basis of Bilateral Trade Relations

As the analysis of the bilateral trade relations via the various standards stocks indicators resulted in the expected significant results in only a very few cases, additional findings are expected from an analysis according to ICS classified groups of the specific significance of standards in technology fields or economic branches. On the basis of the data for the nine countries presented in Table 17.3, 36 bilateral trade relationships can be surveyed, which are grouped according to ICS categories and analyzed separately.

The OLS estimation equation for bilateral trade (see Equation 17.2) on the basis of the 36 bilateral trade relationships in the appropriate ICS categories is defined as follows:

$$TP_{pq} = \alpha + \beta \cdot STD_{pq} + \varepsilon_{pq} \qquad (17.8)$$

Various indicator concepts are applied for the standard stocks STD. Whereas an attempt is made to explain the bilateral trade relations on the one hand by the 'standards stock performance indicator', and on the other hand by the quotient of the per capita standards stocks, for the analysis pertaining to the ICS classification groups the first mentioned indicator will be replaced by the quotient of the relative shares of standards stocks in relation to the respective total stock of national standards $RSTD$:[15]

$$RSTD_{pqs} = (STD_{ps} / \Sigma_s STD_{ps}) / (STD_{qs} / \Sigma_s STD_{qs}) \qquad (17.9)$$

Once again, in order to determine the various effects of the idiosyncratic national standards and the adapted international standards, not only the $RSTD$, but also the $STDPOP$ and the total standards stock STD will be differentiated also, according to the stock of national standards $STDNAT$ and the stock of adapted international standards $STDINT$ in the ICS classification group s.

The results of the simple OLS regression of the basic equation (17.8), differentiated according to the 33 ICS classification groups[16] in Table 17.4 make it clear that the bilateral relation of the national innovation potentials, or the difference of the specialization indexes, can explain the signs of the balance of trade in almost two-thirds of the cases. In less R&D-intensive and not so very innovative branches, such as shipbuilding, the clothing and wood industries, other factors are decisive for the bilateral trade flows, such as factor prices and the availability of raw materials, so that the patent indicators have no explanatory value. This result confirms the findings of the theoretical and empirical literature on international competitiveness (Münt 1996; Wakelin 1997).

In the case of the hypotheses concerning the stock of standards, the results are to be interpreted in a differentiated manner. Although the estimates revealed the expected positive first signs in around one-third of the cases, however there are also ICS classification groups, in which stronger specialization in the standards stocks stimulate imports rather than exports. General differences between the stock of purely national and international standards are not determined; however, in some ICS groups diverging impacts appear. Finally, the indicators are somewhat more conclusive, which do not refer to the standards stock per million inhabitants, but to the relative specialization profiles. Due to the branch- or technology-specific differences, the results are interpreted based on their key features separately according to ICS classification groups.

In trade with products from the medical technology branch, the countries with a higher stock of national standards achieve export surpluses, while the international standards do not represent a significant explanatory factor. Since safety standards, which are also an expression of technological performance, play a special role here, countries with a high stock of standards export more goods to those economies with less standards and possibly less safety requirements. For these reasons the export to countries with many and high quality and safety requirements can be rather difficult.

The cross-sectional area of metrology is determined by very many other factors, so that the stocks of standards on the whole have no explanatory value. The same applies for the cross-sectional areas of fluid systems and components, and energy and heating technology. On the other hand, with mechanical systems, the total and the international standard stock above all have the anticipated positive signs.

In mechanical engineering, the indicators for the relative relation of the national stock of standards display positive signs while the stock of international standards does not play a role. This means that national standards express the high quality of the national products here and this is reflected accordingly in export successes. This conclusion is supported by the very significant value of the patent indicators for technological performance.

In the sector information and communication technology, standards play a special role because of their compatibility characteristics with the corresponding network effects, which are also reflected in the empirical results. Thus export surpluses are explained by above average technological performance, but not, however, by the specialization pattern of the standards stock. Conversely, more standards stimulate the imports in electrical engineering, because the transparency of the local specifications is increased for foreign sellers, and imports made easier for them.

Whereas the effects are clear for the electrical engineering sector, in the case of telecommunications the differences between the operative effects of

Table 17.4   Factors determining bilateral trade in ICS classified groups on the basis of 36 bilateral relations (t-values in brackets)

ICS classification					Indicators			
	RSTD	STD-POP	RSTD-NAT	STDNAT POP	RSTD-INT	STDINTPOP	PAT-POP	ΔRPS
Medical technology	0.64^c	0.03	0.17^c	0.06^c	−0.01	−0.00005	0.17^c	−0.01
	[3.62]	[1.52]	[3.12]	[3.60]	[−0.67]	[−1.02]	[4.90]	[−0.03]
Metrology, measuring, controlling	0.01	0.01	−0.002	0.06^b	−0.11	−0.001	0.02^c	0.53^c
	[0.16]	[0.32]	[−0.44]	[2.40]	[−0.87]	[−0.74]	[2.95]	[5.52]
Mechanical systems	0.62^c	0.002	0.36^a	0.007	0.3^c	0.001^a	0.07^a	0.15
	[4.45]	[0.12]	[1.97]	[0.23]	[7.52]	[1.95]	[2.01]	[0.64]
Fluid systems/components	0.07	0.001	0.04	0.01	−0.03	0.00	0.03^a	0.16
	[0.65]	[0.22]	[0.70]	[0.62]	[−1.03]	[0.33]	[1.97]	[1.04]
Mechanical engineering	0.87^c	0.02	0.91^c	0.10^c	−0.03	0.0002	0.04^b	0.90^c
	[4.59]	[0.29]	[7.41]	[3.179]	[−0.48]	[0.57]	[2.49]	[6.72]
Energy/heating technology	−0.04	−0.01	−0.001	0.01	$−0.03^a$	−0.00004	0.04^a	0.04
	[−0.86]	[−1.00]	[−0.27]	[0.98]	[−1.78]	[−0.62]	[1.77]	[0.16]
Electrical engineering	$−0.53^b$	$−0.03^b$	−0.02	−0.01	0.08	0.0001	0.05^b	0.96^c
	[−2.67]	[−2.42]	[−0.11]	[−0.62]	[1.48]	[0.22]	[2.16]	[4.28]
Electronics	−0.05	$−0.04^b$	0.07^b	0.03	−0.003	−0.0001	0.02^a	0.38^c
	[−0.70]	[−2.55]	[2.09]	[1.29]	[−0.17]	[−0.99]	[1.98]	[3.13]
Telecommunications	$−0.18^c$	$−0.01^c$	0.04^b	−0.001	$−0.12^c$	$−0.0004^a$	0.02	0.68^c
	[−4.20]	[−3.26]	[2.05]	[−0.13]	[−3.21]	[−1.98]	[0.47]	[4.82]
Information/office technology	0.67^c	$−0.08^c$	0.19^c	−0.01	0.28^c	−0.007	0.01	0.73^c
	[3.25]	[−3.07]	[3.96]	[−0.42]	[5.10]	[−1.58]	[0.46]	[7.88]
Reproduction technology	0.05^a	0.03	0.005	0.007	0.05^b	−0.001	0.06^c	0.54^c
	[1.95]	[0.89]	[0.80]	[0.84]	[2.58]	[−0.16]	[2.95]	[5.14]
Precision mechanics, jewelry	0.05^c	0.008^b	0.009	0.002	−0.05	0.0006	0.004	0.26^c
	[3.23]	[2.30]	[1.40]	[1.14]	[−0.32]	[0.84]	[0.27]	[3.85]
Motor vehicles	0.09^c	−0.009	0.04^c	0.002	0.10^c	−0.0002	−0.003	0.56^b
	[4.13]	[−1.09]	[2.80]	[0.19]	[3.00]	[−1.11]	[−0.11]	[2.72]
Railways	0.004	0.001	−0.0001	0.0007	0.21	0.002^a	0.02^b	0.29^c
	[1.19]	[1.26]	[−0.04]	[0.62]	[1.48]	[1.87]	[2.33]	[3.75]

Shipbuilding, marine technology	0.13^c [3.55]	-0.09^a [−1.71]	0.01^c [3.07]	0.02^b [2.65]	0.05 [0.94]	-0.0003^b [−2.05]	0.001 [0.01]	-0.28^a [−1.73]
Aircraft/ spacecraft technology	0.01 [0.36]	0.005 [0.84]	−0.001 [−0.60]	−0.00009 [−0.06]	−0.008 [−0.76]	−0.00005 [−1.18]	0.06^b [2.14]	0.18^a [2.01]
Transportation	−0.05 [−1.51]	0.02^b [2.21]	−0.009 [−0.92]	0.01^a [1.73]	−0.02 [−0.80]	0.0001 [0.86]	0.10^c [4.18]	0.31^c [5.18]
Packaging	−0.12 [−0.84]	0.01 [1.01]	−0.01 [−0.14]	0.04^b [2.34]	−0.01 [−0.31]	−0.00006 [−0.77]	0.08^c [3.32]	0.09 [0.98]
Textile/leather industry	0.04 [0.32]	0.05^c [3.19]	0.004 [0.05]	0.08^b [2.28]	−0.06 [−0.42]	0.003^b [2.54]	0.18^c [4.57]	0.28^c [5.37]
Clothing	−0.02 [−1.19]	0.04 [1.32]	−0.009 [−0.90]	−0.001 [−0.23]	−0.05 [−0.82]	0.0001 [0.22]	−0.02 [−1.00]	0.21 [1.44]
Agriculture	0.11^c [3.70]	0.01^c [2.96]	0.10^c [4.25]	0.02^c [2.92]	0.05^b [2.40]	−0.0003 [−1.21]	0.03 [1.19]	0.24 [1.36]
Food technology	0.05^c [4.49]	0.006^c [4.35]	0.004^b [2.22]	0.009^c [4.79]	0.06^c [5.47]	0.0001 [1.28]	−0.04 [−0.67]	0.54^c [3.619
Chemical processing technology	-0.07^b [−2.17]	0.13 [1.33]	-0.09^b [−2.28]	−0.07 [−0.69]	−0.05 [−1.50]	0.001 [0.91]	0.07^b [2.59]	−0.14 [−0.69]
Mining, minerals; engineering	−0.27 [−1.21]	-0.05^a [−1.82]	-0.55^c [−4.14]	−0.05 [−1.46]	0.16^c [3.12]	−0.003 [−1.18]	0.02^a [1.84]	0.006 [0.04]
Mineral oil technology	0.30^a [2.09]	0.03 [1.40]	0.35 [1.42]	0.03 [0.75]	0.02 [1.23]	0.005 [1.30]	−0.007 [−0.36]	0.44^c [2.80]
Metallurgy	-0.31^c [−2.53]	0.07^c [4.03]	−0.05 [−1.21]	0.02 [1.15]	−0.01 [−0.16]	0.002^b [2.29]	0.06^b [2.40]	0.12^a [1.96]
Wood industry	0.15 [1.01]	0.02^a [2.01]	−0.006 [−0.24]	0.08^c [3.63]	0.10^c [3.45]	0.0005^a [1.93]	0.004 [0.20]	0.17 [1.09]
Glass, ceramics	-0.27^c [−2.99]	0.02 [0.75]	-0.02^c [−3.75]	-0.03^b [−2.43]	0.09^b [2.30]	0.0002^b [2.17]	0.17^a [1.98]	−0.02 [−0.24]
Rubber, synthetics	−0.03 [−0.26]	-0.05^a [−1.76]	−0.006 [−0.18]	−0.03 [−1.06]	0.02 [1.20]	−0.0002 [−0.83]	0.006 [0.94]	0.27^b [2.73]
Paper	-0.20^c [−3.11]	0.02 [0.60]	-0.02^b [−2.38]	−0.008 [−0.62]	-0.10^b [−2.12]	0.0004 [0.20]	0.04^b [2.24]	0.31^a [1.94]
Coatings/ paints	−0.17 [−1.16]	−0.02 [−0.67]	0.12 [0.87]	0.06 [1.28]	0.07 [1.04]	−0.0004 [−0.60]	0.05^a [1.88]	0.85^c [4.60]
Construction/ building materials	0.18^b [2.06]	0.007 [1.36]	0.11^a [1.86]	0.01 [1.35]	0.01^a [1.80]	0.00008 [1.33]	−0.004 [−0.29]	0.28^c [4.07]
Household, entertainment	0.08 [1.29]	0.01^b [2.60]	0.01 [0.46]	0.02^b [2.33]	0.05^b [2.14]	0.0002 [1.33]	0.02 [0.76]	0.10 [1.16]

Notes: ^{a} = a significance level < 0.10; ^{b} = < 0.05; and ^{c} = < 0.01.

national and international standards can be demonstrated in an exemplary manner. National standards, in accordance with the specialization indicator of patents, represent national technological performance and product qualities, which are also manifested in corresponding export surpluses. On the other hand, a large stock of international standards is accompanied by import surpluses, because the interfaces for the various components of telecommunications systems are defined at the international level and foreign suppliers do not have to make any special adaptations for the domestic markets.[17]

This different operative modus of national and international standards is not found in information and office technology, where the three standards' specialization indicators as well as the patent indicators show the expected positive signs.[18] The reason is that network products only represent a part of the products here, many de facto standards exist and the software sector, which is greatly influenced by the question of compatibility, is not included. On the other hand, the results for reproduction technology indicate that in this case the adoption of international standards plays a more important role for competitiveness than the national standards. In the sector of precision engineering and jewelry, only the coefficients of the total standards stock have the expected positive signs.

The results of the different traffic carriers differ very clearly. Whereas the mainly protectionist or monopoly areas of railway technology and air and space technology indicate no significant signs and the export surpluses here depend above all on the relation of the national innovation potentials, in motor vehicle technology and shipbuilding the stocks of standards can explain the foreign trade surpluses.[19] The international standards play no role at all in shipbuilding, whilst automobile production depends very much on the compatibility of the single components of the different suppliers, and, in the meantime, also on international co-operation and thus international standards.

The two cross-sections transportation and packaging combine insignificant signs of the international standards and positive coefficients of quotients of the national standards per inhabitant. While the trade flows in the clothing industry can neither be explained by the innovation potential nor by the stock of standards, in the textile industry the quotients of standards stocks per capita are a positive explanatory factor for international competitiveness.

Above average specialization in the appropriate standards stocks explain export surpluses not only in agriculture, but also in food technology, whereby national standards have a somewhat greater significance than international ones in agriculture, with the reverse being true in the food technology sector.

The results in the chemical industry are surprising, in which a market specialization in the whole stocks of standards is correlated to import rather than export surpluses. Apart from the fact that in chemical process engineering the tendency is to standardize less, which also explains the non-significant results in the mineral oil technology, rubber and synthetics industry and the sector paint and coatings, the results indicate that the signaling of quality and safety requirements through standards makes it easier for exporters to sell their products on foreign markets, and the enterprises in countries with more than average standards are less competitive.

In mining and engineering, the stocks of national and international standards give off different signs. Whilst the stocks of international standards are a positive explanatory factor for foreign trade surpluses, the national standards are a negative one. This means that countries with many international standards tend to realize a foreign trade surplus, while economies with an above-average stock of national standards show an import surplus.[20]

For the area of metallurgy, no clear statements on the role of standards is possible, although the national innovation potential represents a decisive factor for the success in foreign trade.[21] By contrast, for the less technology-intensive woodworking industry the international stocks of standards in particular are an indicator for the explanation of international goods flows.

In the glass and ceramics industries national and international stocks of standards have different significance for the foreign trade performance of the countries studied. Whereas high stocks of international standards correlate positively with the foreign trade surpluses, the national standards are positively correlated with balance of trade deficits. The latter indicates that the competitiveness of domestic producers is not necessarily promoted by a large stock of national standards.

In the paper industry, countries with above-average stocks of national and international standards are less competitive internationally, which is manifested in corresponding balance of trade deficits. The negative effect of the national standards is here even more significant by comparison with the unfavorable impacts of the international standards. On the other hand, in construction and in the cross-section household and entertainment, the number of currently valid standards is positively correlated with balance of trade surpluses.

As a whole, the analysis of the explanation of foreign trade surpluses made it clear, on the one hand, through the relationship between the stocks of standards, and on the other hand, through the relation of the national innovation strength, that standards have different meanings, according to the type of technology or other frame conditions, for the national competitiveness and thus for foreign trade. For despite the principally positive signs, negative coefficients also exist which need to be analyzed in depth in sector-specific studies.

17.4.3 Analysis of the Bilateral Intra-industry Trade of Germany, Austria and Switzerland

Besides the hypotheses derived from the theories of competitive advantage and trade distortions, it is our aim to examine separately the hypothesis derived from the theory of intra-industrial trade of the trade-promoting effect of compatibility and quality standards and of the trade-restricting effect of variety-reducing standards.[22] It is not only foreign trade surpluses as an indicator of international competitiveness that enrich a national economy, by contributing among other things to growth and the securing of jobs, but also pronounced intra-industry trade. Due to the intensive exchange of goods within a branch, positive effects can be recognized for the economy. National industry profits from importing products which cannot be produced domestically, or only at a higher price, or of inferior quality, because the consumers buy more or qualitatively better products with their budget and so increase their utility level. Besides this price and quality aspect, product variety will be enriched because the range of the domestic producers will be supplemented by the foreign range of items.

In order to examine, in a cross-sectional analysis for the three countries, whether the stocks of standards are positively correlated to the volume of intra-industry trade because standards increase the transparency for the consumers and the compatibility with other products, an econometric estimate will be carried out. The OLS equation for estimating the volume of bilateral trade on the basis of the 34 ICS classified groups is defined as follows:

$$TV_{pqs} = \alpha + \beta \cdot STD_{pqs} + \varepsilon_{pqs} \qquad (17.10)$$

whereby TV represents the corrected bilateral trade volume between the exporting country p and the importing country q, defined as:[23]

$$TV_{pqs} = [(EX_{pqs} + EX_{qps}) - (|EX_{pqs} - EX_{qps}|)] / (EX_{pqs} + EX_{qps}) \qquad (17.11)$$

This variable is distinguished by assuming the value 0 when intra-industrial trade is missing, and the value 1 when the export volume to country q and the export volume from country q in the ICS classifications are equally large.

The explanatory variable STD is defined as the sum of both the standards stocks of the countries p an q respectively:

$$STD_{pqs} = STD_{ps} + STD_{qs} \qquad (17.12)$$

In order to be able to determine the different effects of the idiosyncratic national standards and the adapted international standards on intra-industrial trade, we will differentiate, besides the sum of the total standards stocks *STD*, also according to the stocks of national standards *STDNAT* and the stock of adapted international standards *STDINT* in the ICS classified groups *s*.

In order to integrate the theoretically and empirically proved significance of the national innovation potential for foreign trade in the analysis, also for intra-industrial trade, the results are presented for the following OLS regression equation:

$$TV_{pqs} = \alpha + \beta \cdot INN_{pqs} + \varepsilon_{pqs} \qquad (17.13)$$

The structure of the intra-industrial foreign trade is explained by the relative innovation potential *INN* in both countries. Here we fall back on the two indicator concepts already introduced for bilateral trade. The differences of the patent applications per million inhabitants of the exporting and the importing countries $\Delta PATPOP$ are squared, so that higher values point to different absolute national innovation potentials and smaller values to similar structures:

$$(\Delta PATPOP_{pqs})^2 = [(PAT_{ps} / POP_p) - (PAT_{qs} / POP_q)]^2 \qquad (17.14)$$

As in contrast to the standards stocks for patents the worldwide population is known, the relative patent share *RPS* can be calculated. Also, from this specialization indicator, the squaring of the differences between the exporting country *p* and the importing country *q* and similarities of the relative specialization profile can be calculated:

$$(\Delta RPS_{pqs})^2 = (RPS_{ps} - RPS_{qs})^2 \qquad (17.15)$$

For the estimation equation negative coefficients are expected for both innovation indicators, as with great differences in national innovation potentials and in the specialization profiles, the intra-industry trade will be rather weak, in that the one country exports certain categories of goods to the other country and conversely imports other groups of products. On the other hand, with only minor differences, the possibility of intra-industrial trade is more pronounced.

The attempt to try to explain the ICS classification-specific intensity of the bilateral intra-industry trade of Germany, Austria, and Switzerland empirically by the sum of the stocks of standards in both countries respectively has not been successful (compare Table 17.5). Neither the sum of the total standards stocks of the countries surveyed, nor the subsets of national or international standards contribute significantly to the explanation of the sector-specific volume of intra-industry trade. The exceptions are the bilateral

Table 17.5   *Determining factors of the intra-industrial trade of Germany,*
           Austria, and Switzerland with selected countries
           (t-values in brackets)

Indicators	STD	STDNAT	STDINT	$(\Delta PATPOP)^2$	$(\Delta RPS)^2$
Germany					
Germany	–	–	–	–	–
Austria	0.00002 [0.79]	0.0004 [1.25]	−0.00002 [−0.23]	−0.00006 [−0.44]	0.22 [1.30]
Switzerland	0.00003 [1.06]	0.00005 [1.16]	0.00005 [0.59]	0.0001 [0.94]	−0.10 [−0.87]
Spain	−0.000001 [−0.32]	−0.000001 [−0.35]	−0.000001 [−0.17]	0.00004 [0.52]	0.10 [1.68]
France	−0.00001 [−0.43]	−0.00001 [−0.49]	−0.00001 [−0.15]	0.001 [0.75]	−0.19[b] [−2.22]
Netherlands	0.00002 [0.89]	0.00002 [0.74]	0.00004 [0.97]	0.0002 [1.19]	−0.19[a] [−1.96]
United Kingdom	0.00003 [1.23]	0.00003 [1.05]	0.00006 [1.25]	0.00009 [0.48]	−0.14[a] [−1.77]
USA	0.000004 [0.16]	−0.00001 [−0.18]	0.0001 [1.20]	−0.0003[a] [−1.74]	0.02 [0.33]
Japan	0.00003 [0.90]	0.0001 [1.31]	−0.0001 [−0.41]	0.0003 [1.23]	−0.22[b] [−2.74]
Austria					
Germany	–	–	–	–	–
Austria	–	–	–	–	–
Switzerland	0.0001 [0.87]	0.0002 [1.15]	0.0001 [0.52]	0.00001 [0.33]	−0.07 [−1.10]
Spain	0.0001 [0.70]	0.0002 [0.88]	0.0001 [0.44]	0.0001 [0.26]	−0.08 [−1.59]
France	0.00003 [0.54]	0.0001 [1.11]	−0.00004 [−0.33]	0.001 [0.66]	−0.15[b] [−1.76]
Netherlands	0.00005 [0.67]	0.0002 [1.23]	0.00003 [0.31]	−0.0003 [−0.69]	0.12 [1.50]
United Kingdom	0.0001 [1.16]	0.0003[a] [2.01]	0.0001 [0.50]	0.001 [1.56]	−0.13[a] [−1.70]

USA	0.00006 [0.86]	0.00006 [0.69]	0.0002 [1.04]	0.0009a [1.74]	−0.02 [−0.33]
Japan	0.0002 [1.32]	0.0003[a] [1.98]	0.00001 [0.06]	−0.0006 [−1.67]	−0.16[b] [−2.60]
Switzerland					
Germany	−	−	−	−	−
Austria	−	−	−	−	−
Switzerland	−	−	−	−	−
Spain	0.0001 [0.62]	0.0001 [0.32]	0.0001 [0.66]	−0.000007 [−0.25]	−0.05 [−1.16]
France	0.00003 [0.39]	0.00 [0.00]	0.0001 [0.72]	0.00006 [1.35]	−0.04 [−0.47]
Netherlands	0.0001 [0.72]	0.00004 [0.16]	0.0001 [0.78]	0.00008 [1.70]	−0.10 [−1.41]
United Kingdom	0.0001 [1.29]	0.0002 [1.28]	0.0001 [1.12]	0.00004 [1.01]	−0.03 [−0.64]
USA	0.00 [−0.02]	0.00 [−0.03]	0.00 [0.00]	−0.00002 [−0.40]	−0.06 [−0.93]
Japan	0.0002[a] [1.87]	0.0002 [1.22]	0.0005[a] [2.00]	0.00006 [1.56]	−0.13[a] [−1.81]

Notes: [a] = a significance level < 0.10; [b] = a significance level < 0.05; and [c] = a significance level < 0.01.

relations of Austria with Great Britain and Japan and of Switzerland with Japan. In the majority of the bilateral relations examined, however, a positive, but insignificant sign of the stocks of standards is determined for the degree of integration. Systematic differences between the explanatory power of national and international standards were also not found, in contrast to the theoretical hypotheses. Although the empirical results do not admit support for the theoretical hypotheses on the positive influence of standards on intra-industry trade, on the other hand they do not indicate that a negative influence on intra-industry trade emanates from the standards stocks such as the theoretical hypotheses on the impacts of the variety-reducing standards on intra-industry trade and on the generally trade-restrictive effect of standards postulate.

Also, the estimates with the two patent indicators provide only in a few cases the anticipated negative sign for the support of the hypothesis that intra-industry trade between two countries is less in the areas in which the differences in the innovation potentials or specialization profiles are strong, and greater in those areas in which the innovation potentials are similar. On the whole, the results indicate that other factors, such as relative factor prices, cost structures, the degree of product differentiation and the intensity of state regulation, have a stronger explanatory power for the bilateral intra-industry trade of Germany, Austria, and Switzerland.[24] For this reason, and because in the ICS classified groups the composition and the effects of compatibility, quality and variety-reducing standards varies, the intra-industry trade will also be analyzed in depth according to the ICS classifications.

17.4.4 Analysis of the Significance of Standards for the Intra-industry Trade in Technology Fields

On the analogy of the examination of the theories of competitive advantage and trade distortions, the hypothesis derived from the theory of intra-industry trade of the trade-promoting effect of compatibility and quality standards and the trade-reducing effect of variety-reducing standards will also be separately analyzed for the ICS classified groups, since the bilateral country analysis has usually not produced any significant results.

The OLS regression equation for the bilateral trade on the basis of the 36 bilateral trade relations *TV* (see Equation 17.11) of the nine countries in Table 17.5 in the respective ICS classified groups *s* is defined as follows:

$$TV_{pq} = \alpha + \beta \cdot STD_{pq} + \varepsilon_{pq} \qquad (17.16)$$

In order to be able to determine the different effects of the idiosyncratic national standards and the adapted international standards on intra-industrial trade, we will differentiate, in addition to the sum of the total standards stocks *STD*, also according to the stocks of national standards *STDNAT* and the stock of adapted international standards *STDINT* in the ICS classifications *s*. Furthermore, the two patent indicators $(\Delta PATPOP)^2$ and $(\Delta RPS)^2$ will be tested as explanatory variable for their significance, in order to check the hypothesis, if the intra-industry trade turns out to be rather slight if great differences in the innovation potentials and specialization profiles are found.

The cross-sectional analysis on the correlation between the degree of intra-industrial trade and the stocks of standards produced as in Table 17.6 presented the following results. Although in nearly half the cases the total stocks of standards did not correlate significantly with the indicator for intra-industrial trade, in 16 ICS classified groups a significant positive relation can be established. Only in the sector packaging and in the glass and ceramics industries does the sum of standards stocks correlate negatively with the extent of intra-industry trade, so that here there is rather a trade-hindering effect which stems from more standards. The partial stock of international standards even correlates in 19 areas positively and never negatively with the intensity of intra-industrial trade. For the partial stock of national standards in only eight of the ICS classified groups can a positive and in the two above mentioned areas a negative correlation be determined.

These results confirm fundamentally the theoretical hypothesis that international standards contribute more than national standards to promote intra-industry trade.[25] In detail, these facts apply to the following fields: mechanical systems, fluid systems and components, mechanical engineering, electrical engineering, telecommunications, precision engineering and jewelry, motor vehicles, railways, air- and spacecraft, textile and leather industry, clothing, food technology, metallurgy, woodworking, glass and ceramics, paper, coatings and paints and construction. On the other hand, the national stocks of standards, as compared to the international ones, have a closer connection with the indicator for intra-industry trade only in the following fields: medical technology, measuring and controlling, energy and heating technology, electronics, mining and engineering, and mineral oil technology.

In particular, the intra-industry trade in systems technologies with cross-border networks, such as in telecommunications, or technologies which are very dependent on the compatibility of the components from various suppliers, is very closely correlated to the sum of international stocks of standards. In comparison both patent indicators perform worse in explaining intra-industry trade.

17.4.5 Conclusion

In this first section about the impact of standards on trade, data on standards and patents has been combined with data on trade flows in two-digit subgroups of the international classification of standards to explore the effects of standards and innovations on trade. This approach continues the study of Swann et al. (1996) who analyzed the sole effect of standards on British trade before globalization in standardization started.

Still, Swann et al. (1996) found that for each of the British standards variables, the coefficients in the British export and import flows are positive or at least non-negative. Therefore, the United Kingdom's stock of standards is not simply a strategic asset that increases exports and reduces imports; a larger stock of standards may serve to expand trade in general. However, the net effect of British standards on the K trade balance is favorable, but not significant, and the net effect of German standards assumed as 'international best practice' on British trade is unfavorable. Therefore, the effect of standards on net exports is positive as the competitive advantage perspective would indicate. Secondly, idiosyncratic British standards have a stronger positive effect on exports and imports than internationally equivalent standards.

Table 17.6 Determining factors of intra-industry trade in ICS classified groups based on 36 bilateral relationships (t-values in brackets)

ICS classified group	STD	STDNAT	STDINT	$(\Delta PATPOP)^2$	$(\Delta RPS)^2$
Medical technology	0.0002[c] [3.49]	0.0002[c] [3.23]	0.0002 [0.87]	−0.0001[a] [−1.73]	−0.04 [−0.22]
Metrology, measuring, controlling	0.0001[b] [2.69]	0.0001[b] [2.15]	0.0002 [0.99]	−0.0001 [−1.28]	−0.32[c] [−3.63]
Mechanical systems	0.00 [−0.03]	0.00 [−0.70]	0.001[c] [3.56]	−0.002 [−1.04]	−0.18 [−0.88]
Fluid systems/ components	0.00001 [0.46]	0.00 [−0.02]	0.0002[a] [1.96]	−0.0008[a] [−1.87]	−0.29[a] [−1.79]
Mechanical engineering	0.00005 [1.49]	0.00004 [1.02]	0.0002[a] [1.79]	−0.0002 [−1.94]	−0.22 [−0.86]
Energy/ heating technology	0.0002[b] [2.73]	0.0002[b] [2.30]	0.0007[a] [1.91]	−0.00008 [−0.52]	0.10 [0.27]
Electrical engineering	0.00004[a] [1.76]	0.00002 [0.76]	0.0002[c] [3.52]	−0.00002 [−0.21]	0.29 [0.53]
Electronics	0.0001[a] [1.78]	0.0003[b] [2.26]	0.0001 [0.99]	−0.001[b] [−2.24]	−0.19[b] [−2.32]
Telecommunications	0.0001[c] [3.07]	0.0001[a] [1.83]	0.0001[b] [2.33]	0.0002 [0.87]	−0.16 [−0.90]
Information/ office technology	0.0001[a] [1.98]	0.0001 [1.07]	0.0001 [1.57]	−0.0002 [−1.36]	−0.23[c] [−3.00]
Reproduction technology	0.0001 [1.35]	0.0001 [1.00]	0.0007 [1.61]	−0.001[b] [−2.35]	−0.27[c] [−3.50]
Precision mechanics, jewelry	−0.0001 [−0.30]	−0.0002 [−0.47]	0.007[a] [1.89]	−0.002[b] [−2.70]	−0.07[b] [−2.25]
Motor vehicles	0.0003[c] [3.10]	0.0003[b] [2.31]	0.001[c] [3.37]	0.0005[b] [2.09]	−0.04 [−0.27]
Railways	0.0001 [1.01]	0.0001 [0.94]	0.005[b] [2.05]	0.0004 [0.11]	−0.04 [−0.91]
Shipbuilding, marine technology	0.0001 [0.92]	0.0001 [0.71]	0.0007 [0.93]	−0.04 [−0.51]	−0.08 [−0.76]
Aircraft/ spacecraft technology	0.0001[b] [2.21]	0.0001 [1.33]	0.0002[b] [2.69]	−0.07[a] [−1.99]	−0.09[a] [−1.76]
Transportation	0.0002[a] [2.00]	0.0002 [1.34]	0.0005 [1.59]	−0.002[a] [−1.99]	−0.16[c] [−3.46]
Packaging	−0.0002[a] [−1.97]	−0.0002[b] [−2.24]	0.0004 [1.02]	−0.0002[b] [−2.35]	−0.04 [−0.47]

Textile/ leather industries	0.000006 [0.08]	−0.00006 [−0.72]	0.0006[b] [2.59]	−0.0002[b] [−2.05]	−0.12[b] [−2.36]
Clothing industry	0.0001 [0.55]	0.00008 [0.33]	0.02[c] [5.12]	−0.001 [−0.16]	0.13 [1.04]
Agriculture	0.00002 [0.34]	0.000007 [0.11]	0.0004 [1.24]	0.0006[a] [1.85]	−0.05 [−0.32]
Food technology	0.0001[a] [1.70]	0.0001 [1.33]	0.0007[b] [2.34]	0.0003 [0.52]	−0.20 [−1.34]
Chemical processing technology	−0.00003 [−0.79]	−0.00003 [−0.94]	0.0001 [0.70]	0.000002 [0.19]	0.006 [0.02]
Mining, minerals; engineering	0.0002[c] [2.99]	0.0002[b] [2.73]	0.001 [1.63]	−0.002 [−1.66]	−0.25[c] [−2.98]
Mineral oil technology	0.0001[a] [1.98]	0.0001[a] [1.83]	0.0007 [1.56]	−0.002 [−0.13]	−0.02 [−0.17]
Metallurgy	−0.00003 [−1.01]	−0.00004 [−1.64]	0.0004[c] [3.43]	0.0002 [0.21]	0.02 [0.51]
Woodworking	0.0007[b] [2.24]	0.0005 [1.41]	0.002[b] [2.43]	0.17[b] [2.23]	0.08 [0.85]
Glass, ceramics	−0.0005[b] [−2.64]	−0.0006[c] [−3.61]	0.002[c] [2.97]	0.02 [1.22]	0.11 [1.07]
Rubber, synthetics	−0.00003 [−0.86]	−0.00003 [−1.16]	0.0002 [1.44]	0.0001 [1.58]	−0.02 [−0.16]
Paper	0.0004[a] [1.75]	0.0003 [1.00]	0.001[b] [2.18]	−0.003 [−0.32]	−0.07 [−0.57]
Coatings/paints	0.00003 [0.32]	−0.000002 [−0.18]	0.0009[a] [1.98]	−0.0001 [−0.31]	−0.19 [−0.58]
Construction/ building materials	0.00003[a] [2.02]	0.00002 [1.20]	0.0002[c] [3.49]	−0.0001 [−1.46]	−0.14[c] [−2.92]
Household, entertainment	0.00008 [1.66]	0.00007 [1.21]	0.0002 [1.58]	−0.00005 [−0.21]	−0.10 [−1.51]

Notes: [a] = a significance level < 0.10; [b] = a significance level < 0.05; and [c] = a significance level < 0.01.

The comparison between standards and patents as indicators for technological innovativeness underlines that standards are only second best in explaining bilateral trade profiles. The indicator patent application is in many more cases significantly related to trade surplus. Nevertheless, in the analysis separated by technology fields the different functions and therefore impacts on trade become apparent. Therefore, standards are in some technologies indicators for international competitiveness. Furthermore, looking at the intra-industry trade differentiated by technologies, it becomes obvious and supports Swann et al.'s results that particularly international standards are fostering cross-border trade because of identical compatibility, safety and quality standards. The threat of losing product variety by uniform standards cannot be taken seriously. However, the different impacts of standards in the various technologies make further industry- or technology-based in-depth studies necessary.

17.5 THE IMPACT OF INNOVATIONS AND STANDARDS ON TRADE: A PANEL APPROACH ON GERMANY'S WORLD TRADE FLOWS[26]

17.5.1 Introduction

The results of the cross-sectoral and cross-country analyses presented in the previous section have revealed some new insights on the correlation of the stock of standards with international trade flows. However, only time-series analyses allow the detection of causalities between the changes of the stocks of standards and the development of trade volumes. The cross-sectoral analysis was nevertheless necessary, since only for a very limited number of countries and bilateral trade relationships are time series feasible, since very often restricted data availability does not allow a time-series analysis.

Among the feasibility analyses, we concentrate first on the general trade flows of Germany with the rest of the world. Furthermore, the bilateral trade flows with the United Kingdom and France will be analyzed, since these countries are the most important trade partners of Germany and they have significant national standardization activities, which is not the case for most of the smaller countries. Since we have concentrated in our theoretical hypotheses on the differences between idiosyncratic national and adopted international standards, this is an important criteria for this selection.

17.5.2 The Data

The data sources for the trade flows, the patent indicator and the standards are the same as in Section 17.4 (please compare 17.3). For the time-series analyses, we rely on the period between 1980 and 1995 in order to reach a doubling of the period used by Swann et al. (1996), thus making at least a medium-term analysis possible, which is necessary for the research on the more long-term impacts of innovations and standards on trade patterns. The restriction of the time series until the middle of the 1990s allows a differentiation between national and international standards, which loses in importance especially because of the increasing dominance of European Standards.

Table 17.7   Panel estimates for trade equations of Germany versus the world

Model:	1	2	3	4	5	6	7	8	9	10	11	12
Dep. Variables:	ex	ex	ex	ex	im	im	im	im	ex-im	ex-im	ex-im	ex-im
GDPOECD	1.71	1.68	1.89	1.86	–	–	–	–	3.02	3.03	2.75	2.77
	(9.68)	(9.54)	(9.30)	(9.18)					(4.98)	(5.02)	(4.52)	(4.57)
GDPGER	–	–	–	–	2.90	2.87	3.47	3.44	-3.86	-3.85	-3.88	-3.87
					(12.91)	(12.77)	(13.64)	(13.47)	(-6.38)	(-6.39)	(-6.46)	(-6.46)
awr18	-0.52	-0.51	-0.43	-0.42	-0.64	-0.66	-0.37	-0.39	0.09	0.11	-0.05	-0.03
	(-3.11)	(-3.06)	(-2.58)	(-2.49)	(-3.41)	(-3.49)	(-1.98)	(-2.05)	(0.46)	(0.58)	(-0.29)	(-0.17)
PEX / PIM	-0.78	-0.78	-0.82	-0.82	0.33	0.35	0.31	0.32	-1.14	-1.16	-1.15	-1.16
	(-4.36)	(-4.33)	(-4.60)	(-4.59)	(1.59)	(1.68)	(1.52)	(1.57)	(-5.64)	(-5.75)	(-5.87)	(-5.94)
RPSGER	0.06	–	0.06	–	0.06	–	0.05	–	0.01	–	0.01	–
	(2.02)		(2.11)		(1.67)		(1.51)		(0.22)		(0.31)	
RPSGER (–1)	–	0.03	–	0.03	–	0.08	–	0.07	–	-0.06	–	-0.05
		(1.16)		(0.97)		(2.53)		(1.14)		(-1.65)		(-1.42)
STDTGER (–1)	-0.06	-0.03	–	–	-0.01	0.03	–	–	-0.14	-0.16	–	–
	(0.42)	(-0.24)			(-0.05)	(0.22)			(-0.94)	(-1.10)		
STDNGER (–1)	–	–	-0.30	-0.28	–	–	0.17	0.18	–	–	-0.48	-0.48
			(-2.15)	(-2.02)			(1.02)	(1.14)			(-3.25)	(-3.27)
STDIGER (–1)	–	–	-0.03	-0.03	–	–	-0.10	-0.10	–	–	0.05	0.04
			(-1.68)	(-1.59)			(-4.48)	(-4.26)			(2.06)	(1.87)
ρ	0.73	0.73	0.73	0.73	0.78	0.78	0.78	0.78	0.73	0.72	0.70	0.70
adj. R²	0.99	0.99	0.99	0.99	0.99	0.99	0.99	0.99	0.98	0.98	0.98	0.98
RSS	4.05	4.08	3.97	4.01	5.33	5.28	5.08	5.06	5.13	5.10	4.99	4.97
DW statistics	1.82	1.85	1.83	1.87	1.82	1.80	1.81	1.80	1.96	1.95	1.95	1.94

Notes: *t*-values in brackets: $t => 1.96$ represents a level of significance < 0.05; $t => 1.65$ a level of significance < 0.10.

245

17.5.3   Results of the Regression Analysis

In a first step we try to explain Germany's entire foreign trade for the period from 1981 to 1995 by the development of standards stocks and the technological specialization profile, besides other macroeconomic indicators. Basically, this means estimating an export function depending on foreign demand. The demand for a certain good or set of goods is usually determined by the available income, the price of the goods in question, the price for substitute products and diverse quality parameters. The logarithmic form provides the estimated coefficients as elasticities. Three basic models are estimated: an export equation, an import equation and an equation for the export surplus or net export:[27]

$$ln\,(X_{it}) = a_{0i} + a_1 ln\,(GDPOECD_t) + a_2 ln\,(REV18_t) +$$
$$a_3 ln\,(PEX_{it}\,/\,PIM_{it}) + a_4\,RPSGER_i + a_5 ln\,(STDGER_{it-1}) + e_{it} \quad (17.17)$$

$$ln\,(M_{it}) = a_{0i} + a_1 ln\,(GDPGER_t) + a_2 ln\,(REV18_t) +$$
$$a_3 ln\,(PEX_{it}\,/\,PIM_{it}) + a_4\,RPSGER_i + a_5 ln\,(STDGER_{it-1}) + e_{it} \quad (17.18)$$

$$ln\,(X_{it}\,/\,M_{it}) = a_{0i} + a_1 ln\,(GDPOECD_t) + a_2 ln\,(GDPGER_t) +$$
$$a_3 ln\,(REV18_t) + a_4 ln\,(PEX_{it}\,/\,PIM_{it}) + a_4\,RPSGER_i +$$
$$a_5 ln\,(STDGER_{it-1}) + e_{it} \quad (17.19)$$

whereby the variables are defined as follows:

X_{it}, M_{it}	=	Volume of exports and imports[28] in the ICS classification group i at time t at prices in 1980 and the dollar;
a_{0i}	=	Fixed effects of the ICS classified group i;
$GDPOECD_t$	=	Real gross domestic product of the OECD countries at time t;[29]
$GDPGER_t$	=	Real gross domestic product of Germany at time t;
$REV18_t$	=	Weighted real external value of the DM at time t;[30]
$PEX_{it}\,/\,PIM_{it}$	=	Export price index in proportion to the import price index in the ICS classified group i at time t;
$RPSGER_{it\,/\,t-1}$	=	Relative patent share in the ICS classified group i at time t or time $t-1$;
$STDGER_{it-1}$	=	Complete stocks of standards T, or stocks of national standards N or stocks of international standards I of Germany in the ICS classified group i at time $t-1$;
e_{it}	=	$\rho e_{it-1} + u_{it}$;
u_{it}	=	Random disturbance error;
ρ	=	Autocorrelation coefficient of the first order.

The reason for the inclusion of the GDP of the OECD countries in th equation is to encompass the general development of the world inco.... order to depict domestic income, for the imports to Germany the GDP is used for reasons of symmetry instead of the gross value added in the manufacturing sector. The development of the real external value of the DM against the 18 most important currencies, weighted with the corresponding foreign trade volumes, should capture the influence of the exchange rate on Germany's foreign trade. The relation between the export and the import price index reflects the relative prices or the general competitiveness of Germany in price competition, while the relative patent share represents an indicator for the technological performance by comparison to the world, and the stocks of standards a further technology indicator.[32]

Table 17.7 summarizes the results of the panel estimations, which were based on 462 observations. Three blocks were formed, corresponding to the export equation (17.17) (Models 1–4), the import equation (17.18) (Models 5–8) and the export surplus equation (17.19) (Models 9–12).[33]

In the highly significant export equation the development of the GDP of the OECD countries has the expected positive influence on the German exports. Also, the real external value of the DM has the anticipated negative signs, which is consistent with the negative implications of a strong domestic currency for exports. Further the difference between the ex- and import price indexes had a negative effect on exports, as an increase means a lower competitiveness of Germany in price competition.

Due to the increasingly important technology competition, German exports are developing particularly well in those areas in which Germany's relative patent share is above-average. The coefficient certainly always has the desired positive sign, but it is only significant in the model without time delay.

The development of the aggregated sum of national and international standards does not have a significant influence on exports. If one differentiates between national and international standards, it becomes clear that the idiosyncratic national stocks of standards do not have a positive, but rather an unfavorable influence on German exports. But the international standards stocks are also not advantageous for exports. Thus the hypothesis that a competitive export advantage can be created by international and national standards cannot be confirmed. On the contrary, the allegation that a competitive disadvantage for German exporters is caused by national standards receives a certain support. Further, the positive effects derived from the theory of intra-industry trade, especially as a result of international compatibility and quality standards, is not confirmed by the empirical findings.

For imports into Germany the development above all of domestic demand represented by the gross domestic product is decisive. While the sector-specific difference between export and import price indexes has the anticipated positive sign, an increase of the real external value of the DM does not lead to a corresponding increase of cheaper imports.

All in all, Germany imports more in those ICS classified groups in which it displays an above-average technological performance. This means that it imports complementary products in its technologically strong sectors, in order to use them as intermediate products for further processing.

Neither the total national standards stocks nor the subgroups are significant in explaining the import flows, merely the development of the stocks of international standards have a significant negative influence. This means on the one hand, that the German national standards do not present non-tariff barriers to trade and therefore do not lead to trade distortions. On the other hand, the result underlines that the German stock of international standards represents a competitive advantage for the domestic producers and therefore makes imports less attractive.

The export surplus, defined as the difference between the logarithms of export and import flows, is, as expected, explained positively by the development of the GDP of the OECD countries and negatively by the German GDP.[34] The real external value of the DM has no influence any more on the export surplus, while the sector-specific difference between the export and import price indexes has negative effects, as expected.

Due to the fact that Germany is likely to import products in the fields with strong innovative capacities, Germany's relative patent share is not any more significant in the export-surplus equation. While the entire stock of standards has no significance for net trade, a negative impact emanates from the national stock of standards and a positive impact from the international ones. In this way, the hypothesis of the trade-distorting effect or the competitive disadvantage of national standards receives empirical support on the one hand. On the other hand, the German foreign trade surplus is positively influenced by the stock of international standards reflecting an important asset for international competitiveness.

17.5.4   Comparison with the Small Country Switzerland

Although these results on the basis of the entire German foreign trade will be checked by the results on the basis of the German–British and the German–French bilateral trade relations presented in the following sections, for comparison the results for Switzerland (Blind et al. 2000)[35] – a small open economy – are presented in Table 17.8. In contrast to Germany, the results for Switzerland indicate that the Swiss stocks of standards support the trade performance in the sense of the theory of competitive advantage. In particular, the national stock of standards fosters the trade balance. Since Switzer-

represents a small open economy, it is expected from a theoretical point of view that the stock of international standards should be more important. However, for Germany the stock of international standards is more important regarding the trade balance, and the national standards are obviously not positive for the German trade performance. Finally, the technological strength measured by the relative patent share (RPS) is important for the export performance respective the trade balance.

Table 17.8 Impact of standards on German and Swiss trade

	Export Germany	Export Switzerland	Import Germany	Import Switzerland	Trade balance Germany	Trade balance Switzerland
Relative patent share	+	? (+)	+	? (−)	? (+)	+
Total standards	? (−)	+	? (−)	? (+)	? (−)	? (+)
National standards	−	? (+)	? (+)	−	−	−
International standards	−	? (+)	−	? (+)	+	? (+)

Notes: The ? represent insignificant coeffcients with the respective signs in brackets. The +/− indicate significant results at least at a level of significance below 10 per cent.

Since there are obviously some unsolved puzzles among the results, the significance of standards in bilateral trade relations is examined first in the next sections, before a concluding assessment is made regarding the impacts of standards on foreign trade.

17.6   THE IMPACT OF THE STANDARDS STOCK ON THE BILATERAL TRADE FLOWS BETWEEN GERMANY AND THE UNITED KINGDOM

17.6.1   Introduction

After the attempt to explain the entire exports and imports by the stocks of standards for Germany, the bilateral trade relations of Germany with the United Kingdom will now be examined to see how far these are influenced by changes in the stocks of standards in both countries. This approach represents the adequate specification of the models of Swann et al. (1996), who included the German stock of standards, but did not use bilateral trade flows. In addition, they assumed that the German stock of standards 'represent a measure of international best practice' (Swann et al. 1996, p. 1302). This assumption is not proved and it will turn out that in some areas the British standards are more likely to represent best practice. Therefore, the analysis of the bilateral trade between the United Kingdom and Germany is the appropriate approach and will give more insights into the quality of both stocks of standards concerning the trade performance of the two countries.

17.6.2 Descriptive Statistics

In Figure 17.1 the entire standards stock of both countries in the year 1995 are depicted based on the PERINORM database, according to the ICS group classification. In Germany the complete stocks of national and international standards, with the exception of aeronautics and astronautics, and chemical process technology, are higher than in the United Kingdom. If only the stock of international standards (= national standards identical or equivalent to European and international ones) in both countries are considered (Figure 17.2), a more differentiated picture appears. In more than half of the ICS classification groups the United Kingdom has higher stocks of international standards, whereas Germany is leading in only eight fields.

The statistics on the standards stocks in both countries make it clear that significant differences exist, not only between the various ICS group classifications, but also regarding the 'internationality' of the individual stock of standards. The influence of these differences in the stocks of standards generally on the foreign trade of Germany and on the bilateral trade flows of Germany and the United Kingdom should be examined in the following time-series cross-section analyses.

17.6.3 Results of the Regression Analysis

For examining the bilateral trade relations of Germany with the United Kingdom with a special view on the impact of the stocks of standards in both countries, variants of four different models based on equations (17.20) to (17.23) are tested.[36] Besides an export and an import equation, the export surplus and the volume of intra-industry trade will be investigated with respect to their dependence on the stocks of standards:[37]

$$ln\ (X_{it}) = a_{0i} + a_1 ln\ (GVAUK_t) + a_2\ RPSGER_{it-1} +$$
$$a_3\ RPSUK_{it-1} + a_4 ln\ (STDTGER_{it-1}) + a_5 ln\ (STDTUK_{it-1}) +$$
$$a_6 ln\ (STDNGER_{it-1}) + a_7 ln\ (STDNUK_{it-1}) +$$
$$a_8 ln\ (STDIGER_{it-1}) + a_9 ln\ (STDIUK_{it-1}) + e_{it} \qquad (17.20)$$

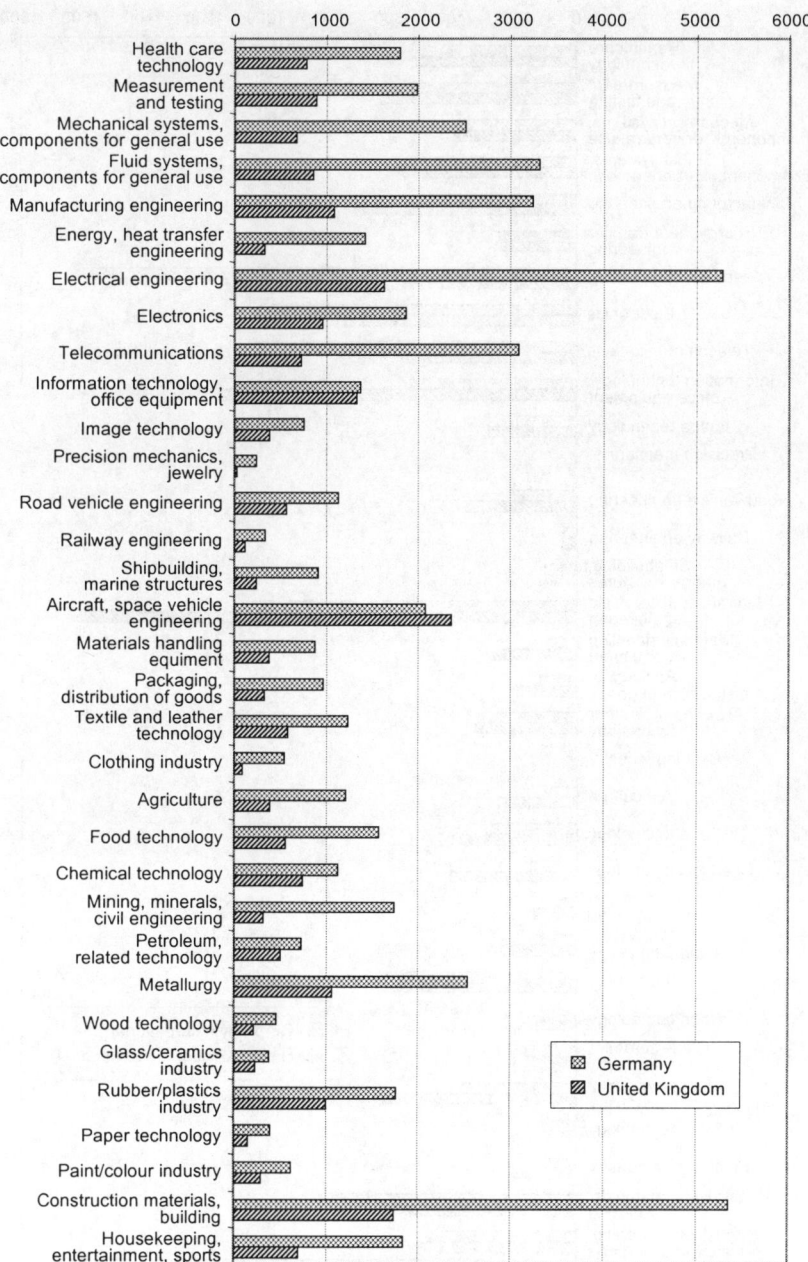

*Figure 17.1   The complete standards stocks in Germany and the United
Kingdom in 1995*

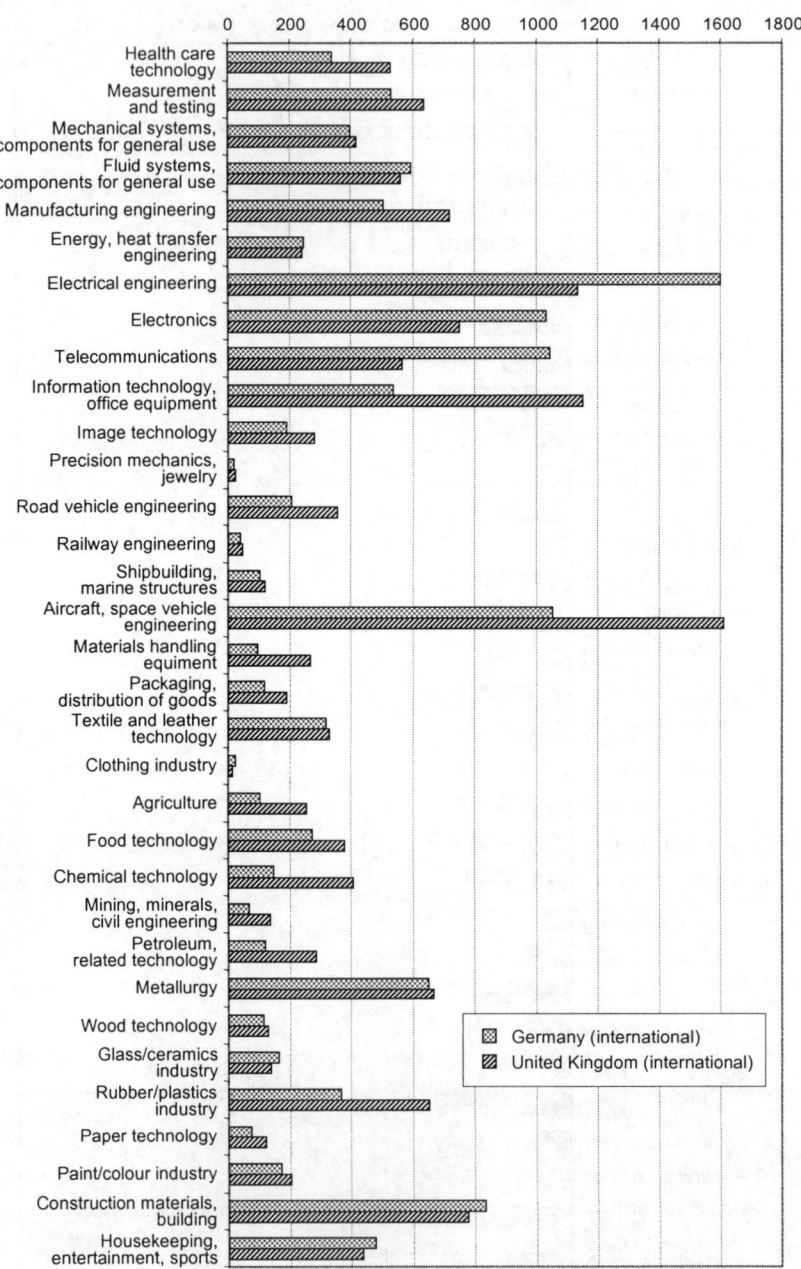

Figure 17.2   Stocks of international standards in Germany and the United Kingdom in 1995

$$ln\ (M_{it}) = a_{0i} + a_1 ln\ (GVAGER_t) + a_2\,RPSGER_{it-1} +$$
$$a_3\,RPSUK_{it-1} + a_4 ln\ (STDTGER_{it-1}) + a_5 ln\ (STDTUK_{it-1}) +$$
$$a_6 ln\ (STDNGER_{it-1}) + a_7 ln\ (STDNUK_{it-1}) +$$
$$a_8 ln\ (STDIGER_{it-1}) + a_9 ln\ (STDIUK_{it-1}) + e_{it} \qquad (17.21)$$

$$ln\ (X_{it}\ /\ M_{it}) = a_{0i} + a_1 ln\ (GVAUK_t) + a_2 ln\ (GVAGER_t) +$$
$$a_3\,RPSGER_{it-1} + a_4\,RPSUK_{it-1} + a_5 ln\ (STDTGER_{it-1}) +$$
$$a_6 ln\ (STDTUK_{it-1}) + a_7 ln\ (STDNGER_{it-1}) + a_8 ln\ (STDNUK_{it-1}) +$$
$$a_9 ln\ (STDIGER_{it-1}) + a_{10} ln\ (STDIUK_{it-1}) + e_{it} \qquad (17.22)$$

$$ln\ [1 - (|X_{it} - M_{it}|)\ /\ (X_{it} + M_{it})] =$$
$$a_{0i} + a_1 ln\ [0.5 \cdot (GVAGER_t + GVAUK_t)] +$$
$$a_2 ln\ (TV_{it-1}\ /\ TVTOT_{t-1}) + a_3\,(RPSGER_{it} - RPSUK_{it}) +$$
$$a_4 ln\ (STDTGER_{it} + STDTUK_{it}) + a_5 ln\ (STDNGER_{it} + STDNUK_{it}) +$$
$$a_6 ln\ (STDIGER_{it} + STDIUK_{it}) + e_{it} \qquad (17.23)$$

whereby the variables are defined as follows:

X_{it}, M_{it}	= Volume of German exports to the UK and the British exports into Germany in the ICS classified group i at time t, deflated by the export or import price index and normed on the dollar rate of 1980;
a_{0i}	= Fixed effects of the ICS classified group i;
$GVAGER_t$	= Real gross value added of the manufacturing sector in Germany at time t;[38]
$GVAUK_t$	= Real gross value added of the manufacturing sector in the UK at time t;
$TV_{it-1}\ /\ TVTOT_{t-1}$	= Quotient of the trade volume (export plus import) in ICS classified group i and the total trade volume in the 33 ICS groups[39]
$RPSGER_{it-1}$	= Germany's relative patent share in the ICS classified group i at time $t-1$;
$RPSUK_{it-1}$	= The UK's relative patent share in the ICS classified group i at time $t-1$;
$STDTGER_{it-1}$	= Complete stock of standards in Germany in the ICS classified group i at time $t-1$;
$STDNGER_{it-1}$	= Stock of national standards in Germany in the ICS classified group i at time $t-1$;
$STDIGER_{it-1}$	= Stock of international standards in Germany in the ICS classified group i at time $t-1$;
$STDTUK_{it-1}$	= Complete stock of standards in the UK in the ICS classified group i at time $t-1$;

$STDNUK_{it-1}$ = Stock of national standards in the UK in the ICS classified group i at time $t-1$;

$STDIUK_{it-1}$ = Stock of international standards in the UK in the ICS classified group i at time $t-1$;

e_{it} = $\rho e_{it-1} + u_{it}$;

u_{it} = Random disturbance error;

ρ = Autocorrelation coefficient of the first order.

As the results in Table 17.9 show, Germany's exports to the United Kingdom are determined above all by the development of gross value added of the manufacturing sector in the United Kingdom and thus the demand. The technological specialization profile represented by the patent applications of both countries at the European Patent Office is a further determining factor. Germany exports in those fields in which it is strong and the United Kingdom weak, as the restricted Model 2 makes especially clear.[40] The entire stocks of standards in both countries do not contribute any significant explanation for Germany's exports to the United Kingdom. By contrast, the differentiation in national and international standards leads to significant results. In the non-restricted form the German international standards have a significant positive effect, whereas not only the German national but also the British international standards have unfavorable effects on German exports.

If it is assumed first of all (Model 4) that the same consequences result from national and international standards in both countries, the results support the hypothesis that the entire national stocks of standards have a positive influence on competitive advantage, in analogy with the national innovation potential.

Instead, when focussing on the different effect dimensions of national and international standards in Model 5, it becomes clear that for German–British trade the respective national stocks of standards have no significance. On the contrary, Germany's international stocks of standards are favorable for promoting German exports to the United Kingdom, while at the same time the British international standards exercise an unfavorable influence. This cannot be so interpreted, however, that the latter represent a barrier to trade for German exports, but that they reflect together with the British innovation potential above all the technological performance and therefore the competitiveness of the United Kingdom in certain ICS classification groups.

As Table 17.10 shows, the imports from the United Kingdom are again dominated by the demand, that is the gross value added of the manufacturing sector in Germany. The technological specialization of both countries has no influence, as Germany obviously also imports goods from those areas in the United Kingdom in which domestic industry has a strong innovation potential, and the British exporters are successful in supplying Germany with goods not only from their areas of innovative and technological strength.

Table 17.9   Panel estimates for the export equation Germany–UK
             (t-values in brackets)

Model:	1	2	3	4	5
Restrictions:					
$a_2 = -a_3$; $a_4 = -a_5$	no	yes	–	–	–
$a_2 = -a_3$; $a_6 = a_7$; $a_8 = a_9$	–	–	no	yes	no
$a_2 = -a_3$; $a_6 = -a_8$; $a_7 = -a_9$	–	–	no	no	yes
a_1 GVAUK	1.60	1.51	1.46	1.43	1.38
	(4.33)	(4.69)	(3.71)	(3.72)	(4.73)
a_2 RPSGER (–1)	0.13	0.12	0.16	0.13	0.14
	(1.36)	(2.30)	(1.66)	(2.46)	(2.48)
a_3 RPSUK (–1)	–0.12	–0.12	–0.11	–0.13	–0.14
	(–1.79)	(–)	(–1.70)	(–)	(–)
a_4 STDTGER (–1)	–0.02	0.14	–	–	–
	(–0.06)	(1.27)			
a_5 STDTUK (–1)	–0.11	–0.14	–	–	–
	(–1.00)	(–)			
a_6 STDNGER (–1)	–	–	–0.41	0.09	–0.01
			(–1.22)	(1.66)	(–0.04)
a_7 STDIGER (–1)	–	–	0.10	0.09	0.10
			(1.80)	(–)	(2.06)
a_8 STDNUK (–1)	–	–	–0.05	–0.10	0.01
			(–0.27)	(–1.99)	(–)
a_9 STDIUK (–1)	–	–	–0.08	–0.10	–0.10
			(–1.45)	(–)	(–)
ρ	0.51	0.50	0.51	0.50	0.50
adj. R^2	0.94	0.94	0.94	0.94	0.94
RSS	34.5	34.5	34.2	34.4	34.3
DW statistics	1.88	1.88	1.89	1.87	1.87
F-test of the restrictions (level of significance in brackets):					
$a_2 = -a_3$	0.00 (0.95)	–	0.13 (0.72)	–	–
$a_4 = -a_5$	0.27 (0.60)	–	–	–	–
$a_2 = -a_3$; $a_4 = -a_5$	0.14 (0.87)	–	–	–	–
$a_6 = a_7$; $a_8 = a_9$	–	–	1.19 (0.30)	–	–
$a_6 = -a_8$; $a_7 = -a_9$	–	–	1.06 (0.35)	–	–
$a_2 = -a_3$; $a_6 = a_7$; $a_8 = a_9$	–	–	0.82 (0.49)	–	–
$a_2 = -a_3$; $a_6 = -a_8$; $a_7 = -a_9$	–	–	0.74 (0.53)	–	–

The entire German stock of standards has principally no negative effects on
the imports from the United Kingdom, whereas the British standards favor the
exports to Germany. However, if one assumes as in the restricted Model 2,

that the relative patent share has no influence and the entire stocks of standards produce opposite effects, then the hypothesis that standards lead to a competitive advantage in trade, is confirmed empirically by the British imports to Germany. The separation into national and international standards in Models 3 and 4 produces no more significant results for the British stocks. Compared with this, the German international standards definitely trigger positive effects for imports from the United Kingdom.

Table 17.10   Panel estimates for the import equation Germany–UK (t-values in brackets)

Model:	1	2	3	4
Restrictions:				
$a_2 = a_3 = 0$; $a_4 = -a_5$	no	yes	–	–
$a_2 = a_3 = 0$; $a_6 = a_7$; $a_8 = a_9$	–	–	no	yes
a_1 GVAGER	1.90	1.90	1.79	1.86
	(6.88)	(6.90)	(6.24)	(6.60)
a_2 RPSGER (–1)	0.04	–	0.04	–
	(0.66)		(0.67)	
a_3 RPSUK (–1)	–0.04	–	–0.05	–
	(–0.93)		(–0.99)	
a_4 STDTGER (–1)	–0.03	–0.20	–	–
	(–0.09)	(–2.04)		
a_5 STDTUK (–1)	0.18	0.20	–	–
	(1.77)	(–)		
a_6 STDNGER (–1)	–	–	–0.38	0.09
			(–1.17)	(2.02)
a_7 STDIGER (–1)	–	–	0.10	0.09
			(2.12)	(–)
a_8 STDNUK (–1)	–	–	–0.14	–0.03
			(–0.73)	(–0.59)
a_9 STDIUK (–1)	–	–	–0.01	–0.03
			(–0.25)	(–)
ρ	0.72	0.72	0.72	0.73
adj. R^2	0.97	0.97	0.97	0.97
RSS	20.4	20.5	20.2	20.5
DW statistics	2.21	2.23	2.19	2.19
F-test of the restrictions (level of significance in brackets):				
$a_2 = a_3 = 0$	0.66 (0.52)	–	0.72 (0.49)	–
$a_4 = -a_5$	0.32 (0.57)	–	–	–
$a_2 = a_3 = 0$; $a_4 = -a_5$	0.52 (0.67)	–	–	–
$a_6 = a_7$; $a_8 = a_9$	–	–	1.58 (0.21)	–
$a_6 = -a_8$; $a_7 = -a_9$	–	–	3.02 (0.05)	–
$a_2 = a_3 = 0$; $a_6 = a_7$; $a_8 = a_9$	–	–	1.12 (0.30)	–

Table 17.11   Panel estimates for the balance of trade equations Germany–UK (t-values in brackets)

Model:	1	2	3	4	5
Restrictions:					
$a_3 = -a_4$; $a_5 = -a_6$	no	yes	–	–	–
$a_3 = -a_4$; $a_7 = a_8$; $a_9 = a_{10}$	–	–	no	yes	no
$a_3 = -a_4$; $a_7 = -a_9$; $a_8 = -a_{10}$	–	–	no	no	yes
a_1 GVAUK	0.83	0.71	0.70	0.70	0.46
	(1.56)	(1.38)	(1.26)	(1.27)	(0.91)
a_1 GVAGER	−1.62	−1.66	−1.69	−1.67	−1.81
	(−3.49)	(−3.59)	(−3.58)	(−3.35)	(−3.95)
a_3 RPSGER (−1)	0.08	0.07	0.10	0.07	0.07
	(0.77)	(1.06)	(0.90)	(1.03)	(1.03)
a_4 RPSUK (−1)	−0.05	−0.07	−0.05	−0.07	−0.07
	(−0.69)	(–)	(−0.61)	(–)	(–)
a_5 STDTGER (−1)	−0.08	0.32	–	–	–
	(−0.17)	(2.24)			
a_6 STDTUK (−1)	−0.27	−0.32	–	–	–
	(−1.78)	(–)			
a_7 STDNGER (−1)	–	–	−0.35	−0.02	−0.13
			(−0.73)	(−0.30)	(−0.46)
a_8 STDIGER (−1)	–	–	−0.01	−0.09	0.02
			(−0.16)	(–)	(0.36)
a_9 STDNUK (−1)	–	–	0.07	−0.05	0.13
			(0.23)	(−0.69)	(–)
a_{10} STDIUK (−1)	–	–	−0.04	−0.05	−0.02
			(−0.60)	(–)	(–)
ρ	0.64	0.63	0.65	0.65	0.65
adj. R^2	0.84	0.84	0.84	0.84	0.84
RSS	50.4	50.5	50.8	50.9	51.1
DW statistics	1.95	1.94	1.94	1.94	1.92

F-test of the restrictions (level of significance in brackets):

$a_3 = -a_4$	0.05 (0.82)	0.14 (0.70)
$a_5 = -a_6$	0.87 (0.35)	–
$a_3 = -a_4$; $a_5 = -a_6$	0.46 (0.63)	–
$a_7 = a_8$; $a_9 = a_{10}$	–	0.26 (0.77)
$a_7 = -a_9$; $a_8 = -a_{10}$	–	0.79 (0.45)
$a_3 = -a_4$; $a_7 = a_8$; $a_9 = a_{10}$	–	0.21 (0.89)
$a_3 = -a_4$; $a_7 = -a_9$; $a_8 = -a_{10}$	–	0.60 (0.61)

As Table 17.11 shows, the trade balance with respect to the export surplus is positively, but not significantly, determined by the demand development in the United Kingdom and negatively by the gross value added in Germany as an indicator for domestic demand.

The indicators for the technological performance of both countries have the expected signs, but do not attain a satisfactory level of significance. In the admissible restricted form (Model 2) the complete standards stocks of Germany and the United Kingdom show the expected significant signs which correspond to the hypotheses from the theory of competitive advantage. Thus the German basic stock of standards favors the German export surplus vis-à-vis the United Kingdom, in that the German exporters realize quality and cost advantages. Correspondingly, the British exporters profit from their stocks of standards, so that they are more competitive in the German market and consequently reduce Germany's export surplus. The partition in national and international standards (Models 3 to 5) does not provide any more significant coefficients. This underlines the multidimensional effects of standards and technical rules in the complex interplay of national and international standards.

Table 17.12   Panel estimates for the intra-industry trade equations Germany–UK (t-values in brackets)

Model:	1	2
a_1 0.5 · (GVAGER + GVAFR)	0.76	0.87
	(2.24)	(1.40)
a_2 TV (–1) / TVTOT (–1)	0.15	0.16
	(2.58)	(2.77)
a_3 abs (RPSGER – RPSFR)	0.05	0.0
	(1.06)	(0.99)
a_4 STDTGER + STDTFR	0.36	–
	(1.96)	
a_5 STDNGER + STDNFR	–	0.65
		(1.99)
a_6 STDIGER + STDIFR	–	0.02
		(0.45)
ρ	0.66	0.67
adj. R^2	0.79	0.79
RSS	21.1	21.0
DW statistics	1.86	1.89

German–British intra-industry trade (Table 17.12) is primarily determined by the development of the gross value added in both countries. The relative trade volume corrected by total trade shows the expected and highly significant positive sign, giving hints to the existence of economies of scale. However, differences in the national innovation capacities do not have a significant but a positive sign, which was not expected, since they should have a negative impact on intra-industry trade due to the strong national specialization, and trade flows in just one direction from the country with absolute advantage into the one with absolute disadvantage should be observed.

The sum of the stocks of standards in total shows the expected positive value. However, the differentiation of standards into national and international does not underline the great importance of international standards for intra-industry trade. On the contrary, the sum of the national standards has a positive impact on intra-industry trade. And the compatibility-ensuring properties of international standards, including their embedded information about consumer preferences and technical specifications, do not foster the German–British trade, because their variety-reducing effect may be stronger than assumed. Quite the opposite, the variety-increasing impact of national standards may be stronger than the intra-industry-enhancing impact of harmonized compatibility and quality standards.

17.6.4 Summary

This section has combined data on patents and standards with data on trade flows in two-digit subgroups of the international classification of standards to explore the effects of innovations and standards on bilateral trade between Germany and the United Kingdom. The approach extends the study of Swann et al. (1996) who analyzed the sole effect of standards on trade in a time period before globalization in standardization took place.

Swann et al. (1996) found that for each of the British standards variables, the coefficients in the British export and import equations are positive or at least non-negative. Therefore, the United Kingdom's stock of standards is not simply a strategic asset that increases exports and reduces imports. Rather, they conclude that a larger stock of standards may serve to expand trade in general. However, the net effect of British standards on the British trade balance is favorable, but not significant, and the net effect of German standards on British trade is unfavorable. Therefore, the effect of standards on net exports is positive as the competitive advantage perspective would indicate. Secondly, idiosyncratic British standards have a stronger positive effect on exports and imports than internationally equivalent standards in their study.

For Germany we found slightly different results. First, the high positive significance of the innovation indicator, the relative patent share, on German exports has to be highlighted. Therefore, the impact of standards is either not significant or even negative. For instance, idiosyncratic national, and at the first glance international, standards are unfavorable for German exports. While the effect of national standards can be explained by the fact that they are suited for national markets and not for preferences abroad, the sign of the international standards can be explained by the institutional change of the harmonization of standards in Europe, which caused an increase in the international stocks of standards which was higher than that of the export flows. This is supported by the negative impact of international standards on imports which represents domestic competitiveness and the blocking of trade. Consequently, the trade balance with the rest of the world is positively influenced by international standards and negatively by national ones.

In the analysis of the German–British trade relations, we observed expected signs of the innovation indicators patents of Germany (positive) and the United Kingdom (negative) in the export equation of Germany. Furthermore, the impact of German standards is positive, while the effect of British standards is negative. Differentiated in the two subgroups of standards, we find a stronger positive impact of international standards. In the import equation, the innovation indicator has no impact, whereas international German standards have a positive impact on the import flows into Germany. In the trade balance equation, only the total stocks of German and British standards have the expected positive respective negative effects.

Although the results of Swann et al. (1996) and our findings cannot be compared directly, it becomes obvious that innovative strength is a very important parameter for the trade performance of highly industrialized countries. We are also aware that standards have different effects in different industrial sectors. Nevertheless, this study is a step further from the pioneering work of Swann et. al. (1996) about the effects of standards on trade patterns and competitiveness.

17.7   THE IMPACT OF THE STANDARDS STOCK ON THE BILATERAL TRADE FLOWS BETWEEN GERMANY AND FRANCE[41]

17.7.1   Introduction

In contrast to the German–British trade relationships, the German–French trade flows are characterized by significantly shorter distances between exporting and importing country and a quasi fixed exchange rate due to the European Monetary System. Both these differences are in general fostering trade, since the transaction costs for the exchange of goods between Germany and France are significantly lower compared to the German–British trade. However, the trade barriers caused by the different languages may be higher in the case of French compared to English. Whether the respective stocks of standards have different impacts on the German exports to and imports from France, this will be analyzed in the following section applying the identical approach as in the German–British analysis.

17.7.2 Descriptive Statistics

In Figure 17.3, the entire standards stock of both countries in the year 1995 are depicted (source: PERINORM), according to the ICS group classification. In Germany the complete stocks of national and international standards, with the exception of railway technology, are higher than in France. If only the stock of international standards (= national standards identical or equivalent to European and international ones) in both countries are considered (Figure 17.4), a more differentiated picture appears. In around one-third of the ICS classification groups France has higher stocks of international standards compared to Germany.

The statistics on the standards stocks in both countries make it clear that significant differences exist, not only between the various ICS group classifications, but also regarding the 'internationality' of the individual stock of standards. The influence of these differences in the stocks of standards generally on the bilateral trade flows between Germany and France should be examined in the following time-series cross-section analyses.

17.7.3 Results of the Regression Analysis

In order to examine the bilateral trade relations between Germany and France with special emphasis on the impact of the stocks of standards in both countries, variants of four different models are tested.[42] Besides an export and an import equation, the export surplus and the volume of intra-industry trade will be investigated with respect to their dependence on the stocks of standards.

First, in addition to other macroeconomic indicators, we try to explain Germany's exports to France for the period from 1981 to 1995 by the development of standards stocks[43] and the technological specialization profile in the same way as in the precious section. Essentially, this means estimating an export- or rather an import-demand function. The demand for a certain good or set of goods is usually determined by the available income, the price of the goods in question, the price for substitute products and diverse quality parameters. The logarithmic form provides the estimated coefficients as elasticities. The indicator for the development of demand is the value added

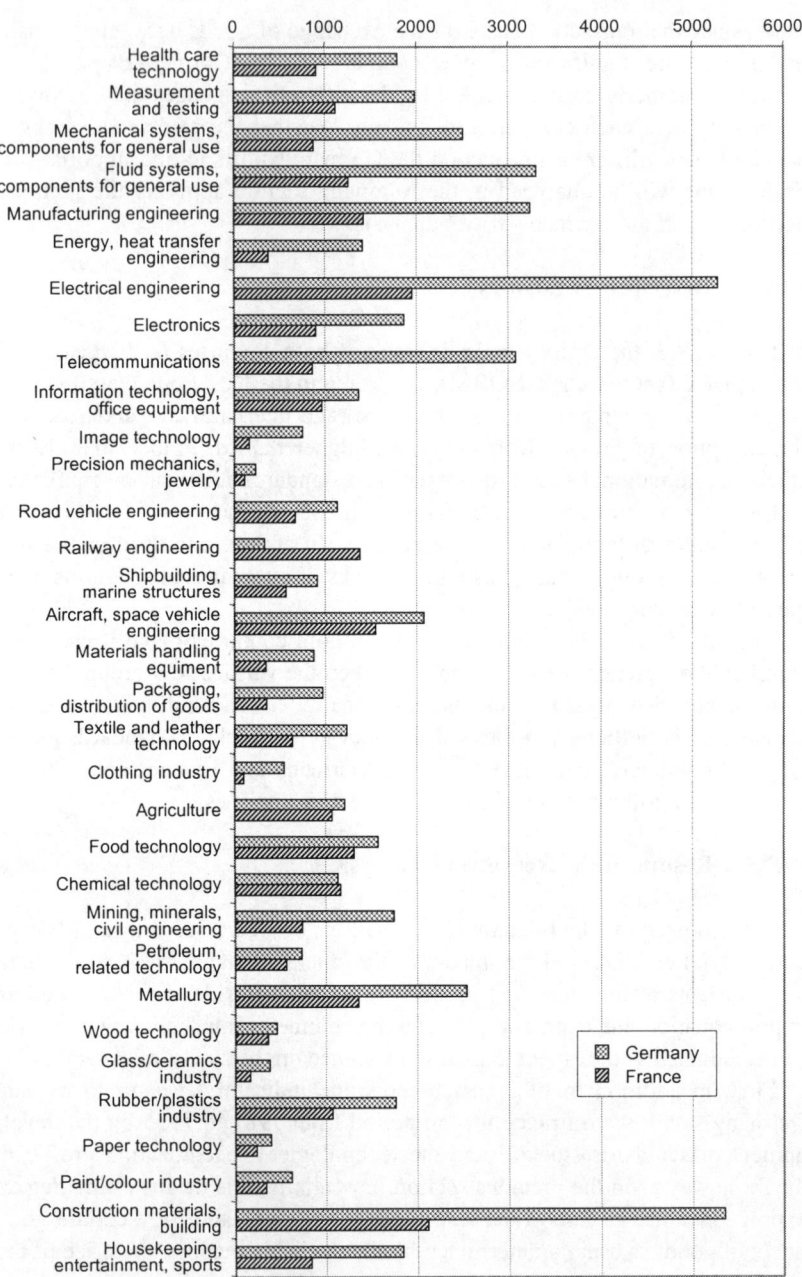

Figure 17.3   The complete standards stocks in Germany and France in 1995

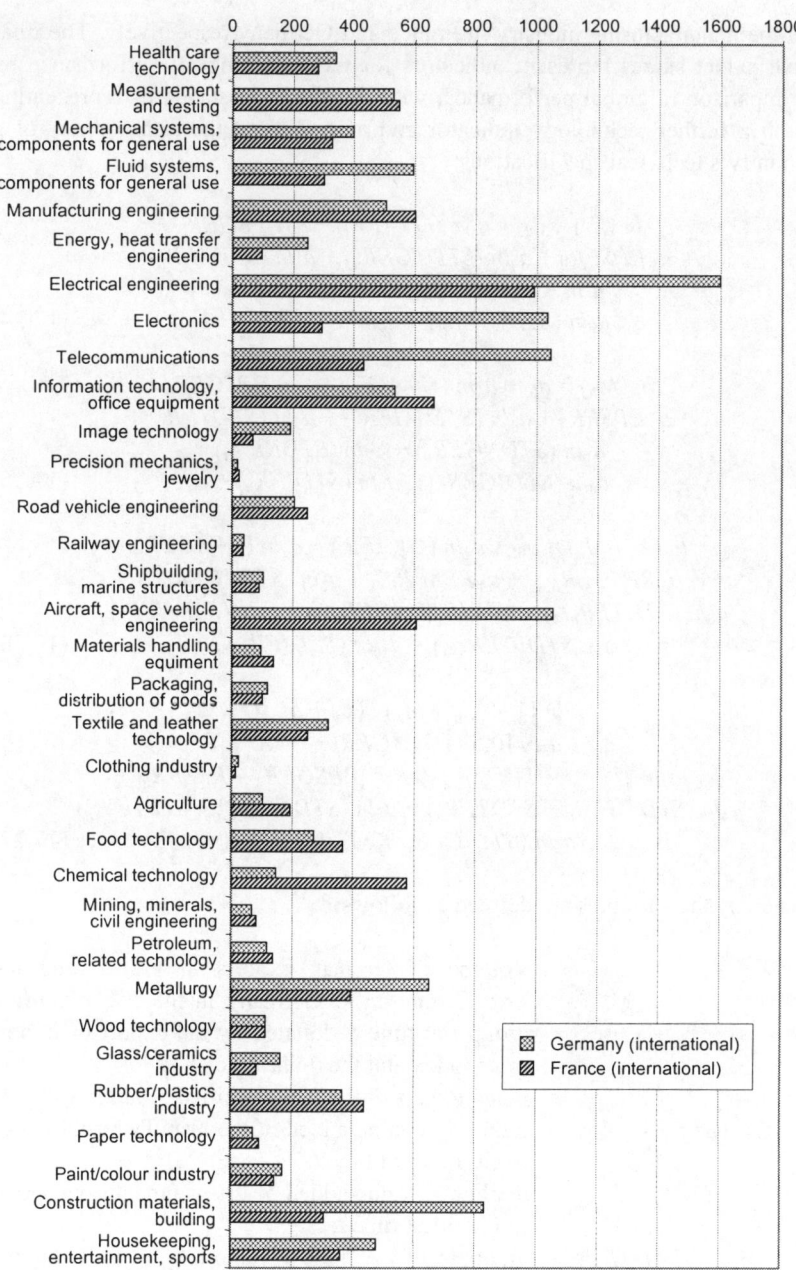

Figure 17.4   Stocks of international standards in Germany and France in 1995

in the manufacturing industry in France and Germany respectively. The relative patent shares represent indicators for the technological performance by comparison to global performance, with the stocks of standards representing both a further technology indicator and a sign about the transparency of a country's technical specifications:

$$ln \, (X_{it}) = a_{0i} + a_1 ln \, (GVAFR_t) + a_2 \, RPSGER_{it} +$$
$$a_3 \, RPSFR_{it} + a_4 ln \, (STDTGER_{it}) + a_5 ln \, (STDTFR_{it}) +$$
$$a_6 ln \, (STDNGER_{it}) + a_7 ln \, (STDNFR_{it}) +$$
$$a_8 ln \, (STDIGER_{it}) + a_9 ln \, (STDIFR_{it}) + e_{it} \qquad (17.24)$$

$$ln \, (M_{it}) = a_{0i} + a_1 ln \, (GVAGER_t) + a_2 \, RPSGER_{it} +$$
$$a_3 \, RPSFR_{it} + a_4 ln \, (STDTGER_{it}) + a_5 ln \, (STDTFR_{it}) +$$
$$a_6 ln \, (STDNGER_{it}) + a_7 ln \, (STDNFR_{it}) +$$
$$a_8 ln \, (STDIGER_{it}) + a_9 ln \, (STDIFR_{it}) + e_{it} \qquad (17.25)$$

$$ln \, (X_{it} \, / \, M_{it}) = a_{0i} + a_1 ln \, (GVAFR_t) + a_2 ln \, (GVAGER_t) +$$
$$a_3 \, RPSGER_{it-1} + a_4 \, RPSFR_{it-1} + a_5 ln \, (STDTGER_{it-1}) +$$
$$a_6 ln \, (STDTFR_{it-1}) + a_7 ln \, (STDNGER_{it-1}) + a_8 ln \, (STDNFR_{it-1}) +$$
$$a_9 ln \, (STDIGER_{it-1}) + a_{10} ln \, (STDIFR_{it-1}) + e_{it} \qquad (17.26)$$

$$ln \, [1 - (X_{it} - M_{it}) \, / \, (X_{it} + M_{it})] =$$
$$a_{0i} + a_1 ln \, [0.5 \cdot (GVAGER_t + GVAFR_t)] +$$
$$a_2 ln \, (TV_{it-1} \, / \, TVTOT_{t-1}) + a_3 \, |RPSGER_{it} - RPSFR_{it}| +$$
$$a_4 ln \, (STDTGER_{it} + STDTFR_{it}) + a_5 ln \, (STDNGER_{it} + STDNFR_{it}) +$$
$$a_6 ln \, (STDIGER_{it} + STDIFR_{it}) + e_{it} \qquad (17.27)$$

whereby the variables are defined as follows:

X_{it}, M_{it}	= Volume of German exports to France and the French imports to Germany in the ICS classified group i at time t, deflated by the export or import price index and the dollar rate of 1980;
a_{0i}	= Fixed effects of the ICS classified group i;
$GVAGER_t$	= Real gross value added of the manufacturing sector in Germany at time t;[44]
$GVAFR_t$	= Real gross value added of the manufacturing sector in France at time t;
$TV_{it-1} \, / \, TVTOT_{t-1}$	= Quotient of the trade volume (export plus import) in ICS classified group i and the total trade volume in the 33 ICS groups;[45]

$RPSGER_{it}$	= Germany's relative patent share in the ICS classified group i at time t;[46]
$RPSFR_{it}$	= France's relative patent share in the ICS classified group i at time t;
$STDTGER_{it}$	= Complete stock of standards in Germany in the ICS classified group i at time t;
$STDNGER_{it}$	= Stock of national standards in Germany in the ICS classified group i at time t;
$STDIGER_{it}$	= Stock of international standards in Germany in the ICS classified group i at time t;
$STDTFR_{it}$	= Complete stock of standards in France in the ICS classified group i at time t;
$STDNFR_{it}$	= Stock of national standards in France in the ICS classified group i at time t;
$STDIFR_{it}$	= Stock of international standards in France in the ICS classified group i at time t;
e_{it}	= $\rho e_{it-1} + u_{it}$;
u_{it}	= Random error term;
ρ	= First-order autocorrelation coefficient.

As the results in Table 17.13 illustrate, Germany's exports to France are determined above all by the development of gross value added of the manufacturing sector in France and thus demand.

The technological specialization profile represented by the patent applications of both countries at the European Patent Office is a further determining factor. Germany exports in those fields in which it is strong.[47]

The entire stocks of standards in both countries do not provide any significant explanation for Germany's exports to France. By contrast, the differentiation in national and international standards leads to significant results (Model 4). In the unrestricted form both the German national and international standards have a negative but insignificant effect, whereas the French international standards have a favorable effect on German exports.

In the permitted restricted models, it is assumed first of all (Model 5) that the same consequences result from national and international standards in both countries. The results support the hypothesis of competitive disadvantage because the entire national stocks of standards have no positive influence on competitive advantage, but the foreign stocks are favorable for their own export activities. The latter underlines the information function of standards that makes consumer and compatibility preferences more transparent for foreign suppliers. When focussing on the effects of different dimensions of national and international standards in Model 6, it is obvious that for German–French trade, the respective national stocks of standards have no significance. On the contrary, Germany's international stocks of standards are unfavorable for promoting German exports to France, while at the same time the French international standards exercise a favorable influence. Thus, together with the technological potential of Germany, the international standards in France that make cross-border compatibility possible specifically foster German exports to France.

Table 17.13　Panel estimates for the export equation Germany–France (t-values in brackets)

Model:	1	2	3	4	5	6
Restrictions:						
$a_2 = -a_3$	no	yes	no	–	–	–
$a_4 = -a_5$	no	no	yes	–	–	–
$a_6 = a_7;\ a_8 = a_9$	–	–	–	no	yes	no
$a_6 = -a_8;\ a_7 = -a_9$	–	–	–	no	no	yes
a_1 GVAFR	1.58	1.57	1.51	1.52	1.54	1.60
	(5.82)	(5.78)	(5.79)	(5.47)	(5.56)	(6.01)
a_2 RPSGER	0.09	0.04	0.09	0.09	0.09	0.09
	(2.00)	(1.31)	(2.00)	(1.90)	(1.89)	(1.90)
a_3 RPSFR	–0.02	–0.04	–0.02	–0.02	–0.02	–0.02
	(–0.54)	(–)	(–0.46)	(–0.61)	(–0.54)	(–0.55)
a_4 STDTGER	–0.24	–0.26	–0.07	–	–	–
	(–1.10)	(–1.20)	(–0.07)			
a_5 STDTFR	0.09	0.72	0.07	–	–	–
	(0.88)	(0.91)	(–)			
a_6 STDNGER	–	–	–	–0.14	–0.04	–0.07
				(–0.74)	(–1.49)	(–0.83)
a_7 STDIGER	–	–	–	–0.04	–0.04	–0.04
				(–1.51)	(–)	(–1.48)
a_8 STDNFR	–	–	–	0.01	0.07	0.07
				(0.14)	(1.71)	(–)
a_9 STDIFR	–	–	–	0.10	0.07	0.04
				(1.80)	(–)	(–)
ρ	0.71	0.72	0.71	0.70	0.70	0.70
adj. R^2	0.98	0.98	0.98	0.98	0.98	0.98
RSS	10.7	10.8	10.7	10.7	10.7	10.7
DW statistics	2.15	2.14	2.13	2.16	2.15	2.16
F-test of the restrictions (level of significance in brackets):						
$a_2 = -a_3$	2.28 (0.13)	–	–	–	–	–
$a_4 = -a_5$	0.74 (0.39)	–	–	–	–	–
$a_6 = a_7;\ a_8 = a_9$	–	–	–	0.44 (0.64)	–	–
$a_6 = -a_8;\ a_7 = -a_9$	–	–	–	0.83 (0.44)	–	–

As Table 17.14 reveals, the imports from France are again dominated by demand, that is the gross value added of the manufacturing sector in Germany. The technological specialization of both countries has a positive influence, as Germany obviously also imports goods from those areas in France in which domestic industry has a strong innovation potential, and the French exporters are successful in supplying Germany with goods especially from their areas of innovative and technological strength.

Table 17.14   *Panel estimates for the import equation Germany–France (t-values in brackets)*

Model:	1	2	3
Restrictions:			
$a_6 = a_7$, $a_8 = a_9$	–	no	yes
a_1 GVAGER	1.73	1.50	1.50
	(6.46)	(5.51)	(5.52)
a_2 RPSGER	0.21	0.22	0.22
	(3.03)	(3.23)	(3.15)
a_3 RPSFR	0.06	0.06	0.06
	(1.14)	(1.25)	(1.32)
a_4 STDTGER	0.37	–	–
	(1.37)		
a_5 STDTFR	0.26	–	–
	(1.88)		
a_6 STDNGER	–	−0.21	0.14
		(−0.85)	(4.04)
a_7 STDIGER	–	0.15	0.14
		(4.09)	(–)
a_8 STDNFR	–	0.09	0.02
		(0.75)	(0.34)
a_9 STDIFR	–	0.01	0.02
		(0.20)	(–)
ρ	0.61	0.59	0.59
adj. R^2	0.97	0.97	0.97
RSS	21.8	21.1	21.1
DW statistics	2.35	2.33	2.34
F-test of the restrictions (level of significance in brackets):			
$a_2 = -a_3$	12.63 (0.00)	–	–
$a_4 = -a_5$	10.75 (0.00)	–	–
$a_6 = a_7$; $a_8 = a_9$	–	1.08 (0.34)	–
$a_6 = -a_8$; $a_7 = -a_9$	–	3.79 (0.03)	–

The entire German stock of standards has a positive, but insignificant effect on imports from France, whereas French standards favor exports to Germany. The separation of standards into both national and international within Models 2 and 3 produces no more significant results for the French stocks. In contrast, the German international standards definitely trigger positive effects for imports from France.

As Table 17.15 shows, the trade balance (export surplus) is positively influenced by demand development in France and negatively by the gross value added in Germany as an indicator for domestic demand. The indicators for the technological performance of France have the expected signs but do not attain a satisfactory level of significance. Because Germany also imports goods in areas with its own strong technological capacity, the signs of the net export equation are ambivalent.

The complete standard stock of Germany shows an unexpected negative value that does not correspond to the hypotheses from the theory of competitive advantage, but rather competitive disadvantage. Thus the German basic stock of standards hinders the German export surplus vis-à-vis France. The partition in national and international standards (Models 2–4) underlines the unfavorable effect of international standards for German net export. On the one hand, they open domestic markets for French importers, and on the other hand, they fail to contribute positively to the competitiveness of German products in France.

German–French intra-industry trade (Table 17.16) is primarily determined by the development of the gross value added in both countries. The relative trade volume corrected by total trade shows the expected positive sign, giving hints to the existence of economies of scale, but this does not reach a critical level of significance. The same is true for the differences in the national innovation capacities, which with increasing dimensions have a negative impact on intra-industry trade; due to the strong national specialization, trade flows in just one direction from the country with absolute advantage into the one with absolute disadvantage should be observed.

The sum of the stocks of standards in total shows the expected positive value. In addition, the differentiation of standards into national and international indeed underlines the great importance of international standards for intra-industry trade. Specifically, their compatibility-ensuring properties, including their embedded information about consumer preferences and technical specifications, foster trade in general. In the context of the other results, they serve to a lesser extent as an indicator for competitive advantage, but more as a measurement for the degree of openness and transparency of an economy.

Table 17.15   *Panel estimates for the balance of trade equations Germany–France (t-values in brackets)*

Model:	1	2	3	4
Restrictions:				
$a_3 = -a_4$	no	–	–	–
$a_7 = a_8$; $a_9 = a_{10}$	–	no	yes	no
$a_7 = -a_9$; $a_8 = -a_{10}$	–	no	no	yes
a_1 GVAFR	0.98	1.18	1.11	1.04
	(1.70)	(2.08)	(1.98)	(1.86)
a_2 GVAGER	−1.06	−0.83	−0.79	−0.97
	(−2.65)	(−2.09)	(−2.01)	(−2.53)
a_3 RPSGER	−0.11	−0.12	−0.11	−0.13
	(−1.28)	(−1.48)	(−1.38)	(−1.58)
a_4 RPSFR	−0.07	−0.08	−0.08	−0.08
	(−1.13)	(−1.36)	(−1.38)	(−1.37)
a_5 STDTGER	−0.44	–	–	–
	(−1.46)			
a_6 STDTFR	−0.20	–	–	–
	(−1.32)			
a_7 STDNGER	–	0.11	−0.20	0.18
		(0.42)	(−5.22)	(1.41)
a_8 STDIGER	–	−0.21	−0.20	−0.22
		(−5.39)	(−)	(−5.58)
a_9 STDNFR	–	−0.08	0.06	−0.18
		(−0.60)	(1.20)	(−)
a_{10} STDIFR	–	0.11	0.06	0.22
		(1.33)	(−)	(−)
ρ	0.56	0.51	0.51	0.51
adj. R^2	0.82	0.83	0.83	0.83
RSS	30.7	29.2	29.3	29.4
DW statistics	2.36	2.32	2.33	2.32
F-test of the restriction (level of significance in brackets):				
$a_3 = -a_4$	3.69 (0.06)	–	–	–
$a_5 = -a_6$	9.47 (0.00)	–	–	–
$a_7 = a_8$; $a_9 = a_{10}$	–	1.08 (0.34)	–	–
$a_7 = -a_9$; $a_8 = -a_{10}$	–	1.43 (0.24)	–	–

Table 17.16   Panel estimates for the intra-industry trade equations
Germany–France (t-values in brackets)

Model:	1	2
$a_1\ 0.5 \cdot (GVAGER + GVAFR)$	0.50	0.43
	(2.04)	(1.71)
$a_2\ TV\ (-1)\ /\ TVTOT\ (-1)$	0.06	0.06
	(1.11)	(1.16)
$a_3\ abs\ (RPSGER - RPSFR)$	−0.05	−0.05
	(−0.99)	(−1.08)
$a_4\ STDTGER + STDTFR$	0.26	–
	(2.75)	
$a_5\ STDNGER + STDNFR$	–	0.10
		(0.76)
$a_6\ STDIGER + STDIFR$	–	0.07
		(2.21)
ρ	0.35	0.35
adj. R^2	0.66	0.66
RSS	12.9	12.9
DW statistics	2.29	2.26

17.7.4   Summary and Comparison

The results of the analysis of the German–French trade are slightly different from the results of the German–British trade relations (see Table 17.17). First, the high positive significance of the innovation indicator – the relative patent share – on German exports has to be highlighted. Therefore, the impact of German standards on exports is not significant or even slightly negative. Furthermore, the impact of French standards on German exports is positive but insignificant. Differentiated in the two subgroups of standards, we find a strong positive impact of French international standards. In the import equation, the German innovation indicator again has a positive impact, as well as international German standards having a positive impact on import flows into Germany. In the trade balance equation with France, the international stocks of German standards have unexpected negative effects. Obviously, German standards open the domestic market for foreign companies, but are not able to support exports. Finally, in the intra-industry trade equation the stocks of international standards are a significant driving force for the increasing trade volumes within industry sectors, whereas in the trade with the United

Kingdom the national standards foster more intra-industry trade.

Table 17.17 makes it obvious both that the impact of standards on trade flows are mostly ambiguous and that there are differences between bilateral trade flows. First, the results confirm that none of the three theoretical perspectives outlined in Sections 4.4 or 17.2 dominates the other two dimensions, therefore the coefficients of the different stocks of standards are often not significant. Second, the impacts of standards on the German–British trade relationships differ from those of the German–French trade. The explanation for these discrepancies is challenging. However, the composition of the trade flows according to the three theoretical perspectives may differ. Consequently, the various impact dimensions of standards gain different weights and the coefficients of the stocks of standards have different signs. Nevertheless, if we recapitulate the results of the cross-country and cross-sector analyses of Section 17.4, then it is more promising to analyze the impact of standards on trade focusing on specific technologies and sectors and not on a specific bilateral relationship between two countries. Consequently, the last time-series analysis in Section 17.8 will rely on the role of standards in one technological area.

Table 17.17   Comparison of the results between German–UK and German–French trade[48]

	Export UK	Export France	Import UK	Import France	Balance UK	Balance France
German relative patent share	+	+	? (+)	+	? (+)	?
Foreign relative patent share	–	? (–)	? (–)	+	? (–)	? (–)
Total German standards	?	? (–)	?	+	?	–
National German standards	? (–)	–	? (–)	? (+)	? (–)	? (–)
International German standards	+	–	+	+	?	–
Total foreign standards	? (–)	? (+)	+	? (+)	–	? (–)
National foreign standards	?	? (+)	? (–)	?	?	?
International foreign standards	? (–)	+	? (–)	? (+)	? (–)	? (+)

Finally, it is obvious that innovative strength is a very important parameter for the trade activities of highly industrialized countries. Nevertheless this study, by extending the approach of Swann et al. (1996) and applying it to the bilateral German–British and German–French trade, takes a further step in an analysis of the effects of standards on trade patterns and competitiveness in the age of globalization, where national standards obviously decrease in importance. This will make empirical analysis more difficult, since national standardization profiles are increasingly dominated by European or international standards, whose origin cannot be traced back to national initiatives in most cases. This development may be also a reason for the differences between the results for the German–British and the German–French trade flows.

17.8   THE IMPACTS OF INNOVATIONS AND STANDARDS ON TRADE OF MEASUREMENT AND TESTING PRODUCTS[49]

17.8.1   Introduction

Before going into the details of measurement and testing and the role of respective standards, the selection of this technological sector for a time-series analysis can be legitimized by the very important role of standards as discussed below, although the results of the cross-sectoral analysis presented in Section 17.4.2 give no reason to expect significant results in the sense that one of the three theoretical perspectives will dominate considerably the other contradicting approaches.

In 1998, the European Commission published a brochure 'Setting the Standard: 25 Years of Quality Measurement'. The description of the 25-year-old tradition of standardization, measurement and testing underlines in an impressive manner the importance of measurements and standards in our lives. The quality of the products and services we consume, the competitiveness of our industries and the quality of our environment depend on being able to make accurate and reliable measurements. However, the economic analysis of measurement and testing has not been well developed until today. Although Kindleberger (1983, p. 377) has stated that standards of measurement can be classified as a public good in the sense that they are available for use by all and that use by any one economic actor does not hinder the use by others, there were no broad follow-up studies about this economic dimension of standards.[50] In addition to their public good character, they produce economies of scale, in that the more economic actors use a given standard, the more everybody gains from use by others caused by increased comparability and interchangeability. Standards of measurement and testing are designed to reduce transaction costs (see also Section 3.5).

The regional dimension of standards is affecting both the economies of scale and the transaction cost aspect of standards. Concerning the first, the larger its economies of scale will be, the wider the regional scope of a measurement and testing standard. Secondly, the reduction of transaction costs will be higher when transnational transactions are considered, because the saving potential is larger compared to international transactions. Consequently, international standards may increase international trade flows more intensively, compared to national standards. Still, the economic impact of measurement and testing on international trade has been not well developed. Furthermore, the system of measurement and testing standards and conformity assessment based upon it represent an important part of the technological infrastructure of a country, which may cause economies of scale for the producers and contributes to their competitive advantage.

In combination with Kindleberger's approach shortly outlined above the technology-based trade theory already discussed in Section 4.4 provides the framework for this section, in which we estimate the importance of differences in technology in explaining the trade volumes of Switzerland's trade of measurement and testing products between 1980 and 1995. We set a very narrow focus in our trade data, because there are close links to measurement and testing standards in these product categories, which allow us to integrate the measurement and testing aspects in our empirical analysis.

The analysis of the correlation between both innovation and standardization and foreign trade should reveal to what extent they exercise a positive or even negative effect considering standards on the bilateral trade flows in measurement and testing products between Switzerland and Germany, France and the United Kingdom. In this study we will differentiate according to different effects of national and international or harmonized European standards, in order to prove the positive trade impacts of international standards according to theoretical considerations.

In Section 17.8.2, the measurement and testing of specific theoretical hypotheses on the mechanisms of standards on trade are presented; the database which formed the basis of the empirical study and some general statistics are examined in 17.8.3. The effects of the innovation profile and the changes in the stock of standards on trade are shown in a time-series analysis pooled for the three countries in Section 17.8.4. The last section summarizes the results.

17.8.2 Theoretical Hypotheses About the Impact of Innovations and Technical Standards on Foreign Trade in Measurement and Testing Equipment

Innovations and standards in measurement and testing have various impacts on exports and imports of measurement and testing products like other standards. However, besides supply-side considerations in the context of the theory of competitive advantage, the generation of innovations and standards facilitate intra-industry trade, and standards especially also have an additional influence on the demand for measurement and testing products. Since we have discussed the supply-side and competition-related dimensions already in Section 17.2, we concentrate on the impacts of standards and technical rules on the demand for measurement and testing equipment.

Besides the supply-side effects, the publication and implementation of new standards also generate a new demand for measurement and testing instruments, in order to fulfill the new, mostly better and more restrictive, measurement and testing standards. In this perspective, a differentiation between the introduction of a national and an international standard allows a distinction of the possible demand (compare Table 17.18). Whereas new national measurement and testing standards have only an impact on national producers and probably on importers of foreign products, international standards generate a worldwide demand for measurement and testing products.

In addition, the introduction of an idiosyncratic national standard, which is not compatible with the standards abroad, may produce a lock-in situation for the national producers. This is especially the case if the development and production capacities of the domestic measurement and testing companies are small and therefore dominated by the demand of national customers for measurement and testing equipment designed for national specifics. The consequence is reduced competitiveness on the international level, especially when the innovative capacity is weak. The latter is a prerequisite for influencing the standard-setting in the international standardization bodies, which may in return be beneficial for the competitiveness of the domestic measurement and testing products suppliers.

From the theoretical perspectives including the general supply-sided considerations in Table 17.2, hypotheses about the impact of innovations and standards can be derived for bilateral trade flows. In total, the influence of formal standards on exports is positive, whereas their consequences for imports are negative. Furthermore, formal standards positively influence exports and imports in the context of intra-industry trade, although the variety-reducing impact of international standards exists. From Table 17.18 it appears that the impact dimensions of the idiosyncratic national and the assumed international standards on foreign trade of measurement and testing products are unambiguously positive.

By aggregating the effects of the different approaches, it can be unambiguously assumed that the domestic innovative capacity positively influences the export of measurement and testing products as for other goods also. Concerning the imports, the negative effects from the viewpoint of competitive advantage and intra-industry trade are also decisive. Standards are also favorable for the export flows. However, the international standards have a slightly stronger effect compared to the national idiosyncratic standards. In contrast, the impacts on imports are not unambiguous, because the negative perspective in the theory of competitive advantage is offset by their positive influence for intra-industry trade and under consideration of the demand-side effects. Finally, we assume symmetric effects of innovations and standards for all countries considered in the empirical analysis.

Table 17.18   Effects of measurement and testing standards: a demand-sided approach

Theoretical approach	Type of standards	Economic effects	Impacts on exports	Impacts on imports
Demand pull	International: possibly stronger in case of strong domestic innovation capacity	Produce demand for measurement and testing products on an international level	? (+)	+
	National	Produce demand for measurement and testing products on a national level	?	+

17.8.3   Descriptive Statistics

The data sources for the trade flows, the patent indicator and the standards are the same as in Section 17.4 (please compare Section 17.3 for the details). For the time-series analyses, we rely on the period between 1980 and 1995 in order to reach a doubling of the period used by Swann et al. (1996), thus making at least a medium-term analysis possible, which is necessary for the research on the more long-term impacts of innovations and standards on trade patterns.

Table 17.19   Concordance standards–patents–trade classification

	ICS classification	IPC classification	SITC classification
Measurement and testing	17; 19	G01B; C; D; F; G; H; J; K; L; M; P; R; S; T; V; W; G04; G12	873; 874

Based on the ICS we have generated a concordance to the international classification of patents (IPC) and the SITC classification (see Table 17.19).

In Figure 17.5, the total exports and imports of Switzerland concerning measurement and testing products are depicted. It is obvious that the exports have been steadily increasing since the middle of the 1980s, whereas the imports have been more or less stagnating since then. Both trends indicate the international competitiveness of the domestic companies of Switzerland.

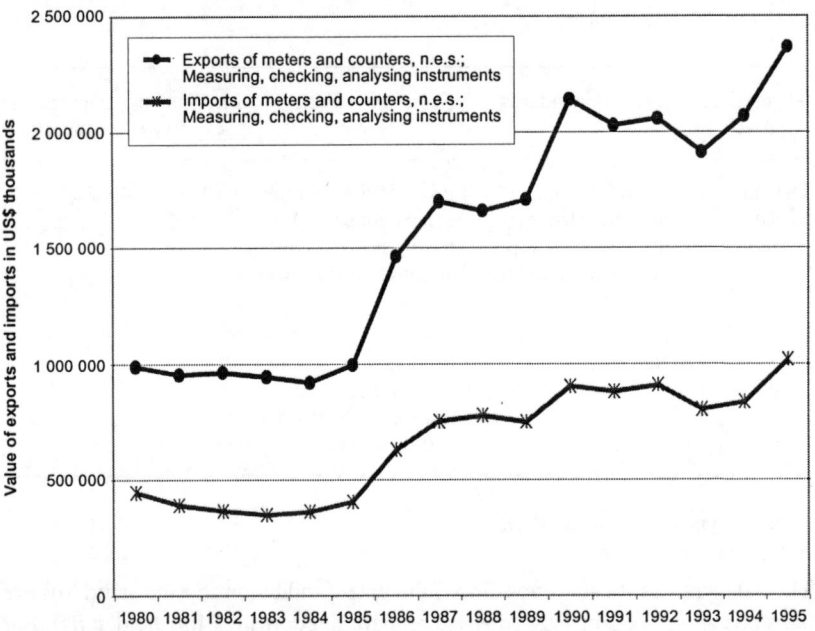

Figure 17.5   Switzerland's worldwide exports and imports in measurement and testing products between 1980 and 1995

In Figure 17.6, the entire standards stocks of the four countries in the years 1980 to 1995 are depicted, according to the ICS group measurement and testing. The statistics on the standards stocks in the four countries make clear that significant differences exist, not only between the various total stocks in the four countries, but also regarding the 'internationality' of the national stock of standards. Whereas Germany holds the largest stock of measurement and testing standards, their share of international standards is relatively low compared with the other three countries. In contrast, Switzerland's total stock of standards is dominated by international standards.

Finally, the technological position of the four countries in measurement and testing technologies is presented in Figure 17.7 by displaying their relative patent shares. Here, in contrast to the standards, Switzerland possesses the strongest position compared to the other three countries. However, they are in general still above averagely active in applying for European patents in the field of measurement and testing.

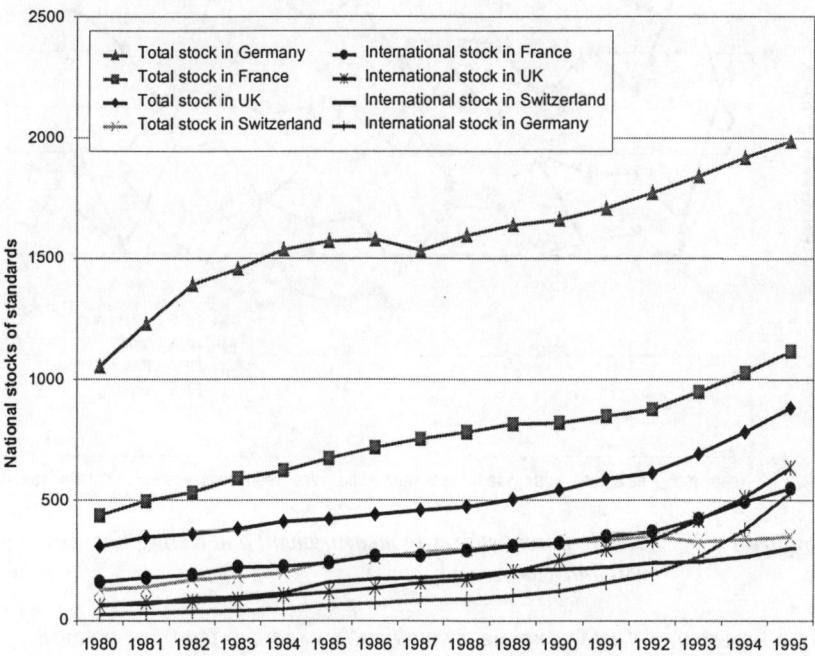

Figure 17.6　The standards stocks in Switzerland, Germany, France and the United Kingdom between 1980 and 1995

The influence of these differences in technological innovativeness and in the stocks of standards on the bilateral trade flows of Switzerland with Germany, France and the United Kingdom will be examined in the following pooled time-series analyses.

17.8.4　Results of the Regression Analysis

For examining the bilateral trade relations between Switzerland and Germany, France and the United Kingdom with a special view to the impact of innovations and the stocks of standards in both countries, variants of four different models are tested.[51] Besides an export and an import equation, the export surplus and the volume of intra-industry trade will be investigated in dependence on the stocks of standards.

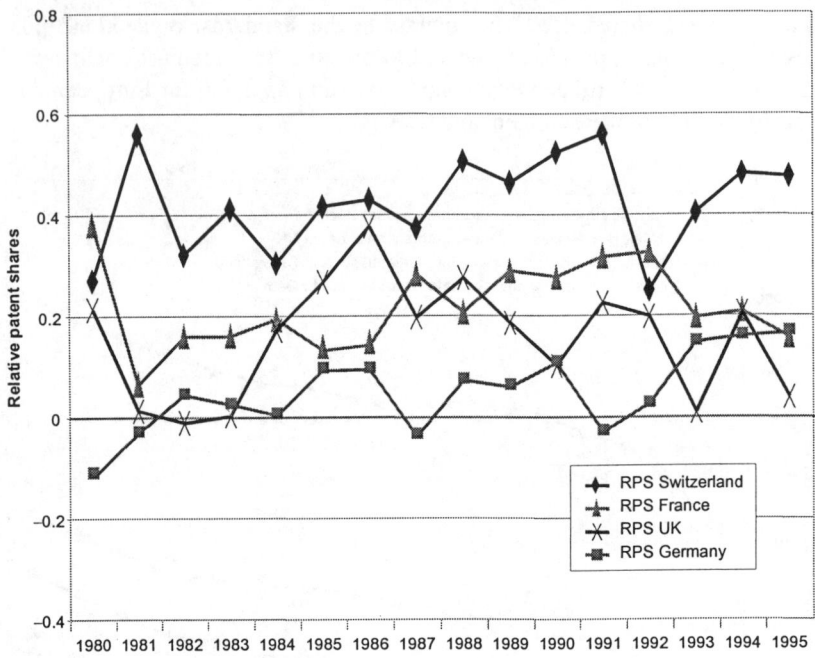

Figure 17.7   Relative patent shares in measurement and testing between 1980 and 1995

In a first step, we try to explain Switzerland's export to the three countries[52] indexed by *i* for the period from 1980 to 1995 by the development of standards stocks and the technological specialization profile, besides other macroeconomic indicators. Basically, this means estimating an export demand function. The demand for a certain good or set of goods is usually determined by the available income, the price of the goods in question, the price for substitute products and diverse quality parameters. The logarithmic form provides the estimated coefficients as elasticities. The indicator for the development of demand is the gross domestic product in the three countries. The relative patent shares represent indicators for the technological performance by comparison with the world, and the stocks of standards both a further technology indicator and a sign of the transparency of a country's technical specifications:

$$ln\ (X_{it}) = a_{0i} + a_1 ln\ (GDP_t) + a_2\ RPSCH_t + $$
$$a_3\ RPS_{it} + a_4 ln\ (STDTCH_t) + a_5 ln\ (STDT_{it}) + $$
$$a_6 ln\ (STDNCH_{it}) + a_7 ln\ (STDN_{it}) + $$
$$a_8 ln\ (STDICH_{it}) + a_9 ln\ (STDI_{it}) + e_{it} \qquad (17.28)$$

$$ln\,(M_{it}) = a_{0i} + a_1 ln\,(GDPCH_t) + a_2\,RPSCH_t +$$
$$a_3\,RPS_{it} + a_4 ln\,(STDTCH_{it}) + a_5 ln\,(STDT_{it}) +$$
$$a_6 ln\,(STDNCH_t) + a_7 ln\,(STDN_{it}) +$$
$$a_8 ln\,(STDICH_t) + a_9 ln\,(STDI_{it}) + e_{it} \qquad (17.29)$$

$$ln\,(X_{it}\,/\,M_{it}) = a_{0i} + a_1 ln\,(GDP_{it}) + a_2 ln\,(GDPCH_t) +$$
$$a_3\,RPSCH_{t-1} + a_4\,RPS_{it-1} + a_5 ln\,(STDTCH_{t-1}) +$$
$$a_6 ln\,(STDT_{it-1}) + a_7 ln\,(STDNCH_{t-1}) + a_8 ln\,(STDN_{it-1}) +$$
$$a_9 ln\,(STDI_{t-1}) + a_{10} ln\,(STDI_{it-1}) + e_{it} \qquad (17.30)$$

$$ln\,[1 - (|X_{it} - M_{it}|)\,/\,(X_{it} + M_{it})] =$$
$$a_{0i} + a_1 ln\,[0.5 \cdot (GDPCH_t + GDP_{it})] +$$
$$a_2 ln\,(TV_{it-1}\,/\,TVTOT_{t-1}) + a_3\,(RPSCH_t - RPS_{it}) +$$
$$a_4 ln\,(STDTCH_t + STDT_{it}) + a_5 ln\,(STDNCH_t + STDN_{it}) +$$
$$a_6 ln\,(STDICH_{it} + STDI_{it}) + e_{it} \qquad (17.31)$$

whereby the variables are defined as follows:

$X_{it},\,M_{it}$	= Volume of Switzerland's exports to the country i and their imports into Switzerland in measurement and testing at time t, deflated by the export or import price index and normed on the dollar rate of 1980;[53]
a_{0i}	= Fixed effects of the country i;
$GDPCH_t$	= Real gross domestic product in Switzerland at time t;[54]
GDP_{it}	= Real gross domestic product in country i at time t;
$TV_{it-1}\,/\,TVTOT_{t-1}$	= Quotient of the trade volume (export plus import) in measurement and testing and the total trade volume respective to country i;[55]
$RPSCH_t$	= Switzerland's relative patent share in the measurement and testing at time t;
RPS_{it}	= Country i's relative patent share in measurement and testing at time t;
$STDTCH_t$	= Complete stock of standards in Switzerland in measurement and testing at time t;
$STDN_{it}$	= Stock of national standards in country i in measurement and testing at time t;
$STDICH_t$	= Stock of international standards in Switzerland in measurement and testing at time t;
$STDT_{it}$	= Complete stock of standards in country i in measurement and testing at time t;

$STDN_{it}$ = Stock of national standards in country i in mea-
                          surement and testing at time t;
$STDI_{it}$ = Stock of international standards in country i in
                          measurement and testing at time t;
e_{it} = $\rho e_{it-1} + u_{it}$;
u_{it} = Random disturbance error;
ρ = Autocorrelation coefficient of the first order.

As the results in Table 17.20 show, Switzerland's exports to the three countries are not determined by the development of gross domestic product in the three destination countries and thus the demand. Measurement and testing equipment is obviously not sensitive to growth cycles in the three export countries.

However, the technological specialization profile of Switzerland represented by its patent applications at the European Patent Office is a determining factor. Switzerland's exports depend on the development of its technological strength in measurement and technology.[56] The specialization profile of the importing countries has the expected negative sign, but is not statistically significant.

The entire stocks of standards of Switzerland have, in accordance with theory, a significant positive impact on the Swiss exports to the three countries, whereas their stocks of standards have a negative impact on Swiss exports. The differentiation in national and international standards (Model 4) confirms the significant results of Switzerland's standards. However, the divided stocks of standards in the three importing countries lose their significance, whereas their innovative capacities gain the expected negative sign. Nevertheless, these results underline the positive role of a country's innovative potential on its success in exporting high-technology products. In the differentiated models both the domestic and the foreign standard indicators support the export of Swiss measurement and testing equipment.

As Table 17.21 shows, the imports of measurement and testing products into Switzerland are again insensitive to demand, that is the gross domestic product of Switzerland. The technological specialization has also no significant influence. The entire Swiss stock of standards has a significant positive effect on the imports from the three countries, whereas the foreign standards do not favor the exports to Switzerland. The separation into national and international standards in Models 3 to 5 produces even more significant results for the international standards of Switzerland. They obviously open up the Swiss market for measurement and testing equipment to foreign suppliers and are generating demand which can be satisfied by companies from abroad. This result confirms the hypothesis of the trade- or even import-fostering effects of international standards, although the direction of the effect of domestic national standard is the same.

Table 17.20 Panel estimates for the export equation (t-values in brackets)

Model:	1	2	3	4	5
Restrictions:					
$a_2 = -a_3$	no	yes	no	–	yes
$a_4 = -a_5$	no	no	yes	–	–
$a_6 = a_7; a_8 = a_9$	–	–	–	no	no
$a_6 = -a_8; a_7 = -a_9$	–	–	–	no	no
a_1 GDP$_i$	0.02	0.04	0.85	0.64	0.57
	(0.05)	(0.07)	(4.65)	(0.92)	(0.84)
a_2 RPSCH	0.16	0.15	0.15	0.14	0.17
	(1.93)	(2.38)	(1.61)	(1.74)	(2.86)
a_3 RPS$_i$	–0.13	–0.15	–0.10	–0.22	–0.17
	(–1.12)	(–)	(–0.76)	(–2.00)	(–)
a_4 STDTCH	1.08	1.08	0.64	–	–
	(4.37)	(4.45)	(5.71)		
a_5 STDT$_i$	–0.49	–0.50	–0.64	–	–
	(–2.96)	(–3.01)	(–)		
a_6 STDNCH	–	–	–	0.34	0.32
				(3.46)	(3.46)
a_7 STDICH	–	–	–	0.91	0.90
				(5.96)	(6.00)
a_8 STDN$_i$	–	–	–	0.55	0.48
				(1.48)	(1.40)
a_9 STDI$_i$	–	–	–	0.03	0.03
				(0.30)	(0.23)
ρ	0.63	0.63	0.41	0.79	0.79
adj. R^2	0.99	0.99	0.99	0.99	0.99
RSS	0.14	0.14	0.15	0.13	0.14
DW statistics	2.08	2.05	1.95	2.12	2.16
F-test of the restrictions (level of significance in brackets):					
$a_2 = -a_3$	0.06 (0.81)	–	–	0.28 (0.60)	–
$a_4 = -a_5$	3.50 (0.07)	–	–	–	–
$a_6 = a_7; a_8 = a_9$	–	–	–	8.04 (0.00)	–
$a_6 = -a_8; a_7 = -a_9$	–	–	–	11.52 (0.00)	–

As Table 17.22 shows, the trade balance with respect to the export surplus is positively determined by the demand development in Switzerland confirming

Table 17.21   Panel estimates for the import equation (t-values in brackets)

Model:	1	2	3	4	5
Restrictions:					
$a_2 = -a_3$	no	yes	no	yes	no
$a_6 = a_7; a_8 = a_9$	–	–	–	no	yes
a_1 GDPCH	−1.51	−1.29	−0.74	−0.60	0.04
	(−1.15)	(−1.04)	(−0.49)	(−0.41)	(0.03)
a_2 RPSCH	0.19	0.12	0.14	0.08	0.02
	(1.09)	(0.88)	(0.72)	(0.52)	(0.10)
a_3 RPS$_i$	−0.02	−0.12	0.01	−0.08	0.10
	(−0.07)	(−)	(0.04)	(−)	(0.40)
a_4 STDTCH	1.68	1.65	–	–	–
	(4.10)	(4.16)			
a_5 STDT$_i$	−0.24	−0.26	–	–	–
	(−1.10)	(−1.23)			
a_6 STDNCH	–	–	0.26	0.25	0.40
			(1.52)	(1.51)	(2.32)
a_7 STDICH	–	–	0.94	0.92	0.40
			(3.22)	(3.25)	(−)
a_8 STDN$_i$	–	–	0.04	0.07	0.19
			(0.11)	(0.18)	(1.45)
a_9 STDI$_i$	–	–	−0.03	−0.04	0.19
			(−0.25)	(−0.26)	(−)
ρ	0.42	0.43	0.46	0.47	0.43
adj. R^2	0.99	0.99	0.98	0.98	0.98
RSS	0.38	0.38	0.43	0.43	0.53
DW statistics	2.08	2.06	1.97	1.96	1.67
F-test of the restrictions (level of significance in brackets):					
$a_2 = -a_3$	0.43 (0.52)	–	0.25 (0.62)	–	–
$a_4 = -a_5$	9.13 (0.00)	–	–	–	–
$a_6 = a_7; a_8 = a_9$	–	–	3.20 (0.05)	–	–
$a_6 = -a_8; a_7 = -a_9$	–	–	4.62 (0.02)	–	–

the unusual role of the overall domestic demand for measurement and testing products. The indicators for the technological performance have neither the expected signs, nor do they attain a satisfactory level of significance, because Switzerland is also importing measurement and testing goods despite its own strong technological capacity. Surprisingly, the total stock of measurement and testing standards in Switzerland has a positive influence on the trade balance due to its import-fostering effect.

Table 17.22 Panel estimates for the balance of trade equations (t-values in brackets)

Model:	1	2	3	4	5	6
Restrictions:						
$a_3 = -a_4$	no	yes	no	yes	no	no
$a_7 = a_8$; $a_9 = a_{10}$	–	–	no	no	yes	no
$a_7 = -a_9$; $a_8 = -a_{10}$	–	–	no	no	no	yes
a_1 GDP_I	−0.81	−0.79	−1.61	−1.60	−2.32	−2.57
	(−0.81)	(−0.80)	(−1.21)	(−1.23)	(−1.68)	(−1.87)
a_2 $GDPCH$	3.36	3.17	2.72	2.68	2.83	1.45
	(2.05)	(2.05)	(1.42)	(1.44)	(1.34)	(0.88)
a_3 $RPSCH$	−0.12	−0.06	−0.08	−0.06	0.01	0.11
	(−0.63)	(−0.41)	(−0.36)	(−0.37)	(0.05)	(0.58)
a_4 RPS_I	−0.03	0.06	0.04	0.06	−0.09	−0.13
	(−0.12)	(−)	(0.14)	(−)	(−0.40)	(−0.57)
a_5 $STDTCH$	−1.07	−1.05	–	–	–	–
	(−2.82)	(−2.90)				
a_6 $STDT_I$	−0.20	−0.17	–	–	–	–
	(−0.80)	(−0.74)				
a_7 $STDNCH$	–	–	−0.13	−0.12	−0.23	0.02
			(−0.78)	(−0.79)	(−1.29)	(0.13)
a_8 $STDICH$	–	–	−0.61	−0.61	−0.23	−0.14
			(−2.50)	(−2.55)	(−)	(−1.14)
a_9 $STDN_I$	–	–	0.02	0.01	−0.05	−0.02
			(0.05)	(0.04)	(−0.34)	(−)
a_{10} $STDI_I$	–	–	0.08	0.09	−0.05	0.14
			(0.61)	(0.63)	(−)	(−)
ρ	0.36	0.35	0.32	0.32	0.49	0.49
adj. R^2	0.93	0.93	0.92	0.92	0.92	0.92
RSS	0.43	0.43	0.46	0.46	0.51	0.51
DW statistics	1.91	1.94	1.93	1.94	1.91	2.06
F-test of the restrictions (level of significance in brackets):						
$a_3 = -a_4$	0.25 (0.62)	–	0.02 (0.90)	–	–	–
$a_5 = -a_6$	7.89 (0.01)	–	–	–	–	–
$a_7 = a_8$; $a_9 = a_{10}$	–	–	2.07 (0.14)	–	–	–
$a_7 = -a_9$; $a_8 = -a_{10}$	–	–	2.23 (0.12)	–	–	–

The partition in national and international standards (Models 3 to 6) under-lines the unfavorable effect of international standards for the Swiss net exports, because they open domestic markets for foreign importers and create additional imports. The stocks of standards in the three foreign countries do not have any significant effect on the trade balance.

Table 17.23   Panel estimates for the intra-industry trade equations (t-values in brackets)

Model:	1	2
$a_1 \, 0.5 \cdot (GDPCH + GDP_i)$	−0.35	−0.57
	(−0.79)	(−1.04)
$a_2 \, TV_i \, / \, TVTOT_i$	0.11	−0.06
	(1.20)	(−0.70)
$a_3 \, abs \, (RPSCH - RPS_i)$	−0.17	−0.13
	(−2.25)	(−1.66)
$a_4 \, STDTCH + STDT_i$	0.33	−
	(1.51)	
$a_5 \, STDNCH + STDN_i$	−	0.61
		(3.35)
$a_6 \, STDICH + STDI_i$	−	0.11
		(1.11)
ρ	0.47	0.38
adj. R^2	0.85	0.88
RSS	0.17	0.14
DW statistics	1.85	1.91

The Swiss intra-industry trade with the three countries (Table 17.23) is not determined by the development of the gross domestic product in the two countries respectively. The rising of the relative trade volume corrected by total trade shows the expected positive sign, which gives hints for the existence of economies of scale, but it does not reach a critical level of significance. However, the differences in the national innovation capacities have a negative impact on intra-industry trade, because due to the increasing differences in the national specialization patterns, trade flows in just one direction from the country with absolute advantage into the one with absolute disadvantage will become stronger.

The sum of the stocks of standards in total shows the expected positive sign. In addition the differentiation in national and international standards underlines – in an unexpectedly strong manner – the strong importance of national standards for intra-industry trade in measurement and testing products. In particular, their compatibility-assuring purpose, including their embedded information about consumer preferences and technical specifications, are trade-fostering in general. In the context of the other results, they represent less an indicator for competitive advantage but more the degree of openness and transparency of an economy.

17.8.5 Summary

The preceding analysis has combined data on patents and standards with data on trade flows in measurement and testing products to explore the effects of innovations and standards on trade in this special area. The approach is based on the approach applied in the previous sections, which extended the study of Swann et al. (1996). First, we found similar results for Germany in a cross-section panel, since the high positive significance of the innovation indicator relative patent share on German exports coincided with its role for Swiss exports in measurement and testing products. Second there are also differences, since the impact of standards is either not significant or even negative for Germany's total exports, which confirms the sector-specific approach applied here, since the Swiss exports are positively influenced by international standards and negatively by national ones.

The analysis in this section confirms Helpman's request for more technologically-oriented trade studies. First, the inclusion of the innovation parameter is a necessary condition for a trade analysis of the high-technology area of measurement and testing products. Innovative capacity is a significant explanatory factor for the export flows of Switzerland. Furthermore, the intra-industry trade is sensitive concerning the differences of the innovative capacities of the respective countries. The wider the technology gap becomes in measurement and trading technology the smaller the intra-industry trade gets, and vice versa. Secondly, the focus on bilateral trade in measurement and testing products calls for a closer look at the specific functions of measurement and testing standards in the theoretical discussion.

Furthermore, the empirical results make it obvious that they have a major impact on the trade flows considered. First, both national and international measurement and testing standards have a positive influence on Swiss exports and imports. Secondly, the intra-industry trade analysis elucidated their positive impact with a particularly strong impact of national standards. The large stock of the latter in the three trade partners of Switzerland is opening up their domestic markets for Swiss companies, whereas for the German, French and British suppliers of measurement and testing equipment the stock of international standards in Switzerland is more important. This observation gives a hint about the importance of country size for the impact strength of national and international standards, which has to be investigated in more depth.

The analysis focused on measurement and testing made it obvious that trade flow analyses have to consider the characteristics of technology, which should include both innovation and diffusion aspects. Furthermore, bilateral relationships are certainly favorable to general export and import flows concerning the whole world, because of the inclusion of country-specific technology indicators and country size as an additional aspect which has to be taken into account.

17.9   COMPARISON AND INTERPRETATION OF SECTORAL AND MICRO-BASED RESULTS[57]

17.9.1   Selection of Data and Methodological Procedure

Analogous to the analysis of the relationship between standardization and technological change, we compare the sector-based results of the trade analysis with the assessment and opinions of companies. Since the trade performance of a whole country in a sector depends also on the framework conditions, such as the national system of innovation, we aim for the separation of the company-specific features. In order to focus just on the role of formal standards and standardization for companies' export behavior, we present some statistics of a company survey.

In this comparison between the sectoral and the microeconomic results, we use the data and follow the methodological approach already outlined in Section 16.5. The micro-analysis is based on the responses of the companies questioned in Germany (see Section 13.3 for further details). Further, the comparison concentrates on the effects of formal standards which are exclusively examined in the sectoral or technology-based analysis in Section 17.8.

As a rule, the depiction is limited to descriptive statistics differentiated according to eleven branches (see Table 13.1). Besides the descriptive statistics, it will be tested to what extent the participation in the standardization process is able to influence export performance in Section 17.9.3.

17.9.2   Comparison of the Sectoral and Micro-based Results

The relationship on the aggregated level turned out to be rather ambiguous, therefore the question remains open, whether this puzzle can be solved by analyzing the behavior of the individual economic units, the companies.

The role of standards for foreign trade is ambivalent, not only from the perspective of the various theoretical approaches, but also in the empirical results of the macroeconomic analysis relying on the trade relations observed. Thus one theoretical hypothesis states that standards increase national competitiveness, in that as a positive signal of quality they improve the sales chances or willingness to pay and by economies of scale make cost savings possible in production. On the other hand, the mere publication of a standard or technical rule on a national level discloses local technical knowledge and preferences, which can be more easily anticipated in the long term by foreign competitors, so that their competitiveness is increased on the whole, and in particular their import efforts regarding the domestic market are supported.

Reputation, quality and transparency effects of company standards
The survey asks about the international reputation of the standards. The majority of those questioned rather agreed with this statement. However, branch-specific differences became apparent. Above all, in the aeronautics and aerospace and in the vehicle construction sectors this statement meets with significantly higher approval, while the enterprises in the chemical industry and in radio, television and telecommunications technology fluctuate between 'so-so' (= 3) and 'applies' (= 4) (Figure 17.8).

The responses to the statement 'Standards increase the quality of our products' correlate significantly with the responses to the question of reputation, whereby the latter had a slightly higher rate of agreement on average. This means that the signaling effect of standards is more highly estimated than their quality-increasing effects. Large enterprises in particular agree with the reputation statement more than the medium-sized enterprises. On the other hand, the small companies support the quality statement more emphatically than the large enterprises. Taken together, the small companies profit more from the quality-promoting effects of standards, while the large ones rather rely on their reputation.

A slim majority of those questioned agreed with the further statement that standards support foreign suppliers, as they reveal technical specifications. This underlines the principally ambivalent impacts of standards. The answers to both statements correlate positively on the basis of the individual responses of the enterprises. In the branch differences it stands out that the companies in domestically-oriented construction do not see an increase in transparency caused by standards. This result is consistent with the (on the whole) critical assessment of the standards stocks in the construction industry.

Two approaches can be chosen to compare these responses with the results of the macroeconomic foreign trade analysis. In a first step, a comparison on an aggregated level is carried out, while in a second step the common features and differences will be worked out at the branch level – as far as possible.

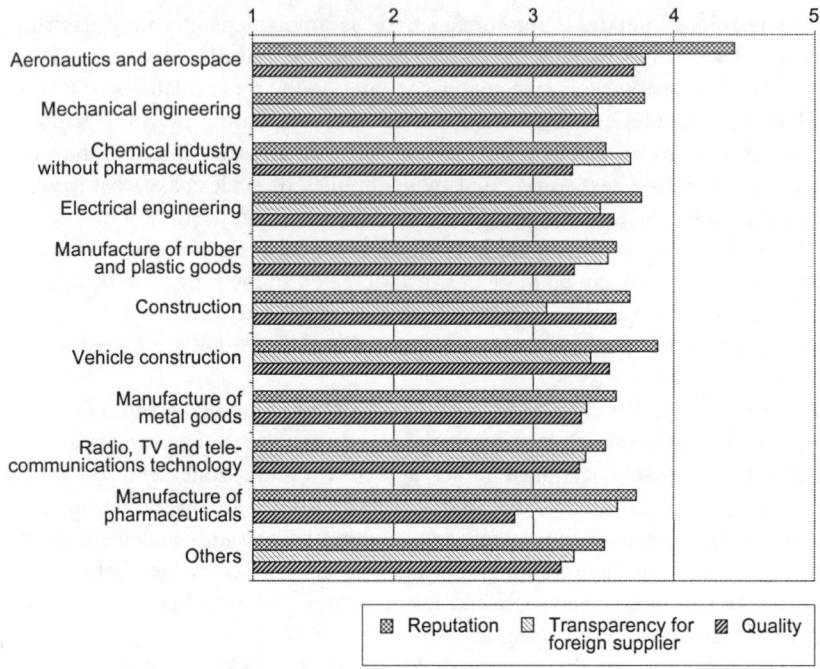

Figure 17.8   Reputation and quality effects of formal standards and their impacts on the transparency of technical specifications for foreign suppliers (disagree absolutely = 1 until agree completely = 5)

From the sectoral analysis of the significance of the standard stocks for foreign trade flows, the following results were obtained for Germany's foreign trade:

- The technological specialization profile is decisive for exports, whereby the national standards stock in particular is not export-promoting.
- The imports are weaker, especially in the areas where Germany has adopted/incorporated very many international and European standards into the national stock. As they cannot be misused as trade barriers because of their internationality, they help the domestic suppliers by improved competitiveness, especially regarding the importers.
- The international standards adopted in Germany therefore positively influence the foreign trade surplus (exports–imports) with the rest of the world, while the originally national standards emanate a negative influence.

As in the questions about the reputation of standards and the increased transparency for foreign suppliers, no differentiation was made into national and international standards, a comparison of the sectoral- and the microeconomic results can only be made in a limited way.

From the basically more positive values in favor of reputation compared with the statement on increased transparency for foreign suppliers, the following can be concluded:

- Standards develop positive impacts for exports.
- Despite the improved transparency for foreign suppliers, standards eventually also lead to an increased competitiveness of domestic producers concerning imports.
- Therefore, the standards are positive for the balance of exports and imports on the whole.

From the sectoral results, an accordance between these results can only be determined for the international standards adopted in Germany. These positive effects for the originally national standards cannot be identified on the sectoral level.

In a second step, the results of the sectoral analysis and enterprise survey should be compared on the branch level. To this end, the company estimations, classified according to ICS categories, on the role of the stocks of standards were referred to, to explain the balance of trade differences. These results are based on 36 bilateral relations between 9 countries. However, the trade relations of Germany, Austria and Switzerland, having a similar standardization system, account for 21 of the 36 relations alone, and on the other hand, these 36 observation points provide a statistically sound data basis, so that this group of countries can be utilized for the comparison.

In order to compare the positive impacts of the reputation effect with the (at least for the domestic competitiveness) negative impacts of the increased transparency, the differences between the answers to both questions are elaborated according to branches. This balance can be compared with the foreign trade results. In Table 17.24 we find the indicators of the standards stocks and their correlations with the foreign trade surpluses in the various ICS categories.

In most of the branches questioned, an agreement between the results of the enterprise survey and the analysis results of the sectoral foreign trade investigation can be noted. Only in the electrical engineering branch is a contradiction apparent between the survey results and the analyses of the foreign trade flows, as the total stocks of standards do not lead to the corresponding export surpluses in the bilateral trade relations observed.

Table 17.24　Determining factors of bilateral trade in selected ICS groups on the basis of 36 bilateral relations, compared to the difference between reputation and transparency effects

ICS Group	Code	RSTD	RSTD-NAT	RSTD-INT	Sectors	Reputation – Transparency[b]
Mechanical systems	21	+ + +	+	+ + +	Manufacture of metal goods	?
Manufacturing engineering	25	+ + +	+ + +	?	Mechanical engineering	
Electrical engineering	29	– –	?	?	Electrical engineering	+
Electronics	31	?	+ +	?	Radio, TV and telecommunications technology	?
Tele-communications	33	– – –	+ +	– – –		
Road vehicle engineering	43	+ + +	+ + +	+ + +	Vehicle construction	+
Aircraft, space vehicle engineering	49	?	?	?	Aeronautics and aerospace	+ +
Chemical technology	71	– –	– –	?	Chemical industry without pharmaceuticals	?
Paint/color industries	87	?	?	?		
Rubber/plastics industries	83	?	?	?	Manufacture of rubber and plastic goods	?
Construction materials, building	91	+ +	+	+	Construction	+ +

Notes:
[a] RSTD = relative difference of the total standards stocks between the export and import country; RSTD-NAT = relative difference of the originally national standards stocks between the export and import country; RSTD-WT = relative difference of the adopted international standards stocks between the export and import country. The question marks mean not significant coefficients; three pluses/minuses a level of significance < 0.01; two pluses/minuses a level of significance < 0.05; and one plus/minus a level of significance < 0.10.

[b] The question marks mean differences less than 0.25 points on the five-point scale between the answers to the reputation and transparency statements; one plus/minus a difference between 0.25 points and 0.5 points; and two pluses/minuses a difference greater than 0.5 points.

Export strategies and standards

In the questions on the companies' export strategies further evidence was supplied on the significance of national and European or international standards. When enterprises are confronted with other standards in foreign markets, almost 80 per cent of the companies react by utilizing European or international standards.[58] Here partly clear differences can be noted among the branches. Whilst almost all the companies surveyed from electrical engineering adopted European or international standards, in building and construction only somewhat over half followed that strategy.

In respect of electrical engineering, this result agrees with a branch-specific explanation of Germany's exports to the rest of the world through the national or the adoption of international standards stocks. For it is one of the few branches in which the European and international standards adopted in Germany statistically significantly support the worldwide exports.

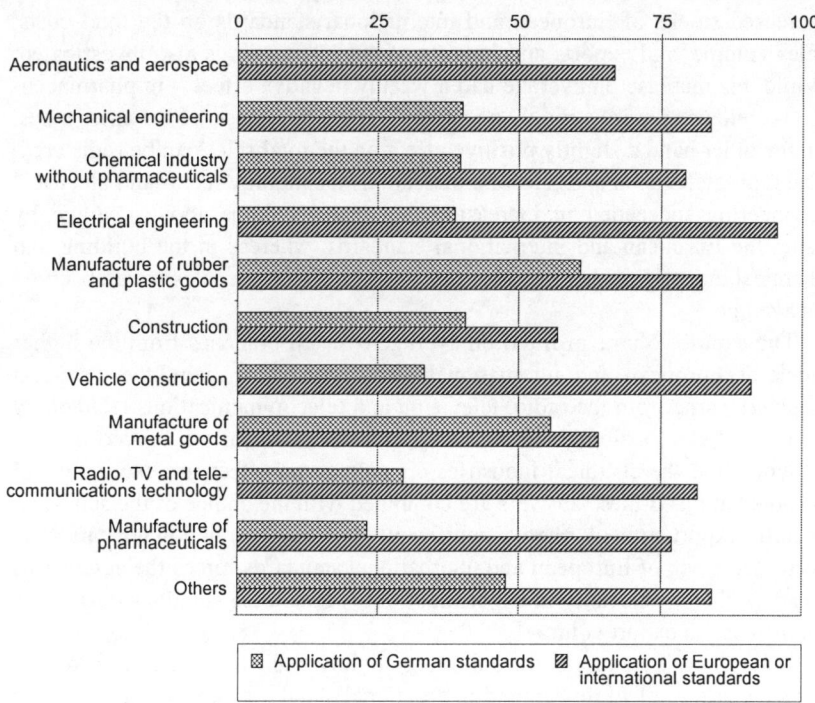

Figure 17.9   *Application of formal German or European/international*
                standards as export strategy
                (share of companies in per cent, multiple answers possible)

The high share of over 42 per cent of enterprises, which assert themselves in export markets with original German standards is surprising.[59] The rubber and plastics industry with 60 per cent and the manufacturers of metal goods with 55 per cent stand out especially here. The optimistic estimations, in particular of these branches, regarding the assertion of German standards in the export markets cannot be supported by the results of the branch-specific econometric analysis.

Barely a quarter of the companies adapt to foreign standards for producing their goods, whereby this strategy is outstanding in radio, television and tele-communications technology with 44 per cent of the enterprises.

Effects of increasing stocks of European and international standards on export and import volumes

As the utilization and adoption of European or international standards – as expected – became the dominant export strategy, the effects of the strongly increased stocks of European and international standards on the total costs, sales volume, and exports and imports of the companies is also investigated. While this increase on average had a weakly negative effect – in pharmaceuticals somewhat more strongly negative – on the total costs of the enterprises, on the other hand a slightly positive effect on the total sales can be registered. The radio, television and telecommunications technology sector and electrical engineering succeeded in increasing their sales volume above average by using the European and international standards, whereas in the building and chemical industry neither negative nor positive effects on sales were observed on average.

The export volume profited on average over all branches from the higher stock of European and international standards. Once again, the strongest impulses came from the radio, television and telecommunications technology sector and electrical engineering, whilst for the companies surveyed in construction and the chemical industries on average no effects on exports could be noted at all. If these answers are combined with the choice of the standard-specific export strategy, then a significantly positive correlation appears. The increased stock of European and international standards forces the enterprises to adopt them, especially in exporting, but rewards them simultaneously with an increase in export volume.

Imports unlike exports are only very slightly positively influenced by the growing stocks of European and international standards, and not at all in the chemical industry, the rubber and plastics industry and in metal goods manufacturing.[60]

If one compares the answers of the companies with the results of the sectoral investigations, which did not show a positive connection between the stocks of European and international standards valid in Germany and imports, at first sight a contradiction seems apparent. This can, however, be partly re-solved because the period investigated lasted only from 1980 until 1995 for data restriction reasons. The effects of the harmonization of standards in Europe on imports, too, were only relevant at their beginning, and the corresponding period compared with the total period under investigation amounted to a mere third and thus had less influence on the estimated results.

Table 17.25   *Influence of the increasing stock of European and international standards on exports and imports and intra-industry trade*

Sectors	Export	Import	Intra-industry trade
Aeronautics and aerospace	10.00	5.00	+ +
Mechanical engineering	3.77	1.47	+
Chemical industry without pharmaceuticals	1.06	0.00	?
Electrical engineering	10.12	2.91	+ + +
Manufacture of rubber and plastic goods	9.00	0.00	?
Construction	1.47	2.94	+ + +
Vehicle construction	5.30	3.91	+ + +
Manufacture of metal goods	8.57	−0.74	+ + +
Radio, TV and telecommunications technology	9.48	3.45	+ +
Manufacture of pharmaceuticals	5.77	0.00	−
Others	8.72	4.38	−

Notes: The scale for the assessment on exports and imports reaches from positive (+25) to negative (−25). Regarding the econometric results, ? means not significant coefficients; three pluses a level of significance < 0.01; two pluses a level of significance < 0.05; and one plus a level of significance < 0.10.

However, the estimates of the surveyed companies regarding the consequences on the sum of exports and imports agree with the results of the connections between European and international standards stocks with the intra-industry trade between nine countries differentiated according to the ICS classified groups. For in none of the 36 ICS groups could a negative correlation be found between the sum of European or international standards stocks and the extent of intra-industry trade defined as the sum of imports and exports in certain product classes.

In Table 17.25 the strength of the correlation between the stocks of European and international standards and the intra-industry trade for the branches is depicted, for which a categorization according to ICS groups is possible for the branches in the company survey (cf. Table 13.1). With the exception of the chemical industry, the influence of the standards on the intra-industry trade is always significantly – or even highly significantly – positive.

Trade-facilitating effects of European and international standards
The trade-facilitating effects achieved by European and international standards are also dealt with in the company survey. More than three-quarters of the companies state that their trading and commercial activities are affected by European and international standards. In electrical engineering it is even over 90 per cent, while conversely less than 60 per cent of the enterprises in the building and construction sector feel any impacts, oriented as they are to the domestic market. It can further be determined that smaller companies are significantly less affected by European and international standards than medium-sized and large enterprises. This depends largely – as will be shown in the following section – on the differing export intensities. Consequently, the trade-facilitating effects are more often determined by increasing company size.[61]

Positive impacts of the European and international standards are especially seen in the increased ease of drawing up contracts on the part of nearly two-thirds of the companies surveyed.[62] But trade barriers are also dismantled by the European and international standards for the majority of enterprises. Once again, electrical engineering stands out positively, with over 80 per cent of the companies questioned, and construction with its domestic orientation at less than 30 per cent stands out negatively from the other branches (Table 17.26).

The savings in logistic costs through European and international standards are of rather less significance. Merely one-quarter of the enterprises realize savings in logistic costs. A somewhat higher percentage is found in aeronautics and aerospace technology, chemicals and vehicle construction, whereas under 10 per cent of the enterprises in the rubber and plastics industries and the pharmaceutical industry are able to cut costs in this field.

Advantages and disadvantages of transferring national standards into European or international standards
Besides the fundamental advantages and disadvantages of European and international standardization, there is the particular situation that the content of national standards is transferred into European and international standards. Such a situation can lead to temporary or lasting cost or competitive advantages, as the foreign competition needs more time to adapt the standards,

Table 17.26   Effects of European and international standards on the intra-industry trade and their trade-facilitating effects

ICS classified group	STD	Sector	No effects	Dismantling trade barriers	Drawing up contracts easier	Lowering logistic costs
Mechanical systems	21	+++	Manufacture of metal goods	22.0%	50.0%	66.7%	22.2%
Manufacturing engineering	25	+	Mechanical engineering	21.0%	62.5%	71.4%	23.2%
Electrical engineering	29	+++	Electrical engineering	8.7%	80.4%	69.6%	19.6%
Electronics	31	?	Radio, TV and tele-communications technology	19.4%	54.8%	64.5%	25.8%
Telecommunications	33	++					
Road vehicle engineering	43	+++	Vehicle construction	18.8%	56.3%	62.5%	31.3%
Aircraft, space vehicle engineering	49	++	Aeronautics and aerospace	14.3%	71.4%	57.1%	42.9%
Chemical technology	71	?	Chemical industry without pharmaceuticals	22.5%	46.9%	67.4%	30.6%
Paint/color industries	87	+					
Rubber/plastics industries	83	?	Manufacture of rubber and plastic goods	26.9%	42.3%	65.4%	7.7%
Construction materials, building	91	+++	Construction	43.9%	29.3%	58.5%	14.6%

Notes: STD = sum of international and European standards stocks. The question marks mean not significant coefficients; three pluses/minuses a level of significance < 0.01; two pluses/ minuses a level of significance < 0.05; and one plus/minus a a level of significance < 0.10.

that is integrate them into the production processes.

In principle, the companies surveyed do not see any chance to realize lasting cost and competitive advantages over competitors by using this strategy. On the other hand, they have temporary advantages to a limited extent, and no permanent disadvantages occur. On the whole, the responses of the enterprises underline that on balance the advantages more than outweigh the disadvantages (Figure 17.10).

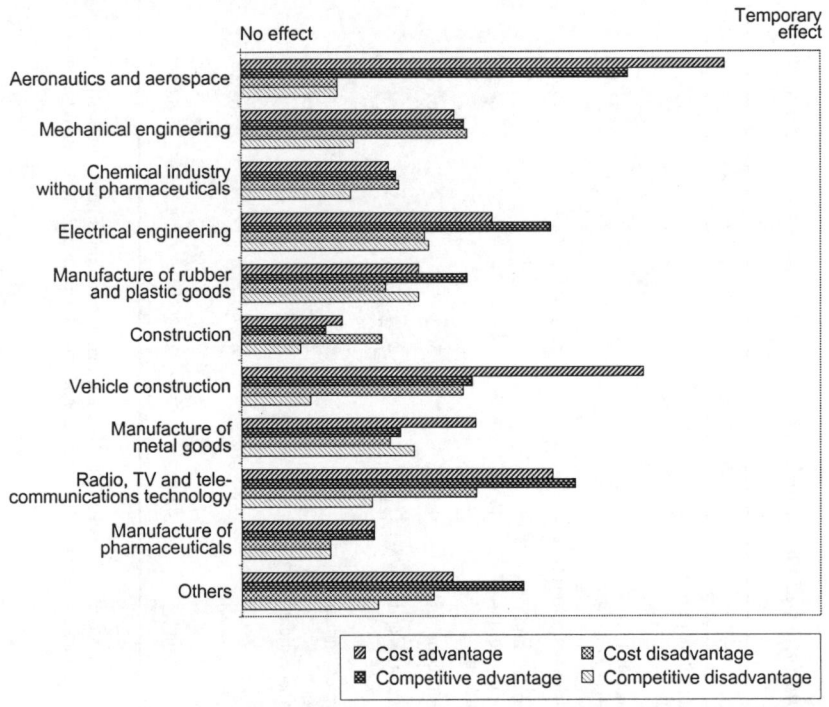

Figure 17.10　Advantages and disadvantages through transferring formal national standards into European and international standards

Surprisingly enough, for mechanical engineering, the chemical industry and building/construction, higher cost disadvantages than advantages occur. This result should be analyzed in more depth. If one differentiates further according to company size, then it can be seen in mechanical engineering that with increasing company size, cost and competitive advantages can be achieved by transferring national standards contents into European and international standards. Large enterprises profit from this procedure more strongly, because they already participate more vigorously in the national standardization processes and exert a greater influence on the contents of the standards than the small firms. Therefore the standardizers more frequently have temporary cost and competitive advantages. Besides the aeronautic and aerospace industry, motor vehicle construction has the highest cost advantages and electrical engineering, radio, TV and telecommunications technology and 'others' have the most enduring competitive advantages.

It was not possible to determine on the basis of the database PERINORM and other possible sources (Blind 2002c) to what extent the transfer of own national standards into supranational standards leads to cost and competitive advantages, because the national origins do not become apparent from the descriptions of the European and international standards. Therefore the data from the company survey are the only ones which permit a conclusive statement on the effectiveness of this standardization strategy. In principle, the responses of the surveyed companies underpin the theoretical considerations, which assume a knowledge advantage and leading time edge in marketing for the actors participating in the standardization process. These advantages are also enjoyed by the companies in a national market, which have been utilizing a national standard for a longer time, as opposed to the foreign enterprises which can apply or must de facto adopt this only after it has been incorporated into a European or international standard.

A disadvantage can even result for the domestic companies if, through the European or international standard, the market power of the national (large) enterprises participating directly in the standardization process is strengthened even further.[63]

17.9.3 Collaboration in the Standardization Process and other Explanatory Factors for Export Activities in Companies

Introduction

In order to conclude the analysis about the role of standards and standardization for trade, a comprehensive, multivariate model to explain the export activities of companies on the microlevel is presented, which may complement the various results at the sector level. In particular, we will examine to what extent (besides other factors) the active collaboration in standardization (*Std*) can explain export performance, since the active involvement in standardization activities allows companies to influence the specifications not only of national standards, but also of international standards. Consequently, export activities can be fostered by trying to define these specifications as close as possible to those used in the own company.

After it becomes clear by means of the Probit estimate of the standardization activity that the majority of the exporting companies are more inclined to engage themselves in standardization, the question arises also whether the participation in standardization increases the capacity and therefore the likelihood to export, which would question the exogenous character of the export intensity in the Probit analyses of Section 13.4 in Part C. If no significant influence of standardization activities on the export success can be observed, the validity of the above tested models can be confirmed.[64]

Besides the standard-related explanatory variable, not included in any other study, the export model includes the most important exogenous variables used already in several other studies. A number of empirical investigations already exist on the influence of innovation activities on the exports of a company, which for the most part show a positive correlation.[65] Therefore, the R&D intensity[66] is taken as a company-specific innovation variable and the patent intensity of the branch as a sector-specific innovation variable (*Inno*)[67] in addition to the collaboration in the standardization process (*Std*).

Derived from the classical foreign trade theories, the relative allocations of capital and labor and the resulting relative factor prices are a determinant for specialization and export flows. As this information is not available on a company level, the labor costs (*LabC*), normalized by the gross output, are taken as an indicator for the relative allocation with labor and the average wage (*AvWag*) as an indicator for the quality of the human capital in the branch.[68] Further, the labor coefficient (*LabInt*) is available as an exogenous variable for each company. As Germany is (comparatively) low in labor supply and high in qualified human resources, it is expected that the labor costs per turnover and the labor coefficient will have a negative effect on the export probability, and the average wage a positive one.[69] In contrast to the labor market, the market for capital is international. Therefore, there should be no large differences in the access to capital for companies in different countries. Consequently, no variable for the availability or the price of capital is included in the equation.

However, as the opening up of export markets initially involves relatively high fixed costs, it can be concluded from this that the larger a company, the more easily the financial means can be raised or the risk taken.[70]

Finally, the intensity of competition (*CompInt*) is an influential factor for the export activities of an enterprise.[71] On the one hand, it is argued that enterprises in highly competitive (domestic) markets are inclined to look for opportunities in export markets (Brouwer and Kleinknecht 1993). On the other hand, companies in markets with low competition have an increased chance to invest their (thus won) profit margins in opening up export markets. After all, with companies which show a high quotient of imports in total turnover (*Imp*), it can conversely be assumed that they have less problems in export activities because of their international business contacts and experience.

Results of Probit and Tobit regressions

In order to determine the influence of the individual explanatory factors on the export activity decision, first the following Probit regression model equation is used:[72]

$$Exp_{ij} = a_0 + a_1 R\&DInt_{ij} + a_2 Inno_j + a_3 LabCo_{ij} +$$
$$a_4 LabC_i + a_5 AvWag_j + a_6 \log Size_{ij} + a_7 (\log Size)_{ij}^2 +$$
$$a_8 CompInt_{ij} + a_9 ImpInt_{ij} + a_{10} StdPart_{ij} + \varepsilon_{ij} \qquad (17.32)$$

The Probit estimate makes it clear that the active participation in standardization has neither positive nor negative impacts on the export behavior of the enterprises questioned.[73] This result confirms observations by Lefebvre and Lefebvre (2002) about the export performance of North American small and medium enterprises, which is only influenced by the implementation of ISO 9000, but not by other standards.[74] However, in particular the company-specific R&D intensity and the innovation intensity of the branch as a whole are crystallizing out as the driving forces for the export activities of the company. Against expectation, companies in branches with high average wages show a lower likelihood to export, while the tendency to export increases with rising labor costs. The company size is significant for the export decision neither in its linear nor in its quadratic dimension. Finally, companies with few competitors export clearly more than those with very many competitors, so that the former can leverage their market-dominating position also successfully into export activities. The same applies for companies which fall back especially on imports. On the one hand, the growing worldwide specialization, as well as the increasingly important intra-industry trade, foster the procurement of cheap and high-quality intermediate products abroad and thus strengthen own competitiveness. On the other hand, companies like these are more integrated in international markets, so that their sales efforts abroad are more successful.

In addition to the Probit model, a Tobit model was estimated in order to analyze the influence of the exogenous variables on their export intensity:[75]

$$ExpInt_{ij} = a_0 + a_1 R\&DInt_{ij} + a_2 Inno_j + a_3 LabCo_{ij} +$$
$$a_4 LabC_i + a_5 AvWag_j + a_6 \log Size_{ij} + a_7 (\log Size)_{ij}^2 +$$
$$a_8 CompInt_{ij} + a_9 ImpInt_{ij} + a_{10} StdPart_{ij} + \varepsilon_{ij} \qquad (17.33)$$

In general, the results of the Tobit estimation confirm the signs of the coefficients of the Probit estimation. However, the R&D intensity does not explain the export intensity, whereas it was significant for the overall export decision. On the other hand, the export intensity increases with company size.

Summary

First of all, the participation in standardization cannot exert a positive influence on either the probability for export or export intensities. Consequently, the assumption of export activities as an exogenous variable for the determination of the probability of joining standardization processes is confirmed. Companies are obviously not able to influence their success in exports by joining standardization processes and influencing the technical specifications of standards. The temporary advantage in respect of the companies not joining the standardization processes is not great or long enough that it results in a competitive advantage. The further important result is the great significance of R&D and innovations for the export success of companies.

Table 17.27   Regression results of Probit and Tobit estimation to explain export behavior

Explained variable: export behavior	Probit estimation Coefficient	z-value	Tobit estimation Coefficient	t-value
R&D intensity	0.072^c	4.02	0.012^b	2.45
Innovation intensity of the branch	$4.28e + 07^a$	1.82	$8.68e + 06$	1.29
Labor coefficient	−0.537	−1.05	−0.010	−0.17
Labor costs of the branch	2.328	1.23	1.524^c	2.80
Average wage of the branch	$−0.000^a$	−1.81	0.000	−1.02
Company size (log)	−0.275	−0.46	0.050	0.27
Company size $(\log)^2$	0.005	0.36	−0.001	−0.34
Competition intensity	$−0.427^b$	−2.49	$−0.194^c$	−3.76
Import intensity	158.258^c	2.74	3.252^c	8.55
Active collaboration in standardization	0.000	0.10	0.000	0.24
Constant	3.634	0.63	−0.411	−0.23
Log likelihood	−158.258		−204.225	
Pseudo R^2	0.43		0.23	
Number of observations	417		417	

Notes: Innovation intensity is defined as the quotient between the number of German patent applications at the German patent office and the production value of the industry branch (OECD 1997d).
^{a} = a significance level < 0.10; ^{b} = a significance level < 0.05; and ^{c} = a significance level < 0.01.

17.9.4 Conclusion

The ambivalence of technical standards for foreign trade, not only in the economic foreign trade theory, but also in the econometric analyses based on sectoral data, is only partly confirmed in the results of the enterprise survey.

The double-edged effects of formal standards for national competitiveness result because they have, on the one hand, a high reputation internationally and thus contribute to an improved competitiveness. On the other hand, they also support the foreign suppliers, because they reveal technical specifications. Therefore more than a third of the companies find themselves confronted with an increased competitive pressure, especially from the European and international standardization.

On account of the positive impacts on exports, the large majority of the companies apply European and international standards, without attempting to assert originally German standards in the export markets. Consequently, the increasing stock of European and international standards also leads to increased export and import volumes. This corresponds too with the positive correlation between the extent of intra-industrial trade and the stocks of rules identical to the European and international standards. This can be seen concretely in the facilitation of contracts and the breaking down of trade barriers, which is felt by a great majority to foster trade.

This special strategy, of transporting the substance of national standards into European or international standards, is not a permanent or lasting success. Some companies can thereby achieve merely temporary cost and competitive advantages. Only a few enterprises suffer cost and competitive disadvantages in the short term.

The results of the multivariate estimation model show that neither active participation in standardization work nor the extent of the actively utilized standard stock affect the export success of companies. Own R&D efforts and thus the innovation strength are decisive for this, because by these means the enterprises can achieve a crucial competitive edge, which can be supported in parallel by active work in standardization or by utilizing above all the European and international stocks of standards.

Coming back to the results of the sectoral-based analyses, the step-by-step approach towards a more differentiated analysis based on single industries or technologies seems to be more promising than remaining on a very aggregated level, since the different economic impacts of formal standards are obviously not equally distributed among sectors or technologies. Therefore, the assumption that standards have the same impact across sectors is restrictive and should not be made in the application of sector- or technology-related analyses.

NOTES

[1] Lecraw (1987) finds trade-hampering impacts of Japanese standards based on the assessment of a very limited number of companies exporting to Japan.

[2] It has even been observed that despite the recent efforts towards European harmonization there still exists a significant and not decreasing number of national technical regulations (cf. Europäische Kommission 1998a, p. 11).

[3] Blind (2002c) investigated the feasibility of identyfing national initiatives leading to European or international standards, but ended up with no feasible approach.

[4] Ganslandt and Markusen (2001) are able just to model the cost-raising impact of standards and to simulate it using numerical values.

[5] Casella (2001) raises in her model the question whether the uncontrolled production of international standards through private coalitions of very few large companies may lead to anti-trust problems. If this is the case, then all kinds of harmonized international industry standards may be negative for the development for intra-industry trade.

[6] The data also contain technical rules published by sectoral standardization bodies.

[7] Only the standards data for Japan are taken out of the Japanese Industrial Standards database (Japanese Industrial Standards Committee 1997, 1998).

[8] This Section 17.4 refers to Blind (2000).

[9] Here the situation in 1995 is referred to, as a direct application of new standards in production processes is assumed. In addition, it could be argued that the stocks of standards are endogenous and positively influenced by the development of trade flows. However, the development of standards is a very complex process, co-ordinating the preferences of various interest groups. Consequently, the time span between proposing a new standardization process and the final publication of a standard could last more than five years – at least in the past and in the case of controversial international standards. In addition, standardization is primarily driven by technological change which has an impact on trade, but again only with a significant time lag.

[10] Hereby the sum of patents from the years 1993 to 1995 is used, on the one hand to compensate for random deviations and to account for the fact that a certain time lag occurs before an innovation is transformed into a marketable product. This procedure is based on the results of Münt (1996), and Wakelin (1997).

[11] Due to the different data collecting methods of the statistical offices, only the export figures are used. ε represents the error term.

[12] The stocks of standards were divided by millions of inhabitants (for the population data, cf. Institut der Deutschen Wirtschaft 1997, p. 139). In some cases, no stock of standards is given for the export or import countries under scrutiny. In order to integrate these ICS classified groups into the analysis, the existence of a single norm is assumed.

[13] An international standard is present in the national stock of standards, if a reference to the equivalence or identity with a European or international standard document is contained in the PERINORM database. In the following, for purposes of simplification, the term international standard is used to describe all national documents with positive references to European or international standards.

[14] This can be explained, because Austria has adapted many German DIN standards.

[15] This corresponds to Wakelin's procedure (1997) to explain the bilateral OECD trade by the innovation indicator patent applications.

[16] The ICS classification 'military technology' cannot be separately analyzed, because too many countries do not have standards for military technology.

[17] The result of the simple regression analysis is also based on the fact that in the Member States of the European Union through the numerous standards of the European Standards Institute (ETSI) the stock of international or European standards is very high, while in Japan and the USA, on the one hand, a much smaller number of international telecommunications standards exist, and on the other hand, these countries are more likely to realize export surpluses with telecommunications products.

[18] The surprising differences between the signs of the specialization indicator for the total

standards stock RSTD and the correlation of the standards per million inhabitants stems from the fact that although Japan has specialized very much in this area, the standards stock of Japanese industrial standards per million inhabitants is however clearly below the average of the other countries.

[19] See Note 78 on the discrepancies between the coefficients of RSTD and STDPOP for shipbuilding.

[20] This result can be so interpreted, according to the hypothesis of the 'competitive disadvantage', that the domestic producers are too much hampered by the national standards system and therefore lose out on international competitiveness.

[21] Cf. Note 78 on the discrepancies between the coefficients of RSTD and STDPOP.

[22] See Caves (1981) on further definitive features of intra-industry trade.

[23] This formula was utilized by Hesse (1974) for his analysis of the growth of the intra-industrial trade and is known as the Grubel–Lloyd Index (Grubel and Lloyd 1975).

[24] Cf. on other factors influencing intra-industry trade Greenaway and Milner (1984) for United Kingdom's intra-industry trade.

[25] Weissinger (1995) has already found some evidence for this hypothesis regarding international ISO standards based on simple correlation analysis.

[26] Sections 17.5 and 17.6 refers to Blind and Jungmittag (2001). Blind et al. (2000) performed a similar analysis for Switzerland, which results will be referred to later.

[27] It could be argued that the stocks of standards are endogenous and positively influenced by the development of trade flows. However, the development of standards is a very complex process, co-ordinating the preferences of various interest groups. Consequently, the time span between proposing a new standardization process and the final publication of a standard could last more than five years – at least in the past and in case of controversial international standards. In addition, standardization is primarily driven by technological change which has an impact on trade, but again only with a significant time lag.

[28] Source: OECD (1998a, 1998b).

[29] Cf. OECD (1997b).

[30] Cf. Deutsche Bundesbank (1996, p. 74) on the external value of the DM.

[31] A more exact indicator for the development of demand is the value added in the manufacturing industry of the OECD countries. As this is not available, the more comprehensive indicator of the gross domestic product is applied.

[32] Due to unavailable data, the German stock of standards cannot be compared with the worldwide stock of standards.

[33] In Table 17.7 and the following tables the logarithmic variables are given as small-case letters. Further, RSS stands for the squared sum of the residuals.

[34] The export surplus equation is theoretically the difference between the export and the import equation. However, the parameter ρ differs between both equations, so that the identity is not quite given.

[35] Thierstein et al. (2000) also include the results of a survey among Swiss companies regarding the role of standards for their export behavior.

[36] The export and import price indexes cannot be taken into consideration, because they are not available for bilateral trade relations. At the same time, the real exchange rates have proved to be not significant in the bilateral estimations. This may be based on the fact that the development of the DM against the British pound was subject to relatively short-term fluctuations and the long-term development was relatively well anticipated and could be hedged by corresponding contracts.

[37] It could be argued that the stocks of standards are endogenous and positively influenced by the development of trade flows. However, the development of standards is a very complex process, co-ordinating the preferences of various interest groups. Consequently, the time span between proposing a new standardization process and the final publication of a standard could last more than five years – at least in the past and in the case of controversial international standards. In addition, standardization is primarily driven by technological change which has an impact on trade, but again only with a significant time lag.

[38] The gross value added was taken from OECD (1997b).

[39] This variable represents the presence of economies of scale. The higher this ratio, the more

likely economies of scale exist.
40 In the following panel estimates various versions were tested, if restrictions regarding the symmetrical effects of national and international standards or regarding the opposite effects of domestic and foreign standards are statistically admissible.
41 Section 17.7 refers to Blind and Jungmittag (2002).
42 The export and import price indexes cannot be taken into consideration, because they are not available for bilateral trade relations. At the same time, the real exchange rates have proved insignificant in the bilateral estimations. This may be based on the fact that the development of the DM against the French franc was subject to relatively short-term fluctuations in the first part of the time period analyzed and the long-term development was relatively stable due to the European Monetary System.
43 Regarding the question of the endogeneity of the stock of standards compare Note 59.
44 The gross value added was taken from the OECD (1997b).
45 This variable represents the presence of economies of scale. The higher this ratio, the more likely economies of scale exist.
46 In contrast to the German–British approach, the indicator variable has no time lag, since the estimations with time-lags resulted in insignificant coefficients.
47 In the following panel estimates, various versions were tested, if restrictions regarding the symmetrical effects of national and international standards or regarding the opposite effects of domestic and foreign standards are statistically admissible.
48 We refer only to the non-restricted models. The ? represent insignificant coeffcients with the respective signs in brackets. The +/– indicate significant results at least at a level of significance below 10 per cent.
49 Section 17.8 refers to Blind (2001b) and is reprinted from *Information Economics and Policy*, vol. 13, Blind, K. (2001),'The impacts of innovations and standards on trade of measurement and testing products', pp. 439–60, with permission from Elsevier.
50 This evaluation is confirmed by the extensive survey of David and Greenstein (1990). However, recently Swann (1999) discussed the economics of measurement in general and Temple and Williams (2002) investigated the impact on the United Kingdom of the measurement activity that provides the basic underpinning technology, or infra-technology, for a diverse set of economic activities.
51 The export and import price indexes cannot be taken into consideration, because they are not available for bilateral trade relations. At the same time, the exchange rates have proved to be not significant in the estimations.
52 This pool approach allows the regression analysis to rely on 45 and not 48 observations, because of the integration of the autoregressive term.
53 The export and import price indices were delivered by the Eidgenössische Oberzolldirektion. The dollar exchange rate was taken from OECD (1997a).
54 The real gross domestic products were taken from OECD (1997b), except for Switzerland which come from Bundesamt für Statistik (1996).
55 Total trade volume is defined as the sum of trade in 33 product groups based on the ICS classification. This variable represents the presence of economies of scale. The higher this ratio, the more likely is the existence of economies of scale.
56 In the following panel estimates various versions were tested, if restrictions regarding the symmetrical effects of national and international standards or regarding the opposite effects of domestic and foreign standards are statistically admissible.
57 The following presentations refer partly to part 5 of Blind and Grupp (2000).
58 Over 87 per cent of the standardizing companies apply this strategy.
59 Small- and medium-sized enterprises apply this strategy surprisingly more often than large enterprises. On the other hand, there are no differences between 'standardizers' and 'non-standardizers'.
60 The standardizing companies see greater effects on sales volume, export and import volumes than those not involved in standardization work.
61 Further, the standardizing enterprises more often anticipate the trade-facilitating effects of European and international standards. Moreover, the companies conducting R&D are much more positively affected in drawing up contracts than the enterprises not engaged in R&D.

[62] Clemens and Hauser (1993) found this ambivalence in their study also based on a company survey. In total, the advantages outweigh the disadvantages of European standards respective trade.

[63] This unexpected result can perhaps also be explained by the fact that the question was so interpreted that the transfer also of other national and foreign standards into European and international standards was included.

[64] A more adequate approach is to estimate a set of export and standardization equations simultaneously. However, the available data set does not allow this more sophisticated approach.

[65] Cf. for example Wakelin (1998a) and Ebling and Janz (1999).

[66] Cf. Hirsch and Bijaoui (1985) on this explanatory factor.

[67] Source: German Patent Office and OECD (1997d).

[68] Source: OECD (1997d).

[69] Cf. on this Ebling and Janz (1999, p. 7), among others.

[70] Equally, companies which are already a part of a multinational enterprise or employees abroad find it easier to be successful in export markets. On the other hand, these organizational connections can lead to products and services being produced on the spot and exports are then superfluous. Because of too many missing values, these variables cannot be considered in the Probit model.

[71] The answers on the number of competitors are on three-point scale ranging from one, over some to many.

[72] In contrast to the first approach, the variable export behavior takes only two values, with yes = 1 and no = 0.

[73] In addition, three further models are estimated with dummies for the standardization activity at the national, European and international level. Since the results are very similar to the general model, they are not reported here. However, it has to be mentioned that the impact of being active in standardization on the likelihood to export decreases with the internationalization of the standardization bodies.

[74] Additionally, the actively utilized stock of standards (log$StdStock$) may be used as a substitute for own R&D or may be an indicator for one part of the company's codified knowledge base which is able to support its export efforts. Therefore, both the collaboration in standardization and the active use of standards should have positive impacts on a company's export performance. Since both variables are highly correlated, they have to be integrated in separated equations, but the results show no differences and remain insignificant.

[75] A Tobit and not a normal OLS model is used, since the majority of the observations is censored at zero.

18. The Macroeconomic Impacts of Innovation and Standardization

18.1 INTRODUCTION

Innovation systems not only have the task of stimulating innovation, they also have to ensure its efficient diffusion. According to Rogers (1995, p. 5) the diffusion of knowledge is 'the process by which an innovation is communicated through certain channels over time among the members of a social system'. IFO (1998) listed the most important channels of diffusion:

- Diffusion by imitation;
- Diffusion by licensing;
- Diffusion by R&D co-operation;
- Diffusion by marketing of new products.

Standardization is only mentioned indirectly by the authors. For the diffusion of new ideas, products and technologies, besides private channels, standardization is suitable by state-recognized standard development organizations (SDOs), like DIN and BSI on the national level, CEN on the European, and ISO on the international level. Thus, the development of standards by state-authorized bodies constitutes a fundamental element of a country's technical and economic infrastructure and influences significantly the macroeconomic development like other infrastructures.[1]

The purpose of this final chapter on the impacts of standards is to assess for the first time the influence of the national stock of standards – as a basic element of the technical and economic infrastructure – on macroeconomic development. In the following section, the role of standards in a macroeconomic production function is assessed. Since we find significant impacts, Section 18.3 reports the results of the survey among German companies, which allow if not a comparison, at least a check of the results of the macroeconomic model.

18.2   THE ROLE OF STANDARDS IN THE MACROECONOMIC PRODUCTION FUNCTION[2]

18.2.1   Technical Progress in Macroeconomic Production Functions

The significance of technological activities as an essential determinant of the economic performance of industrialized economies is generally acknowledged today. It is also undisputed in the meantime that technical standards are very important for the fast diffusion of new technologies. In clear contradiction to the theoretical insights and economic relevance, however, is the consideration of the level of technology with regard to technological progress, and the role of standardization in macroeconometric production models. So when estimating production functions (for example a Cobb–Douglas production function), technological progress is commonly approximated only by a linear time trend. This procedure reveals a series of weaknesses. On the one hand, the inclusion of a time trend does not provide an explanation for technical changes, that is the causes or sources underlying technical progress are not distinguishable. At most, the order of magnitude of the technical progress can be estimated. On the other hand, no changes in the rate of technical progress can be identified, rather technical progress grows uniformly, as if dropping from heaven. Only a few authors have attempted to take technical progress into account by using more appropriate indicator variables (cf. Budd and Hobbis 1989a, 1989b; Coe and Moghadam 1993). A formal record of the influence of standardization in macroeconometric production functions by means of appropriate indicator variables is – to our knowledge – completely missing.

In the following analysis alternative sources of technical progress will be identified and approximated by means of indicator variables, which will then be considered when estimating long-term production functions for the business sector of the Federal Republic of Germany, without agriculture, forestry, and fishing and without flat rental, from 1960 until 1996. The theoretical reference model for technical progress – since a true theory of innovation is still lacking – is borrowed from Grupp (1998, chapter 1). Along these lines of empirical operationalization, we shall distinguish between technical progress which is the result of own domestic inventive achievements, and the import of technological know-how through licensing agreements. The first source of technical progress will be approximated through the time-lagged stock of patents at the German Patent Office (Deutsches Patentamt), the second by the real fees for licenses captured in the balance of payments of the Federal Republic of Germany. In addition, the important role of standardization for facilitating the diffusion of technologies will be integrated in the long-term production function. It will be approximated by the indicator variable – the stock of effective technical standards.

For estimating the long-term production functions, the concept of the cointegration of time series introduced by Engle and Granger (1987) will be used. This concept allows the differentiation between actual long-term relations and merely spurious regressions if time series are trending. As in this study only the long-term relations and not the short-term dynamics between the output, the usual production factors and the indicator variables for technical progress, as well as for the role of standardization, are being considered, first of all the first step of Engle and Granger's two-step procedure will be applied, in which existing long-term relations are identified and estimated without specifying the short-term dynamics. However, the distribution of the estimators of the cointegrating vector provided by such a static regression is generally non-normal and so inference cannot be drawn about the significance of the individual parameters by using the standard 't' tests. For this reason the three-step procedure, proposed by Engle and Yoo (1991), is subsequently used to remedy this shortcoming. Their third step, added to the Engle–Granger two-step procedure, provides a correction to the parameter estimates of the first stage static regression which makes them asymptotically equivalent to full information maximum likelihood (FIML) estimates and provides a set of standard errors which allows the valid calculation of standard t tests. The superior long-term production function will then be used to at least roughly assess the effects of the technical progress approximated by the indicator variables and of the role of standardization, approximated by the stock of technical standards, as well as the impact of the usual production factors on economic growth from 1961 until 1990.[3]

This section is structured as follows. In the next sub-section we will discuss how technology innovation and diffusion can be integrated by means of appropriate indicator variables in a conventional Cobb–Douglas production function. The data used here will also be described. The empirical results will be presented in 18.2.3. The main results will be summarized in the 18.2.4.

18.2.2 Technological Innovation and Standardization in Macroeconomic Production Functions

The starting point for the following illustration is given by the usual Cobb–Douglas production function:

$$Y_t = A \cdot K_t^{\alpha} \cdot L_t^{\beta} \cdot e^{\lambda - t} \qquad (18.1)$$

where Y_t represents the output, K_t the capital employed and L_t the amount of labor. The parameters α and β represent the partial production elasticities of the factors capital and labor. Their sum results in the degree of homogeneity or scale elasticity of the production function. The parameter A is called the efficiency parameter. Its influence corresponds to that of the degree of homogeneity. Here too a change of this parameter, while all other parameters remain constant, leads to a uniform proportional change of output for each factor input combination. Technical progress is usually taken into account in the Cobb–Douglas production function in a disembodied and neutral form, with the efficiency parameter formed by the equation $A(t) = A \cdot e^{\lambda \cdot t}$ as time-dependent, where t is a linear time trend. In logarithmic form the production function can then be written as:

$$y_t = a + \alpha \cdot k_t + \beta \cdot l_t + \lambda \cdot t \tag{18.2}$$

where lower case letters represent the variables in logarithmics.

At a first glance the use of a linear time trend to record technical progress appears to be an admissible simplification. However, this procedure reveals a series of weaknesses.[4] On the one hand, the inclusion of a time trend does not provide an explanation for technological changes, that is the causes or sources underlying technical progress are not distinguishable. At most, the order of magnitude of technical progress can be estimated. On the other hand, when using a time trend, no changes in the rate of technical progress can be identified, rather technical progress grows uniformly, as if dropping from heaven like manna. These weaknesses can be remedied if the status of technology or of technical progress could be approximated by appropriate indicators. To this end it is useful to distinguish alternative sources of technical progress (cf. Grupp 1998).

A central possibility to attain technical progress is represented by research and development (R&D) activities. It does not appear promising to include the R&D expenditures directly in a production model. As Kennedy and Thirlwall (1972) already emphasized in an overall survey, the immense growth of expenditure on R&D appears to have only small effects on the aggregated growth rates on a country level. This is not surprising, however, since as Griliches (1980) states, R&D is an investment flow, the output of the enterprises on the other hand is affected by the accumulated stock of earlier results of such investments and of other knowledge sources apart from explicit R&D activities. Further we have to note that R&D comprises basic (academic) and defense research as well as experimental development in industry, the productive effects of which, respectively, are quite diverse (cf. the differentiated reference scheme in Grupp 1998, pp. 18 ff.). Therefore, apart from data-technical problems, the inclusion of a stock of R&D capital in the production function, as done by Coe and Moghadam (1993), does not provide a suitable approximation for technical progress. A large stock of

R&D capital is surely the pre-condition for numerous technological innovations, but it does not guarantee that technological innovations are indeed created. Thus it is necessary to find an appropriate indicator for the stock of results of R&D activities. In this study the mean stock of patents in the German Patent Office is used as such an indicator. This patent stock at year's end, pat_t^{end}, is defined as:

$$pat_t^{end} = pat_{t-1}^{end} + pat\ (basic)_t^g + pat\ (add)_t^g - pat_t^c - pat_t^l \qquad (18.3)$$

where $pat\ (basic)_t^g$ represents the number of granted (basic) patents, pat $(add)_t^g$ the number of granted additional patents, pat_t^c the number of cancelled patents and pat_t^l the number of lapsed patents.[5] The mean stock of patents, pat_t, is then calculated as the average of the patent stocks at previous year's end and at current year's end. As a certain period will elapse between the granting of a patent and the full implementation of the respective innovation in the companies, this indicator is to be taken into account in the empirical investigations with an appropriate time lag. In concordance with other empirical investigations (Griliches and Lichtenberg 1984; Griliches and Mairesse 1984; Geroski 1991; Münt 1996; Grupp and Jungmittag 1999), our empirical examinations showed for the mean stock of patents that a time lag of three years elapses before production is affected.

A further possibility to utilize technological innovations in production is presented by licensing agreements with foreign companies. This import of technological know-how will always be worthwhile if it is cheaper and/or faster than the own development of corresponding technologies (cf. Budd and Hobbis 1989b, p. 5). The expenditure for licenses and patents, lex_t, from the balance of payments of the Federal Republic of Germany will be taken as an indicator for this source of technical progress.[6] Although these payments are mainly transacted between affiliated firms and so influenced by transfer price settings, they give quite general evidence about the trends of technology transfer mainly due to foreign direct investment (cf. Beise and Belitz 1996, p. 60). As this data is only available with respect to prices, it was deflated with the price index for gross fixed capital formation on the basis of 1991. Although this price index will only reflect the price development for expenditure on licenses relatively imprecisely, it should be the most adequate among the available price indexes (cf. Budd and Hobbis 1989a, p. 15).

Besides patents, technical standards are also an appropriate indicator for the stock of results of research and development activities. Traditionally, technical standards have three different main economic functions (compare Chapter 3 and Section 4.5). First, as compatibility standards, they allow products or components of products to work together. Secondly, they define a certain level of product or process quality in the form of minimum quality standards. Thirdly, they reduce the number of variants in a product range – a variety reduction standard. From this it follows that technical standards are also an indicator for the technological capability of an economy, because technological innovations with a market relevance are brought into the standardization process to promote their diffusion by network effects through compatibility standards and by gaining acceptance and reducing risk through quality and safety standards. The variety-reduction-type standard can lead to scale effects and thus fosters diffusion – which is certainly more relevant to overall productivity than (initial) innovation. Furthermore, technical standards claim to represent the state of the art in the sense of the developed stage of technical capability at a given time as regards products, processes and services, based on consolidated findings of science, technology and experience.

The PERINORM contains the complete data base of all technical standards[7] in Germany among other European nations, including technical regulations since the mid 1970s. The stock of standards at a year's end, std_t^{end}, is defined as:

$$std_t^{end} = stdt_{t-1}^{end} + std_t^p - std_t^w \tag{18.4}$$

where std_t^p represents new technical standards published and std_t^w technical standards withdrawn. The mean stock of patents, std_t, is also calculated as the average of the stocks of standards at previous year's end and at current year's end. Because of the years which elapse between the beginning of a standardization process and preliminary publication as a pre-standard and the final publication of the document, the companies in general do not have a time lag in getting aware and implementing the results of the standards. Thus, no time lag seems to be required for this variable.

With the technological innovations and the role of standardization explicitly taken into consideration, the extended Cobb–Douglas production function is now in logarithmic form:

$$y_t = a + \alpha \cdot k_t + \beta \cdot l_t + \gamma \cdot pat_{t-3} + \delta \cdot lex_t + \varepsilon \cdot std_t + u_t \tag{18.5}$$

An error term u_t was added to this equation which fulfills the usual assumptions. In the course of the empirical analysis various variants of this function for the business sector without agriculture, forestry, fishing and without flat rentals were estimated, whereby real gross value-added of this sector in 1991 prices served as the endogenous variable. The capital stock for this sector was determined in the usual way as in the annual average employed gross fixed assets in prices of 1991, that is the average from the gross fixed assets at the beginning of the actual year and at the beginning of the following year was computed. The number of employees in this sector of the economy was taken as labor input variable. Other input variables, which take the number of hours worked into consideration, were not available for the complete period reviewed.

While the usual time series which are used to estimate the production functions refer to the business sector without the atypical fields of agriculture, forestry, fishing and flat rentals, the selected indicators for the technological innovations and standardization encompass the economy as a whole. For these indicators there are no time series available which refer to the individual economic sectors. They are likewise not easily established by on-line patent statistics, as the concordance problem between patent classification and sector definition, because of heterogeneity within sectors, is very difficult to solve (cf. Grupp 1998, pp. 162 onwards). As these atypical economic sectors have in any case benefited very little from technological innovations, we assume that the distortion is tolerable.

It is often assumed in empirical investigations that the scale elasticity of the factors capital and labor is equal to unity, that is $\alpha = 1 - \beta$. This restriction can be very simply realized if the initial logarithmic production functions (18.1) and (18.5) are written as:

$$y_t - l_t = a + \alpha \cdot (k_t - l_t) + \lambda \cdot t + u_t \qquad (18.6)$$

or

$$y_t - l_t = a + \alpha \cdot (k_t - l_t) + \gamma \cdot pat_{t-3} + \delta \cdot lex_t + \varepsilon \cdot std_t + u_t \qquad (18.7)$$

The admissibility of such a restriction of the scale elasticity can be tested by means of an F-test.

18.2.3 Empirical Results

The starting point of the empirical investigation is given by the univariate analysis of the time series under consideration which are displayed in Figure 18.1. Nearly all time series show permanent growth over time. The only exception is the time series of the log of employees which is strongly influenced by business cycles. Furthermore, a strong increase in employment can be observed in the second half of the 1980s. Additionally, the time series of real gross value added as well as of the production factors capital and labor show a jump in 1990–91 due to German unification. Also obvious is the strong increase of the real license expenditures in the first half of the 1980s and again in 1995–96.

In order to check whether the time series are integrated of order one (that is the series are characterized by unit roots) or whether they are following deterministic trends, ADF tests are carried out (Dickey and Fuller 1979).

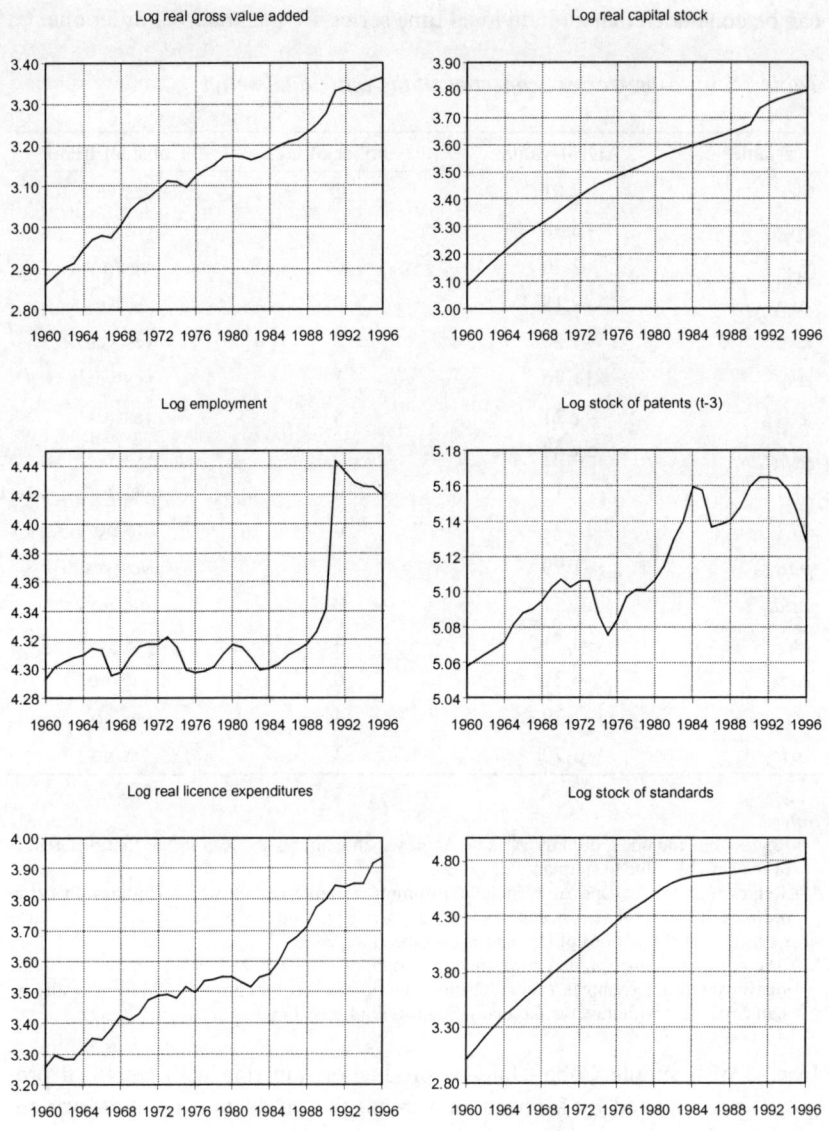

Figure 18.1   Variables used for the unrestricted production function from 1960 to 1996

As Table 18.1 indicates, the null hypothesis of a unit root in favor of stationary cannot be rejected for all eight variables. For the first differences the null hypothesis must be rejected at a significance level of at least 5 per cent. So, it

The Economic Impacts of Standards

can be concluded that all individual time series are integrated of order one.

Table 18.1   Time-series properties of the individual series

Variable	ADF-t-value	Number of lags	Constant/trend
y_t	−2.01	2	yes/yes[a]
Δy_t	−6.38	1	yes/yes[a]
$y_t - l_t$	−2.70	0	yes/no[a]
$\Delta(y_t - l_t)$	−9.18	0	yes/yes[a]
k_t	−3.20	2	yes/yes[a]
Δk_t	−11.46	1	yes/yes[a]
$k_t - l_t$	−1.21	6	yes/yes[a]
$\Delta(k_t - l_t)$	−8.04	6	yes/yes[a]
l_t	−2.19	0	yes/no[a]
Δl_t	−15.71	0	no/no[a]
pat_{t-3}	−1.99	1	yes/yes
Δpat_{t-3}	−3.57	0	no/no
lex_t	−3.51	4	yes/yes[b]
Δlex_t	−6.21	0	yes/no
std_t	−1.39	4	yes/no[c]
Δstd_t	−3.70	3	yes/no[c]

Notes:
[a] = Regressions include a dummy variable Δdgu which captures the leap in the first differences of the variables due to German unification.
[b] = Regressions for lex_t and Δlex_t include a dummy variable D9596 which captures an extra-ordinary increase of real license expenditures in 1995 and 1996, mainly caused by large mergers and acquisitions of US pharmaceutical enterprises.
[c] = The regressions for std_t and Δstd_t include a dummy variable D81 which captures a highly significant structural break (policy change) due to the fact that prior to 1980 practically no standards were withdrawn in Germany (refer to Figure 18.1).

Instead of a simple Cobb–Douglas production function with technical progress approximated by a linear time trend, the production function was re-specified in such a manner that the time trend was substituted by the three indicator variables. Besides the capital stock and employment, this function includes the dummy variables *DOC* which captures the first West German depression in 1967 and the first oil crisis, as well as *D80* and *D81* which capture a structural break starting in 1980 and reinforced in 1981 following the second oil crisis. Furthermore, a dummy variable DGU is added to the equation to catch the effects of German unification.[8]

Table 18.2   Estimation results for the augmented production function

Variable	First step of Engle/Granger		Third step of Engle/Yoo	
	unrestricted	$\hat{\alpha} + \hat{\beta} = 1$	unrestricted	$\hat{\alpha} + \hat{\beta} = 1$
Constant	−2.3992	−2.3038	−2.6262	−2.4840
	(−4.7392)[a]	(−5.8305)	(−8.9754)	(−8.7991)
k_t	0.3871	0.3607	0.4161	0.3599
	(2.9334)	(3.6419)	(5.1754)	(6.1000)
l_t	0.6620	0.6393	0.6517	0.6401
	(5.3206)	−	(10.9346)	−
pat_{t-3}	0.1194	0.1268	0.1660	0.1602
	(1.4925)	(1.6889)	(3.3603)	(3.2694)
lex_t	0.1292	0.1373	0.1206	0.1393
	(3.6923)	(6.0173)	(5.5321)	(8.8165)
std_t	0.0634	0.0700	0.0546	0.0708
	(1.7387)	(2.3955)	(2.4053)	(4.0000)
DOC	−0.0171	−0.0172	−0.0160	−0.0158
	(−4.0261)	(−4.1141)	(−6.1538)	(−6.3200)
D80	−0.0107	−0.0108	−0.0099	−0.0119
	(−1.7556)	(−1.8021)	(−2.5385)	(−3.1316)
D81	−0.0165	−0.0168	−0.0183	−0.0192
	(−2.5966)	(−2.7299)	(−4.8158)	(−5.0526)
DGU	−0.0383	−0.0343	−0.0385	−0.0340
	(−2.7649)	(−6.4176)	(−8.7500)	(−9.1892)
R^2	0.9990	0.9984	0.9990	0.9983
R_{adj}^2	0.9987	0.9980	0.9986	0.9979
DW test	2.1436	2.1840	−	−
EG test	(36.6)[b]	(36.5)	−	−
	−6.9645	−7.0260	−	−
	(0.0043)[c]	(0.0019)	−	−
F-test of the restriction	−		1.3500	
	−		(0.2554)[c]	

Notes:
[a] = Empirical *t* values in brackets but statistical conclusions on the base of usual *t* tests are only permitted if the third step of the Engle/Yoo procedure has been applied.
[b] = Number of observations available after forming lags and first differences and number of I(1) variables in brackets.
[c] = Significance levels in brackets.

The estimation results for the unrestricted and restricted version of this long-term production function is reported in Table 18.2. A view of the *t* values calculated for the estimates of the third step of the Engle and Yoo procedure shows that all coefficients of the unrestricted as well as the restricted estimation are unequal to zero at least at a significance level of 5 per cent. Therefore, all three indicator variables have a highly significant power of explanation. Furthermore, the magnitudes of their coefficients verify that the factors approximated by the indicator variables make contributions to real gross value added that cannot be neglected.

The estimates of the coefficients of the factors capital and labor also seem to be very reliable. They are rather similar to the estimates in Schröer and Stahlecker (1996) where a long-term Cobb–Douglas production function is estimated using quarterly data from 1970 until 1994. Schröer and Stahlecker (1996) introduced, after a data mining process, dummy variables which change the slope of the time trend to approximate changes of technical progress.

The R^2 values of 0.9990 and 0.9984 respectively for the first step of the Engle and Granger procedure and 0.9990 and 0.9983 respectively for the third step of the Engle and Yoo procedure indicate again a very good fitting of the models to the observed data. The *DW* test statistics suggest that the presence of first order autocorrelation can be excluded. Turning to the *EG* test statistics it can be seen that the unrestricted as well as the restricted production function forms a cointegration relation. Therefore both models can be interpreted as long-term production functions. Additionally, the restriction of the sum of the partial production elasticities is now permitted beyond all usual significance levels as the *F*-test shows. Based on these estimation and testing results, the restricted product function containing all three indicator variables is superior to other augmented production functions with only one or two indicator variables which had been considered during the empirical investigation, but are not reported here due to the limitation of space.[9]

Due to the approximation of different sources and causes of technical progress and of standardization by means of appropriate indicator variables it is now possible to assess, at least roughly, the effects of these variables as well as of the usual production factors on the growth of real gross value-added. The results of the ex-post forecasts of average annual growth rates for the whole observation period before German unification as well as for different subperiods before and after German unification are reported in Table 18.3. The comparison of the realized total and the forecasted total growth rates of real gross value-added in the business sector without agriculture, forestry, and fishing and without flat rental shows a good fitting of the model to the observed data. Only in two subperiods (from 1961–65 and from 1971–75) the model overestimates the growth rates by 0.6 and 0.4 percentage points respectively. In two further subperiods (from 1966–70 and from 1992–96) it underestimates the growth rates by 0.3 and 0.5 percentage points respectively. However, in three of these subperiods economic growth is strongly affected by exogenous influences, which are not fully captured by the dummy variables.

Table 18.3   Sources of growth in the business sector from 1961 to 1996

Source	Average annual percentage changes							
	61–90	61–65	66–70	71–75	76–80	81–85	86–90	92–96
k_t	1.6	2.6	2.0	1.7	1.3	0.8	1.1	1.1
l_t	0.2	0.6	0.1	−0.6	0.5	−0.6	1.1	−0.7
pat_{t-3}	0.1	0.2	0.2	−0.4	0.3	0.2	0.0	−0.3
lex_t	0.5	0.6	0.5	0.4	0.2	0.1	1.3	0.6
std_t	0.9	1.5	1.2	0.9	1.1	0.4	0.2	0.3
Total:								
fitted	3.3	5.7	4.1	2.1	3.5	1.0	3.7	1.0
realized	3.3	5.2	4.4	1.7	3.6	1.1	3.8	1.5

Notes: Differences between the sums of the individual components of the growth rates and the fitted total growth rates are caused by rounding and by joint effects.

Turning to the individual factors, it can be seen that the development of the capital stock has the greatest impact on the growth rates of gross value added in most cases, accounting for 0.8 up to 2.6 percentage points. This result is in accord with the results for other countries (cf. Budd and Hobbis 1989a, 1989b; Coe and Moghadam 1993). The role of standardization is in second position, accounting for 0.2 up to 1.5 percentage points of the average annual growth rates. However, coinciding with the reduction of growth of the stock of standards at the beginning of the 1980s, the impact of standards on economic growth moves to a new lower level.

The impact of the factor labor on economic growth is strongly influenced by fluctuations of the number of employees due to business cycles. In particular the reductions in the number of employees after the first and second oil price crisis had negative impacts on economic growth. On the other hand, the strong increase of the number of employees in the second half of the 1980s has fostered economic growth.

The technical progress in total over the whole time range attributes almost 50 per cent to the macroeconomic growth in Germany and empirically supports the hypotheses of the endogenous growth theories, which highlight the role of knowledge for economic growth. Table 18.3 differentiates also between the three sources of technical progress. It turns out that the role of standardization, responsible for the effective diffusion of innovations, is very important, accounting for 0.2 up to 1.5 percentage points of the average annual growth rates. However, coinciding with decreasing growth of the stock of official formal standards at the beginning of the 1980s due to the increasing informal private industry standardization by industry consortia because of shorter product life cycles, the impact of the former standards on economic growth moves to a new lower level. The stock of patents and the real license expenditures had a moderate influence on growth in most periods. Nevertheless, these two sources of technical progress account for slightly more than 18 per cent of the total increase of gross value added in the period from 1961 until 1990. Their share increases even to 35 per cent in the second half of the 1980s, but only due to the strong increase of real license expenditures, which represents the increase in technology imports, even for a high-technology country like Germany.

Altogether, when the three factors are compared, it became evident that standards were at least as important for technical innovation as patents. It is clear that the innovation potential is not the only deciding factor in economic development, but that it must also be broadly disseminated by means of standards and technical rules. However, the decreasing contribution of formal standards to economic growth since the 1980s hints at the increasing importance of de facto standards in times of shorter product cycles.

18.2.4 Summary

In this section cointegration analysis is applied to estimate a long-term production function for the German business sector covering a period from 1960 until 1996. In extension and refinement to most other empirical studies technical progress is not approximated by a summarizing linear time trend, but the alternative sources of technical progress are taken into account by more specific indicator variables: the results of own inventive achievements by the lagged stock of patents plus the import of technological know-how by the real license expenditures. In addition, the role of standardization with its important functions, for example for the diffusion of technological innovations, is integrated in the production function.

This restricted production function is then used to assess, at least roughly, the effects of the influences approximated by the indicator variables as well as the impact of the usual production factors on growth of real gross value added. Here, it became clear that the development of the capital stock has the greatest impact on economic growth. But facilitating technology diffusion is also very important and it can be expected that this role will further increase with the growth of network industries. Furthermore, it is obvious that the influence of various aspects of technical progress may not be neglected in growth models.

18.3 COMPARISON AND INTERPRETATION OF MACROECONOMIC AND MICRO-BASED RESULTS[10]

In this last section, the results of the macroeconomic analysis of the significance of standards and technical rules for the growth of the economy as a whole should be compared with the estimates of the surveyed companies. This approach is analogous to the analysis of the impact dimensions technological change and trade.

Although an individual enterprise cannot make a statement valid for the overall economic significance of formal standards, it is still possible for it to estimate their influence on the development of the own company, or even of the own branch. From these single responses conclusions can reasonably be drawn about the general impacts of standards on the economy as a whole.

18.3.1 The Significance of Standards for the Economic Development of Branches and Enterprises

In the company survey, the impacts of standards on the development of the own branch as well as on the own company were the subject of questioning. On average, their influence on the own branch was estimated as slightly positive. In the effects on branches, radio, television and telecommunications technology stand out positively, whilst the majority of the companies questioned in the pharmaceutical industry did not perceive positive impulses for the development of their branch due to standards. However, in view of the very few pharmaceutical standards, this estimation is not surprising.

The estimates of the effects standards have on the branch correlate positively with the growth effects for the own enterprise; however, the average estimate of the latter is no longer slightly positive, but only ambivalent.[11] In the construction and building sector, which is less influenced by systems technologies, and in the remaining group 'Others', the skepticism concerning the company-specific effects of standards on growth is greatest.

These estimates underline that standards represent a knowledge stock which develops more positive impacts for the own branch as a whole than for the own enterprise. Thus the benefit for the sum of companies – that is the branch – is greater than for the individual enterprise, because the standards are a public good whose total benefit is composed of the sum of the individual benefits of the enterprises. If this thought is further developed and the branch benefit aggregated to an overall economic benefit in the sense of a total benefit for all enterprises, then the agreement of the companies surveyed would be even more clear and unambiguous. This result tallies with the result of the positive contribution of standards and technical rules towards economic growth.

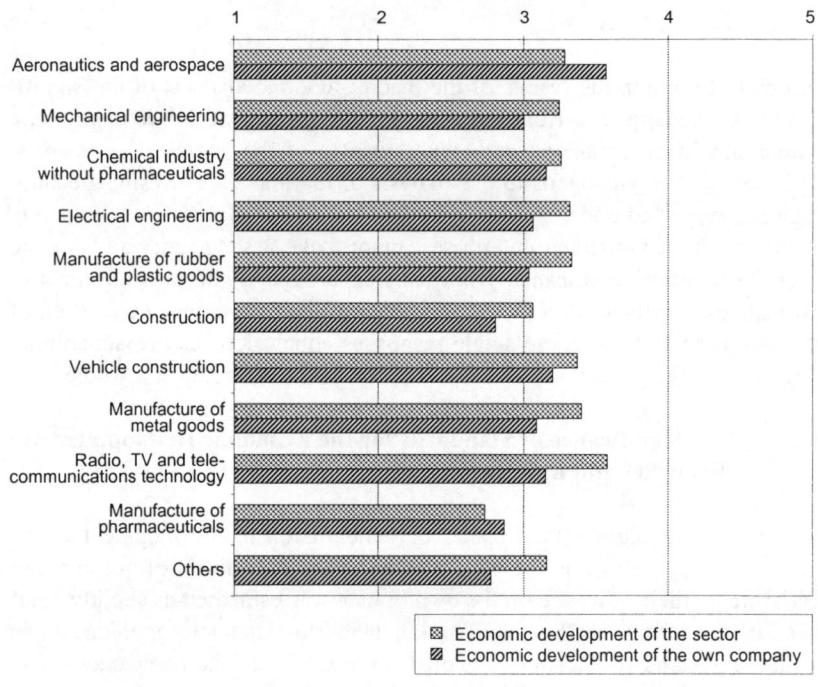

Notes: In this graphic on the questions about opinions, the numbers on the axis represent the following: 1 = does not apply at all; 2 = does not apply; 3 = partly; 4 = applies; 5 = applies entirely.

Figure 18.2   Significance of standards for the economic development of branches and enterprises

18.3.2   The Significance of Standards for the Development of Turnover and Value Added

The question of the impacts of standards on turnover and in-house value added is closely bound up with the above question. The answers to both questions correlate therefore significantly with the responses to the question

about the influence of standards on the development of branches and companies. The radio, television and telecommunications technology industry stand
out significantly positive to the question of effects on turnover, while the
remaining group 'Others' can discern only neutral impacts on the part of
standards. The effects on the in-house value added are somewhat smaller on
the whole.[12]

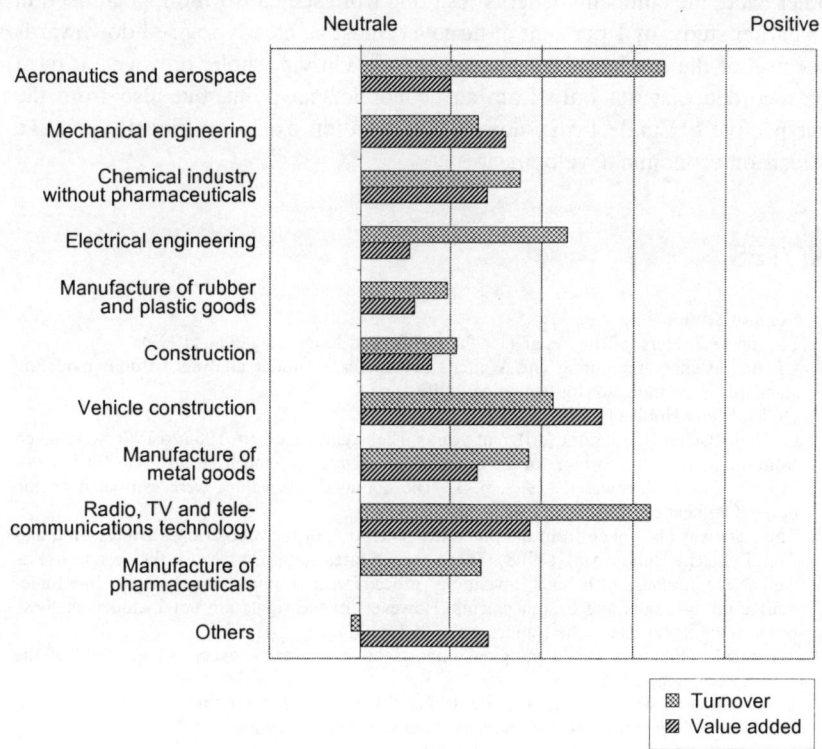

*Figure 18.3   Influence of standards on the development of turnover and
value added*

18.4   CONCLUSION

If these results are seen in the context of the Cecchini study on the introduction of the European Single Market, which arrives at an economic benefit of
standardization[13] amounting to DM20 billion for the old Federal Republic of
Germany, or 1 per cent of the GNP, and a study quoted by Collazzo (1979)
with a company benefit of 1 per cent of the turnover,[14] then quite a consistent

picture of the economic benefits of standardization emerges.

The available results of the macroeconomic analysis can basically confirm the results arrived at in the past, using other methodological approaches, on the economic benefits of standardization to the value of approximately 1 per cent of GNP. However, the share of standardization in the overall economic growth has shrunk somewhat in the last years in favor of globalization. On the other hand, the company benefits resulting from standardization, postulated in an earlier study, of 1 per cent of turnover, must be clearly revised downwards because of the results of the present survey. On the whole, however, it must be recorded that not only from an economic viewpoint, but also from the perspective of single firms, does standardization exert significantly positive effects on economic development.

NOTES

[1] See also Section 4.5.
[2] Section 18.2 refers to Blind et al. (1999a, 1999b) and Jungmittag et al. (1999).
[3] Cf. for instance Jungmittag and Welfens (1996) for a similar analysis without modeling standards as an indicator for technology diffusion.
[4] Cf. Budd and Hobbis (1989a, p. 2).
[5] Cf. Statistisches Bundesamt (different years). The patent data from 1963 to 1996 were taken from the Statistical Yearbook of the Statistisches Bundesamt and from the Blatt für Patent-, Muster- und Zeichenwesen (1998, p. 3). The non-available values were estimated on the basis of a linear trend.
[6] This data was also taken from the Statistical Yearbook of the Statistisches Bundesamt, and from Deutsche Bundesbank (1998). These expenditures comprise fees for the current use as well as the purchase of patents, inventions, procedures and other property rights like trade-marks, utility-model and design patents. However, movie rights are not included in these property rights (cf. Deutsche Bundesbank 1998, p. 21).
[7] Technical regulations are of marginal importance in Germany, except in the field of the environment and safety.
[8] In concordance with Schröer and Stahlecker (1996), we find no shifts of the production elasticities of the factors capital and labor after German unification.
[9] Compare for details Jungmittag et al. (1999).
[10] The micro-analysis is based on the responses of the companies questioned in Germany. See Section 13.3 and Table 13.1 for further details. The following presentations refer partly to Part 5 of Blind and Grupp (2000).
[11] The small enterprises differ from the others in that they regard standards on the whole as slightly positive for the development of the own company. The standardizing enterprises assess in particular the branch effects of standards more positively compared with the firms which are not active in standardization.
[12] Further, the surveyed companies assess the cost–benefit ratio of standardization on average as slightly positive. The large enterprises and the standardizing companies stand out with extraordinarily affirmative estimates. In vehicle construction the cost–benefit ratio is also above-average good. On the other hand, the manufacturers of pharmaceuticals see the ratio in a slightly negative light.
[13] Cf. DIN (1998b, p. 10). This figure was indirectly determined for the economic effects of the completely barrier-free Single Market in Europe (cf. Cecchini 1988, p. 122 and 133; European Commission 1988).
[14] The benefit of standardization to single companies amounting to 1 per cent of turnover

cannot be directly compared with the results of the enterprise survey. However, the additional costs of the enterprises per turnover, which would arise if DIN were no longer available as a platform, represent a comparable basis. This quantitative value can be calculated for only 75 companies. It is true that an average value of 1.7 per cent is arrived at. However, because of the high variance of the data, the margin of fluctuation between the first quartile at 0.01 per cent and the third quartile at 0.3 per cent must be considered further. This statistical range lies clearly below the figure calculated by Collaz (1979). It must, however, be considered that the opportunity costs arising for the companies when abolishing DIN are very difficult to determine, also because of indirect modes of action and thus are rather underestimated. The difference compared to the results of the economic investigative approach can be partly explained in that in the economic consideration on the one hand, the considerable implicit and intangible effects of standardization, for example on the safety of employees and consumers; and on the other hand, the positive external effects of standards on economic activity, which is difficult to quantify, have been generally included. These aspects would be neglected in a purely cost-oriented company perspective, or could not have been calculated from a single-company viewpoint at all.

19. Overall Summary, Future Research and Policy Considerations

As a final step, the results of the sectoral and macroeconomic analyses on the basis of official statistics and the responses to the company surveys will be summarized and respective policy considerations derived. The responses to the surveys corresponded in the main with the analyses based on economic statistics. There are only minor differences in matters of detail, which can however be explained by the insufficient comparability of the questions put in the survey and the statistical analyses. However, some research questions have still not yet been adequately answered or even all addressed. These future challenges are also included in this concluding chapter.

19.1 STANDARDS AND TECHNOLOGICAL CHANGE

19.1.1 Main Results

The sectoral and technology-based analyses revealed that standards have a positive and not a negative effect on technological change and innovation. It was also shown that the standards collection adapts to the rate of technological change. The German companies surveyed did not regard the standards collection as out-of-date, but as being too large in some sectors. It is also evident that neither industry-wide standards nor private industry standards, when seen in the context of other barriers to innovation, are significant hurdles.

19.1.2 Future Research

Although technical standards are no serious obstacles for innovation, the unanswered question is whether standards are able to channel technological change, and in which direction they will lead. Besides the driving factor of technological change there are other influencing factors for standards, such as changing user requirement or even new business models. The influence of these parameters has also to be investigated in a more systematic way, and also its interplay with technological change. Finally, standards are becoming a more important issue for the service sector. Consequently, it has to be asked what factors are responsible for the development of new service standards, especially since besides new technological opportunities consumer needs are crucial for the success of new services.

19.1.3   Policy Consideration

The growing importance and speed of technological change and innovations for the competitiveness of highly industrialized countries in the age of globalization should be taken into account in the standardization strategies by setting priorities on innovative areas and by adjusting the existing stock of standards to keep up with the state-of-the-art of science and technology. In order to keep abreast of the higher speed of technical change, the standardization process needs to be closer to the development of science and technology. In this context, with the standardization at the R&D stage and pre-standards, a first step is being made in the right direction.

Because of the close link between innovation and its diffusion by means of standards, the selection of subjects for standardization must be more closely guided by technological change and the current state of science and technology. In sectors characterized by very short product lives and development cycles, standardization organizations should systematically withdraw standards which have outlived their scientific and technical significance. Additionally, effective mechanisms for the conversion of the faster and more flexible industry standards into the standardization processes of the standards development organizations should be elaborated.

19.2   STANDARDS AND RESEARCH AND DEVELOPMENT

19.2.1   Main Results

As opposed to the econometric results on sector and technology level, the results of the German survey show that the effect of standards on research and development is contradictory, and even negative in some sectors. Nevertheless, most businesses benefit from participating in standards work, because they gain access to the research results of other businesses. The responses given by the companies surveyed do not provide an answer to the question of how this advantage weighs against the disadvantage of revealing their own R&D results in standardization processes. However, it was shown that non-involvement in standards work generally increases the costs of R&D.

Companies with little or no own R&D activities are more likely to participate in standardization in order to access the R&D results of the other participants. There are even indications from the survey of the European companies, that those with strong patent portfolios stay away from formal standardization processes.

19.2.2   Future Research

Since the results about the R&D and IPR characteristics of companies and their likelihood of joining formal standardization processes are still rather ambivalent, further investigations have to be carried out which especially take into account the role of strategies to protect intellectual property. Furthermore, service companies are developing an increasing demand for standards. The main characteristics of those which actively participate in standardization processes have to be investigated, especially since most of them do not perform traditional R&D activities.

Due to the increasing importance of standardization within industry consortia, it has to be checked whether the results about company characteristics for formal standardization processes are also valid for informal standardization activities and what the reasons for differences are. A side aspect of this research path is a systematic comparison between standardization activities and the current Open Source movement in the software industry.

The user side was rather neglected in the study. Therefore, one has to ask what role standards play for the knowledge management of companies, in the sense of whether they are an important source of information for innovation activities and, if this is the case, whether they promote innovation by giving the companies a more stable framework for future-oriented and risky activities. It is also interesting to know-how strong standards are in channeling the innovation activities of companies.

19.2.3   Policy Consideration

Standardization is a form of technology transfer. Therefore, it is particularly important to involve companies which are leaders in their sectors in new standards projects. In particular, possessors of proprietary de facto standards should be given incentives to integrate their technical specifications into official standards in order to promote the diffusion of their technical knowledge. Further, all participating companies must be convinced that the benefits of sharing the R&D results of other companies are greater than the risk involved in revealing their own results. In order to reach these targets, it is appropriate to shape the IPR rules for standardization processes in a way that they represent both an incentive for IPR holders and not a threat for companies not using IPR to contribute their technological know-how to the process. Therefore adequate licensing and protection schemes have to be developed.

Because of the low propensity to standardize in sectors with a small market concentration on the one hand, and the higher demand for compatibility in these sectors, strategies should be elaborated which improve the access to standardization in general for the small and medium-sized enterprises of these sectors and facilitate the standardization process. Furthermore, although de facto standardization cannot be prevented, at least incentives should be provided to make the details of the technical specifications public in order to reduce information asymmetries and wasted R&D invested in incompatible solutions by other companies which are not members of the consortium. A first approach is the new opportunity to publish so-called publicly available specifications (PAS).

19.3   STANDARDS AND FOREIGN TRADE

19.3.1   Main Results

The contradictory effects of technical standards on exports, predicted by both economic foreign trade theory and the econometric analyses carried out on the basis of that theory, are only partly corroborated by the results of the company survey.

Standards affect competitiveness in two ways: on the one hand, their high international reputation leads to greater competitiveness when they are used; on the other hand, they also benefit foreign suppliers, because they make technical specifications more transparent.

Most businesses in the German survey make use of European and International Standards because of their positive effect on exports. In consequence, an increase in the number of these standards leads to an increase in export and import volumes. This corresponds to the positive link between intra-industry trade and standards which are identical to European and international standards. A large majority of the surveyed companies identified positive effects as including a simplification of contractual affairs and a lowering of trade barriers.

19.3.2   Future Research

As already seen in the study, analyses must be performed on a sectoral level to receive clearer results about the impact of standards on trade, because in some sectors specific types of standards dominate. A more sophisticated approach would require the differentiation of the total stock of standards into different types, like compatibility and quality standards. Furthermore, a more comprehensive trade model has to take into account governmental regulations, which have probably a more negative impact on trade flows than standards.

Another aspect which is gaining more importance is trade in services. Although there are several models to explain the increasing trade volumes in services, the role of standards has not yet been considered. Because we are only at the beginning of standardization in service industries, we will have to wait for some time in order to perform similar statistical analyses.

19.3.3   Policy Consideration

From the macroeconomic analyses and the company survey we can conclude that European and international standards have a much more positive effect on exports than national standards do. Companies should therefore be encouraged to take an even more active role in European and international standards work. Furthermore, European and international standards should be quickly incorporated into the national standards collections. A prerequisite for international involvement is active participation in standards work at a national level, and businesses must be convinced of the benefits of this as an effective export strategy.

International as well as national standards can provide support for technology transfer from technological leaders to developing nations. Although this is conducive to development policy, these nations may present a threat to our own competitiveness, because standards enable them to imitate our products and production processes. Technology transfer also reveals the preferences of domestic consumers, making them transparent to competitors worldwide. These threats should be countered by concentrating national standardization activities in those sectors in which the national innovation potential is greater than elsewhere in the world. Further, the interested parties should decide whether a well thought-out standards proposal should be presented at the international level in order to improve the chances of establishing an advantage for their own technology. In general, export-intensive sectors and companies should be supported to start European or international standardization projects, in order to transform or integrate their R&D results into common technical standards and to provide them with at least temporary cost and quality advantages compared to their competitors abroad, without being non-tariff trade barriers.

Considering strategic aspects, interested companies have to decide whether to initiate a standardization project first on the national, or immediately on an international, level. If a country has a significant lead in the technological development respective to other countries and there are no serious alternative specifications available, then a standardization project should be started on European or international level. In case of competing technologies, bilateral negotiations and consensus with the most important trading partners may prepare the ground for a success on a European or international level.

A further challenge for policymakers is the completion of the Single Market for service in the European Union by intensively using the instrument of common European standards, taking into account that the demand for services is even more heterogeneous than for material goods, which requires either additional national standards or a new kind of European standard which gives national priorities a greater leeway. However, the national standardization initiatives in services should not be misused to build up additional barriers for service trade.

19.4 STANDARDS AND MICRO- AND MACROECONOMIC BENEFITS

19.4.1 Main Results

The results of our macroeconomic analyses basically confirm those of previous analyses using other methods. Our study shows the economic benefits of standardization as being about 1 per cent of the gross national product. However, the assessment by an earlier study that the benefits of standardization were 1 per cent of business sales must be corrected downwards.

19.4.2 Future Research

The result of assessment of the macroeconomic impact of the stock of standards should at first be replicated for other countries with significant national standardization activities, taking into account possible implementation costs. Then it would also be possible to perform analyses focusing on specific sectors which takes into account the sector-specific functions and impacts of standards. Another aspect which has not yet been considered is the general shift from national to European and international standardization activities and its impact on growth which may be ambivalent because stronger economies of scales have to be weighted against higher adaptation costs during the implementation of standards and variety-reducing effects. Second, the approach has to be extended by including the increasing activities of industry consortia in producing informal standards, taking into account that new emerging industries use more informal ways to standardize than the mature industries in the manufacturing sector. Third, it might be interesting to segment the total stock of standards into differentiating the types by economic functions. Then it would be possible to compare their different macroeconomic impacts. Furthermore, the approach should also be applied to the service sector, although not even a comprehensive growth model in general has been developed yet, because a simple transfer of the models elaborated for the manufacturing sector does not reflect the specifics of the service sector. At the microeconomic level, it is still rather vague what impact the involvement in standardization and the use of formal standards have on the success and therefore growth of companies.

19.4.3   Policy Consideration

The positive macroeconomic effects, which far exceed the sum of individual benefits for the economy, and the relief of the state through technical standards, justify public financial support for standards work and give standardization a firm place in economic policy and research and innovation policies. In particular, the latter should take a more integral approach, taking full account of the relationship between innovation and its diffusion by means of standards.

In the service sector, policymakers face the challenge of developing a new comprehensive standardization policy approach which has to take into account that the innovation process is rather different from the R&D-dominated one in the manufacturing sector, requiring an even stronger involvement of users and rather different trade patterns.

Bibliography

Adams, M. (1996), 'Norms, standards, rights', *European Journal of Political Economy*, **12**, 363–75.

Adolphi, H. (1997), 'Die Stellung der Normenabteilungen', in Hesser, W. (ed.), *Innerbetriebliche Normung: Organisation, Europäische Richtlinien und Strategien*, Forschungsbericht Nr. 3, Hamburg: Universität der Bundeswehr, pp. 1–21.

Akerlof, G. (1970), 'The market for lemons', *Quarterly Journal of Economics*, **84** (3), 488–500.

Allen, R.H. and R.D. Sriram (2000), 'The role of standards in innovation', *Technological Forecasting and Social Change*, **64**, 171–81.

Amable, B. and B. Verspagen (1995), 'The role of technology in market shares dynamics', *Applied Economics*, **27**, 197–204.

Anderton, B. (1999), 'Innovation, product quality, variety, and trade performance: an empirical analysis of Germany and the UK', *Oxford Economic Papers*, **51**, 152–67.

Antonelli, C. (1994), 'Localized technological change and the evolution of standards as economic institutions', *Information Economics and Policy*, **6**, 195–216.

Arthur, W.B. (1989), 'Competing technologies, increasing returns, and lock-in by historical events', *The Economic Journal*, **99**, 116–31.

Arundel, A. (2001), 'The relative effectiveness of patents and secrecy for appropriation', *Research Policy*, **30**, 611–24.

Baldwin, R. (2000), *Regulatory Protectionism, Developing Nations and a Two-tier World Trade System*, World Bank Working Paper, Washington, DC: World Bank.

Barras, R. (1986), 'Towards a theory of innovation in services', *Research Policy*, **15**, 161–73.

Barrett, Ch. and Y.-N. Yang (2001), 'Rational incompatibility with international product standards', *Journal of International Economics*, **54**, 171–91.

Barrett, S. (1994), 'Strategic environmental policy and international trade', *Journal of Public Economics*, **54**, 25–38.

Bartsch, M. (1987), *Das DIN – Deutsches Institut für Normung e.V. – als marktbeherrschendes Unternehmen*, i.S.v. §22 GWB, Münster: University Dissertation.

Baskin, E., K. Krechmer and H.M. Sherif (1998), 'The Six Dimensions of Standards: Contribution Towards a Theory of Standardization', in Lefebvre, L.A., R.M. Mason and T. Khalil (eds), *Management of Technology, Sustainable Development and Eco-Efficiency*, selected papers from the Seventh International Conference on Management of Technology, UK: Elsevier Science Ltd., pp. 53–62.

Bauer, C.-O. (1980), 'Technische Normen – ihre Methoden und Wirkungen als Gegenstand von Forschung und Lehre', *DIN-Mitteilungen*, **59**, 329.

Baumol W.J. and W.F. Oates, (1971), 'The use of standards and prices for protection of the environment', *Swedish Journal of Economics*, **73** (1), 42–54.

Beck, H. (1995), 'Patent issues in establishing technological standards', *Computer Law*, **12** (3), 1 and 3.

Belleflamme, P. (2002), 'Coordination on formal versus de facto standards: a dynamic approach', *European Journal of Political Economy*, **18** (1), 153–76.

Beise, M. and H. Belitz (1996), *Internationalisierung von Forschung und Entwicklung in multinationalen Unternehmen*, Materialien zur Berichterstattung zur technologischen Leistungsfähigkeit Deutschlands 1996, Berlin and Mannheim: ZEW.

Beitz, W. (1986), 'Normung und Innovation – ein Spannungsfeld?', *DIN-Mitteilungen*, **65**, 86–9.

Benezech, D., G. Lambert, B. Lanoux, C. Lerch and J. Loos-Baroin (2001), 'Completion of knowledge codification: an illustration through the ISO 9000 standards implementation process', *Research Policy*, **30**, 1395–407.

Berg, B. (1998), 'Europäische Normen zur Biotechnik', *transkript*, **2** (4), 8–11.

Berger, H. (1998), 'Regulation in Germany: some stylized facts about its time path, causes and consequences', *Zeitschrift für Wirtschafts- und Sozialwissenschaften*, **118**, 185–220.

Berger, W.G. and R. Clement (1990), 'Ökonomische Bedeutung der Normung auf dem Gebiet der Informationstechnik: Einige Eckwerte für die Bundesrepublik Deutschland', *DIN-Mitteilungen*, **69**, 490–91.

Berry, L.L., V.A. Zeithaml and A. Parasuraman (1992), 'Five Imperatives for Improving Service Quality', in Lovelock, C.H. (ed.), *Managing Services – Marketing, Operations, and Human Resources*, New Jersey: Prentice-Hall, pp. 224–35.

Besen, S.M. and J. Farrell (1994), 'Choosing how to compete: strategies and tactics in standardization', *Journal of Economic Perspectives*, **8** (2), 117–31.

Besen, S.M. and L.L. Johnson (1986), *Compatibility Standards, Competition, and Innovation in the Broadcasting Industry*, Santa Monica, US: RAND Corporation.

Blankart, C.B. and G. Knieps (1992), 'Netzökonomik', in Herder-Dornreich, P. (ed.), *Jahrbuch für Neue Politische Ökonomie*, 11, Tübingen: Mohr-Siebeck, pp. 73–87.

Blankart, C.B. and G. Knieps (1993a), 'Network Evolution', in Wagener, H.-J. (ed.), *On the Theory and Policy of Systemic Change*, Heidelberg: Physica Verlag, pp. 43–50.

Blankart, C.B. and G. Knieps (1993b), 'State and standards', *Public Choice*, **77**, 39–52.

Blind, K. (2000), *The Impact of Technical Standards and Innovative Capacity on Bilateral Trade Flows*, working paper, Karlsruhe: Fraunhofer Institute for Systems and Innovation Research ISI.

Blind, K. (2001a), 'Standardisation, R&D and Export Activities: Empirical Evidence at Firm Level', in *Proceedings of the Third Interdisciplinary Workshop on Standardization Research at the University of the German Federal Armed Forces*, Hamburg: Universität der Bundeswehr, pp. 165–86.

Blind, K. (2001b), 'The impacts of innovations and standards on trade of measurement and testing products: empirical results of Switzerland's bilateral trade flows with Germany, France and the United Kingdom', *Information Economics and Policy*, **13**, 439–60.

Blind, K. (2002a), 'Driving forces for standardisation at standardisation development organisations', *Applied Economics*, **34** (16), 1985–98, http://www.tandf.co.uk/journals/routledge/00036846.html.

Blind, K. (2002b), *Literature Survey, Standards in the Service Sectors: An Explorative Study*, First Interim Report for DG Enterprise of the European Commission, Karlsruhe: Fraunhofer Institute for Systems and Innovation Research ISI.

Blind, K. (2002c), *Normen als Indikatoren für die Diffusion neuer Technologien*, Endbericht für das Bundesministerium für Bildung und Forschung im Rahmen der

Untersuchung 'Zur Technologischen Leistungsfähigkeit Deutschlands' zum Schwerpunkt 'Methodische Erweiterungen des Indikatorensystems', Karlsruhe: Fraunhofer Institute for Systems and Innovation Research ISI.

Blind, K. (2002d), *The Interrelationship Between Standardisation, R&D and Export Activities: Empirical Evidence at Firm Level*, paper presented at the Fourth Annual Conference of the European Trade Study Group, Kiel: ETSG.

Blind, K. (2003), 'The Impact of Patent Rights on the Propensity to Standardise at Standardisation Development Organisations: an International Cross-Section Analysis', in *EURAS Yearbook of Standardization*, München: Accedo-Verlags-Gesellschaft.

Blind, K. and H. Grupp (2000), *Gesamtwirtschaftlicher Nutzen der Normung. Volkswirtschaftlicher Nutzen: Zusammenhang zwischen Normung und technischem Wandel, ihr Einfluss auf die Gesamtwirtschaft und den Aussenhandel der Bundesrepublik Deutschland*, Berlin: Beuth Verlag.

Blind, K. and C. Hipp (2003), 'Driving forces for the introduction of quality standards in innovative service companies: an empirical analysis for Germany', in *Technological Forecasting and Social Change*, New York, NY: American Elsevier Publications Co., pp. 653–69.

Blind, K. and A. Jungmittag (2001), 'The Impacts of Innovations and Standards on German Trade in General and on Trade with the UK in Particular', working paper, in *Proceedings of the Annual Conference of the European Association of Research of Industrial Economics (EARIE)*, CDROM, Dublin: EARIE.

Blind, K. and A. Jungmittag (2002), *The Impacts of Innovations and Standards on German–French Trade Flows*, working paper, Karlsruhe: Fraunhofer Institute for Systems and Innovation Research ISI.

Blind, K. and N. Thumm (2002), 'Survey of the Relationship Between IPR and Standardisation and Contractual Problems in RTD Projects', in Blind, K., R. Bierhals, E. Iversen, K. Hossain, B. Rixius, N. Thumm and R. van Reekum, *Study on the Interaction Between Standardisation and Intellectual Property Rights*, Karlsruhe: Fraunhofer Institute for Systems and Innovation Research ISI, pp. 59–82.

Blind, K. and N. Thumm (2003), 'Interdependencies Between Intellectual Property Protection and Standardisation Strategies', in Jakobs, K. (ed.), *Euras Proceedings 2003*, proceedings of the 8th EURAS Workshop, 11–12 July 2003, Aachener Beiträge zur Informatik, Band 33, Aachen: Wissenschaftsverlag Mainz in Aachen, pp. 88–107.

Blind, K., H. Grupp, A. Hullmann and A. Jungmittag (1999a), *Der Zusammenhang zwischen Normung und technischem Wandel*, Endbericht an das Bundesministerium für Wirtschaft und Technologie und das Deutsche Institut für Normung, ISI-Research-Report B-44-99, Karlsruhe: Fraunhofer Institute for Systems and Innovation Research ISI.

Blind, K., H. Grupp and A. Jungmittag (1999b), 'The Influence of Innovation and Standardisation on Macroeconomic Development: the Case of Germany', in Jakobs, K. and R. Williams (eds), *Proceedings of the 1st IEEE Conference on Standardisation and Innovation in Information Technology*, Piscataway, US: IEEE Service Center, pp. 125–31.

Blind, K., H. Grupp and A. Jungmittag (2000), 'Der Zusammenhang zwischen technischem Wandel und Normung und ihr Einfluß auf den Außenhandel in der Schweiz', in Institut für Öffentliche Dienstleistungen und Tourismus (ed.), *Der gesamtwirtschaftliche Nutzen der Normung*, Bericht 1, St. Gallen: Universität St. Gallen.

Blind, K., R. Bierhals, E. Iversen, K. Hossain, B. Rixius, N. Thumm and R. van

Reekum (2002), *Study on the Interaction Between Standardisation and Intellectual Property Rights*, final report for the Generaldirektion Forschung der Europäischen Kommission (EC Contract No G6MA-CT-2000-02001), Karlsruhe: Fraunhofer Institute for Systems and Innovation Research ISI.

Blind, S. and S. Bühring (1996), 'Die ökonomische Theorie der Standards und ihre Anwendung auf den Medienbereich', in *Homo Oeconomicus*, **13**, 515–60.

Blum, U., A. Töpfer, G. Eickhoff and I. Junginger (2000), *Gesamtwirtschaftlicher Nutzen der Normung: Unternehmerischer Nutzen*, 2 Bände, Berlin: Beuth Verlag.

Böhm, E., H. Conrad, H. Hiessl, T. Hillenbrand, V. Kühn, K. Lützner and R. Walz (1998), *Effektivität und Effizienz technischer Normen und Standards für kommunale Umweltschutzaufgaben am Beispiel der kommunalen Abwasserentsorgung*, Studie im Auftrag des Umweltbundesamtes, Bonn: Umweltbundesamt.

Boom, A. (1995), 'Asymmetric international minimum quality standards and vertical differentiation', *Journal of Industrial Economics*, **43**, 101–19.

Brander, J.A. and B. Spencer (1981), 'Tariffs and the extraction of monopoly rents under potential entry', *Canadian Journal of Economics*, **14**, 371–89.

Brander, J.A. and B. Spencer (1985), 'Export subsidies and international market share rivalry', *Journal of International Economics*, **18**, 83–100.

Brouwer, E. and A. Kleinknecht (1993), 'Technology and a firm's export intensity: the need for adequate innovation measurement', *Konjunkturpolitik*, **39**, 315–25.

Brouwer E. and A. Kleinknecht (1996), *Alternative Innovation Indicators and Determinants of Innovation*, EU Report 16963, Brussels: European Commission.

Budd, A. and S. Hobbis (1989a), *Output Growth and Measure of Technology*, discussion paper, London: London Business School, Centre for Economic Forecasting (CEF).

Budd, A. and S. Hobbis (1989b), *Cointegration, Technology and the Log-run Production Function*, discussion paper, London: London Business School, Centre for Economic Forecasting (CEF).

Bundesamt für Statistik (1996), *Statisches Jahrbuch für die Schweiz 1997*, Zürich: Verlag Neue Zürcher Zeitung.

Casella, A. (2001), 'Product standards and international trade. Harmonization through private coalitions?', *Kyklos*, **54**, 243–64.

Caves, R.E. (1981), 'Intra-industry trade and market structure in the industrial countries', *Oxford Economic Papers*, **33**, 203–23.

Cecchini, P. (1988), *Europa '92. Der Vorteil des Binnenmarktes* (Cecchini–Bericht), Baden-Baden: Nomos.

CEN (1996), *Service and Standardization*, Brussels: CEN.

Chiesa, V., R. Manzini and G. Toletti (2002), 'Standard-setting processes: evidence from two case studies', *R&D Management*, **32**, 431–49.

Choi, J.P. (1997), 'Herd behavior, the "penguin effect" and the suppression of informational diffusion: an analysis of informational externalities and payoff interdependency', *RAND Journal of Economics*, **28**, 407–25.

Christoph, H. (1980), 'Internationale Normen in der Schweißtechnik – Nutzen und Kosten des Normungsaufwands', *DIN-Mitteilungen*, **59**, 528–33.

Chu, P.-Y. and H.-J. Wang (2001), 'Benefits, critical process factors, and optimum strategies of successful ISO 9000 implementation in the public sector: an empirical examination of public sector services in Taiwan', *Public Performance and Management Review*, **25** (1), 105–21.

Church J. and N. Gandal (1993), 'Complementary network externalities and technological adoption', *International Journal of Industrial Organization*, **11**, 239–60.

Clemens, R. and H.-E. Hauser (1993), *Die Harmonisierung technischer Normen in der EG und ihre Auswirkungen auf den industriellen Mittelstand*, Stuttgart: Schäffer-Pöschel.

Coe, D.T. and R. Moghadam (1993), 'Capital and trade as engines of growth in France', *IMF Staff Papers*, **40**, Washington, DC, 542–66.

Cohen, W., R.R. Nelson and J. Walsh (2000), *Appropriability Conditions and Why Patent and Why They Do Not*, National Bureau of Economic Research, Working Paper 7552, Washington, DC: NBER.

Cohendet, P. and W.E. Steinmueller (2000), 'The codification of knowledge: a conceptual and empirical exploration', *Industrial and Corporate Change*, **9** (2), 195–209.

Collazzo, C. (1979), 'Einflüsse der Normung auf die Wirtschaft und die Verbraucher', in DIN (ed.), *Wirtschaftlichkeit der Normung*, Berlin: Beuth Verlag, pp. 41–79.

Coursey, B.M. and A.N. Link (1998), 'Evaluating technology-based public institutions: the case of radiopharmaceutical standards research at the National Institute of Standards and Technology', *Research Evaluation*, **7**, 147–57.

Cowan, R.A. (1990), 'Nuclear power reactors: a study in technological lock-in', *Journal of Economic History*, **50**, 541–67.

Cowan, R. and J.H. Miller (1998), 'Technological standards with local externalities and decentralized behaviour', *Journal of Evolutionary Economics*, **8**, 285–96.

Crampes, C. and M. Wolkowicz (1995), *Standards and Industrial Property*, working paper, Toulouse Cedex: Université des Sciences Sociales.

Darsie, B. (1990), 'Financial Services', in Toth, R.B. (ed.), *Standard Management – a Handbook for Profits*, American National Standards Institute, New York: ANSI, pp. 351–9.

Dasgupta, P. (1995), 'Non-tariff barriers to trade: the issue of technical standards', *Foreign Trade Review*, **30**, 46–66.

David, P.A. (1985), 'Clio and the economics of QWERTY', *American Economic Review*, **75**, 332–6.

David P.A. (1987), 'Some New Standards for the Economics of Standardisation in the Information Age', in Dasgupta, P. and P. Stoneman (eds), *Economic Policy and Technological Performance*, Cambridge: Cambridge University Press.

David, P.A. and S. Greenstein (1990), 'The economics of compatibility standards: an introduction to recent research', *Economics of Innovation and New Technology*, **1**, 3–41.

David, P.A. and H.K. Monroe (1994), *Standards Development Strategies Under Incomplete Information – Isn't the 'Battle of the Sexes' Really a Revelation Game?*, paper at the Telecommunications Policy Research Conference, 1–3 October 1994, Solomon's Island, Maryland: TPRC.

David P.A. and W.E. Steinmueller (1994), 'Economics of compatibility standards and competition in telecommunications networks', *Information Economics and Policy*, **6**, 217–41.

David, P.A. and M. Shurmer (1996), 'Formal standard-setting for global telecommunications and information services', *Telecommunications Policy*, **20**, 789–815.

Davis, L. (1997), *Quality Assurance: ISO 9000 As a Management Tool*, D(26), Copenhagen: Copenhagen Business School Press.

Deregulierungskommission (1991), *Marktöffnung und Wettbewerb*, Stuttgart: Schäffer-Poeschel.

Deutsche Bundesbank (1996), *Monatsbericht Dezember 1996*, Frankfurt/Main: Bundesbank.

Deutsche Bundesbank (1998), *Technologische Dienstleistungen in der Zahlungs-bilanz*, Statistische Sonderveröffentlichung 12, Frankfurt/Main: Bundesbank.

Deutsche Elektrotechnische Kommission (1996), *Jahresbericht 1996*, Frankfurt/ Main: DKE.

Deutsches Institut für Wirtschaftsforschung (1995), *Produktionsvolumen und -potential. Produktionsfaktoren des Bergbaus und des Verarbeitenden Gewerbes – Bundesrepublik Deutschland ohne Beitrittsgebiet – Statistische Kennziffern*, 37. Folge 1970–94, Berlin: DIW.

Deutsches Patent- und Markenamt (1998), *Blatt für Patent-, Muster- und Zeichenwesen*, 3, München, Köln, Berlin: Heymann.

Dickey, D.A and W.A. Fuller (1979), 'Distribution of the estimators for autoregressive time series with a unit root', *Journal of the American Statistical Association*, **74**, 427–31.

DIN (1995), *Grundlagen der Normungsarbeit des DIN*, 6th edition, Berlin et al.: Beuth Verlag.

DIN (ed.) (1996a), *Europäische Normung: Ein Leitfaden des DIN*, Berlin: Deutsches Institut für Normung e.V.

DIN (ed.) (1996b), *DIN Geschäftsbericht 1995–96*, Berlin: Deutsches Institut für Normung e.V.

DIN (1998a), *DIN Geschäftsbericht 1997–98*, Berlin: Deutsches Institut für Normung e.V.

DIN (1998b), *Etwas über DIN*, Berlin: Deutsches Institut für Normung e.V.

DIN (2000a), *DIN Geschäftsbericht 2000*, Berlin: Deutsches Institut für Normung e.V.

DIN (2000b), *Gesamtwirtschaftlicher Nutzen der Normung*, 3 Bände, Berlin: Beuth Verlag.

DIN (2002), *Standardisierung in der deutschen Dienstleistungswirtschaft – Potenziale und Handlungsbedarf*, DIN-Fachbericht 116, Berlin: Beuth Verlag.

Dixit, A. and J. Stiglitz (1977), 'Monopolistic competition and optimum product diversity', *American Economic Review*, **67**, 297–308.

Docking, D.S. and R.J. Dowen (1999), 'Market interpretation of ISO 9000 registration', *Journal of Financial Research*, **22** (2), 147–160.

Dosi, G. (1982), 'Technological paradigms and technological trajectories: a suggested interpretation of the determinants and directions of technical change', *Research Policy*, **12**, 147–162.

Ebling, G. and N. Janz (1999), *Export and Innovation Activities in the German Service Sector: Empirical Evidence at the Firm Level*, discussion paper no. 99–53, Center for European Economic Research, Mannheim: ZEW.

Economides, N. (1996), 'The economics of networks', *International Journal of Industrial Organization*, **14**, 673–99.

Egan, M. (2001), *Construction of a European Market*, Oxford: Oxford University Press.

Eichener, V. and H. Voelzkow (1993), 'Entwicklungsbegleitende Normung: Integration von Forschung & Entwicklung, Normung und Technikfolgenabschätzung', *DIN-Mitteilungen*, **72**, 764–8.

Engle, R.F. and C.W.J. Granger (1987), 'Co-integration and error correction: representation, estimation, and testing', *Econometrics*, **55**, 251–76.

Engle, R.F. and B.S. Yoo (1991), 'Cointegrated Economic Time Series: an Overview with New Results', in Engle, R.F. and C.W.J. Granger (eds), *Long-run Economic Relationships – Readings in Cointegration*, Oxford et al.: Oxford University Press.

Europäische Kommission (1990), *Grünbuch der EG-Kommission zur Entwicklung der europäischen Normung*, Brussels: EU.

Europäische Kommission (1998a), *Weniger Gesetzgebung für besseres Handeln: Die Fakten*, Brussels: EU.

Europäische Kommission (1998b), *Effizienz und Verantwortlichkeit in der europäischen Normung im Rahmen des neuen Konzepts*, Brussels: EU.

European Commission (1988), *Research on the 'Cost of Non-Europe'. Basic Findings*, 6, Brussels: EU.

European Commission (1992), *Communication from the Commission 'Intellectual Property Rights and Standardization'*, COM (92), Brussels: EU.

European Commission (1998), *Setting the Standard: 25 Years of Quality Measurements*, Luxembourg: Office for Official Publications of the European Communities.

European Commission (2000a), *Guide to the Implementation of Directives Based On the New Approach and the Global Approach*, Luxembourg: Office for Official Publications of the European Communities.

Farrell, J. (1989), 'Standardization and intellectual property', *Jurimetrics Journal*, 35–50.

Farrell, J. and G. Saloner (1985), 'Standardization, compatibility, and innovation', *RAND Journal of Economics*, **16**, 70–83.

Farrell, J. and G. Saloner (1986), 'Installed base and compatibility: innovation, product preannouncements and predation', *American Economic Review*, **76**, 943–54.

Farrell, J. and G. Saloner (1987), 'Competition, Compatibility and Standards: the Economics of Horses, Penguins and Lemmings', in Gabel, H.L. (ed.), *Product Standardization and Competitive Strategy*, Amsterdam: North-Holland, pp. 1–21.

Farrell J. and G. Saloner (1992), 'Converters, compatibility, and the control of interfaces', *Journal of Industrial Economics*, **40** (1), 9–36.

Farrell J. and C. Shapiro (1988), 'Dynamic competition with switching costs', *RAND Journal of Economics*, **19**, 123–37.

Fischer, R. and P. Serra (2000), 'Standards and protection', *Journal of International Economics*, **52**, 377–400.

Flam, H. (1992), 'Product markets and 1992: full integration, large gains', *Journal of Economic Perspectives*, **6**, 7–30.

Foss, K. (1996), *A Transaction Cost Perspective on the Influence of Standards on Product Development: Examples from the Fruit and Vegetable Market*, DRUID Working Paper, 96–9, Copenhagen, Aalborg: DRUID.

Fredebeul-Krein, M. (1997), 'Veränderte Anforderungen an eine europäische Standardisierungspolitik in der Telekommunikation', *Zeitschrift für Wirtschaftspolitik*, **46**, 140–66.

Frenkel, M. and H.-R. Hemmer (1999), *Grundlagen der Wachstumstheorie*, München: Vahlen.

Fritsch, M., T. Wein, H.-J. Ewers (1996), *Marktversagen und Wirtschaftspolitik. Mikroökonomische Grundlagen staatlichen Handelns*, 2nd edition, München: Vahlen.

Funke, M. and H. Strulik (2000), 'On endogenous growth with physical capital, human capital and product variety', *European Economic Review*, **44** (3), 491–515.

Gabel, H.L. (1993), *Produktstandardisierung als Wettbewerbsstrategie*, London et al.: McGraw-Hill.

Gandal, N. (1994), 'Hedonic price indexes for spreadsheets and an empirical test for network externalities', *RAND Journal of Economics*, **25**, 160–70.

Gandal, N. (1995), 'Competing compatibility standards and network externalities in the PC software market', *The Review of Economics and Statistics*, **77**, 599–608.

Gandal, N. (2002), 'Compatibility, standardization and network effects: some policy implications', *Oxford Review of Economic Policy*, **18** (1), 80–91.

Gandal, N. and O. Shy (2001), 'Standardization policy and international trade', *Journal of International Economics*, **53**, 363–83.

Ganslandt, M. and J.R. Markusen (2001), *Standards and Related Regulations in International Trade: a Modeling Approach*, NBER Working Paper 8346, Cambridge, MA: NBER.

Geroski, P. (1991), 'Innovation and sectoral sources of UK productivity growth', *Economic Journal*, **101**, 1438–51.

Geroski, P. (2000), 'Models of technology diffusion', *Research Policy*, **29**, 603–25.

GEWIPLAN (1988), 'The "Cost of Non-Europe": Some Case Studies on Technical Barriers', in European Commission (1988), *Research on the 'Cost of Non-Europe'. Basic Findings*, pp. 39–233.

Gilpin, S. and S.P. Kalafatis (1995), 'Issues of product standardisation in the leisure industry', *The Service Industries Journal*, **15** (2), 186–203.

Glanz, A. (1990), 'Kann der Anwender mit Standards rechnen?', *online*, **5**, 24–30.

Goerke, L. and M.J. Holler (1995), 'Voting on standardisation', *Public Choice*, **83**, 337–51.

Graßmuck, J. and W. Heller (1986), 'Inner- und überbetriebliche Normung – Gegensatz oder gegenseitige Ergänzung?', *DIN-Mitteilungen*, **65**, 561–5.

Greenaway, D. and Ch. Milner (1984), 'A cross section analysis of intra-industry trade in the UK', *European Economic Review*, **25**, 319–44.

Greenaway, D. and Ch. Milner (1986), *The Economics of Intra-industry Trade*, Oxford et al.: Blackwell.

Greenhalgh, C. (1990), 'Innovation and trade performance in the United Kingdom', *Economic Journal*, **1000**, 105–18.

Greenstein S.M. (1993), 'Did installed base give an incumbent any (measurable) advantages in federal computer procurement?', *RAND Journal of Economics*, **24** (1), 19–39.

Greenstein, S.M. (1997), 'Lock-in and the cost of switching mainframe computer vendors: what do buyers see?', *Industrial and Corporate Change*, **6**, 247–74.

Griliches, Z. (1980), 'R&D and the productivity slowdown', *American Economic Review*, **70**, 343–8.

Griliches, Z. and F. Lichtenberg (1984), 'Interindustry technology flows and productivity growth: a reexamination', *Review of Economics and Statistics*, **61**, 324–9.

Griliches, Z. and J. Mairesse (1984), 'Productivity and R&D at the Firm Level', in Griliches, Z. (ed.), *R&D, Patents and Productivity*, Chicago et al.: University of Chicago Press.

Grossman, G. and E. Helpman (1991), *Innovation and Growth in the Global Economy*, Cambridge, MA: MIT Press.

Grubel, H.G. and P.J. Lloyd (1975), *Intra-industry Trade: the Theory and Measurement of International Trade in Differentiated Products*, New York: Wiley.

Gruber, H. (2000), 'The evolution of market structure in semiconductors: the role of product standards', *Research Policy*, **29**, 725–40.

Grupp, H. (1997), *Messung und Erklärung des Technischen Wandels: Grundzüge einer empirischen Innovationsökonomik*, Berlin: Springer.

Grupp, H. (1998), *The Foundations of the Economics of Innovation: Theory, Measurement and Practice*, Cheltenham, UK and Northhampton, MA, USA: Edward Elgar.

Grupp, H. and A. Jungmittag (1999), 'Convergence in global high technology? A decomposition and specialisation analysis for advanced countries', *Jahrbücher für Nationalökonomie und Statistik*, **218**, 552–73.

Gustavson, E. (2000), *Selling Practices and Standardization – a Study on Interaction Practices Between Seller and Buyer on the Industrial Market*, contribution to the 14th Nordic Conference on Business Studies, School of Economics and Commercial Law, Gothenburg: University of Gothenburg.

Händel, S. (1980), 'Gedanken über die Wirtschaftlichkeit der Normung', *DIN-Mitteilungen*, **59**, 659–65.

Harhoff, D. (1997), 'Innovationsanreize in einem strukturellen Oligopolmodell', *Zeitschrift für Wirtschafts- und Sozialwissenschaften*, **117**, 333–64.

Harhoff, D. and D. Moch (1997), 'Price indexes for PC database software and the value of code compatibility', *Research Policy*, **26**, 509–20.

Hartlieb, B. (1993), 'Entwicklungsbegleitende Normung (EBN) – Geschichtliche Entwicklung der Normung. Gründung eines Sonderausschusses des DIN-Präsidiums', *DIN-Mitteilungen*, **72**, 332–9.

Hartlieb, B. and H. Behrens (1996), 'Dienstleistung und Normung – Ergebnisse von der Arbeitsgruppe "Dienstleistung und Regelsetzung" im BMBF-Vorhaben "Dienstleistung 2000plus"', *DIN-Mitteilungen*, **75**, 746–51.

Hartman, R.S. and D.J. Teece (1990), 'Product emulation strategies in the presence of reputation effects and network externalities: some evidence from minicomputer industry', *Economics of Innovation and New Technology*, **1**, 157–82.

Hauser, H. (1979), 'Qualitätsinformationen und Marktstrukturen', *KYKLOS*, **32**, 739–63.

Hawkins, R. (1996), *Determining the Significance of Industrial Standards as Indicators of Technical Change*, Science Policy Research Unit, final report, Brighton: University of Sussex.

Hawkins, R, R. Mansell and J. Skea (1995) (eds), *Standards, Innovation and Competitiveness*, Camberley, UK and Brookfield, US: Edward Elgar.

Helbig, J. and J. Volkert (1998), 'Potentiale und Grenzen freiwilliger Standards im Umweltschutz', *IAW-Mitteilungen*, **4**, 4–15.

Helpman, E. (1998), 'Explaining the structure of foreign trade: where do we stand?', *Weltwirtschaftliches Archiv*, **134**, 573–89.

Helpman, E. (1999), 'The structure of foreign trade', *Journal of Economic Perspectives*, **13**, 121–144.

Hemenway, D. (1975), *Industrywide Voluntary Product Standards*, Cambridge, MA: Ballinger.

Heß, G. (1993), *Kampf um den Standard! Erfolgreiche und gescheiterte Standardisierungsprozesse – Fallstudien aus der Praxis*, Stuttgart: Schäffer-Pöschel.

Hesse, H. (1974), 'Hypotheses for the Explanation of Trade Between Industrial Countries, 1953–70', in Giersch, H. (ed.), *The International Division of Labour: Problems and Perspectives*, Tübingen: Mohr, pp. 39–59.

Hesser, W. and A. Inklaar (eds) (1997), *An Introduction to Standards and Standardization*, DIN, Berlin: Beuth-Verlag.

Hesser, W. and R. Meyer (1993), 'Parameter der Wirtschaftlichkeit von Normungsvorhaben – Das Wachstum der Typenvielfalt als überbetrieblich verwendbarer Parameter', *DIN-Mitteilungen*, **72**, 349–54.

Hesser, W., C. Herb and R. Hildebrandt (1994), 'Standards and Law', in Hesser, W. (ed.), *Different Aspects of Research on Standardization*, Hamburg: Universität der Bundeswehr, pp. 1–23.

Hildebrandt, R. (1995), 'The Significance of the EC Directives Policy for Company Standardization', in Hesser, W. (ed.), *From Company Standardization to*

European Standardization, Hamburg: Universität der Bundeswehr, pp. 87–99.

Hipp, C., B.S. Tether and I. Miles (2000), 'The incidence and effects of innovation in services – evidence from Germany', *International Journal of Innovation Management*, **4** (4), 417–53.

Hirsch, S. and I. Bijaoui (1985), 'R&D intensity and export performance: a micro view', *Weltwirtschaftliches Archiv*, **121**, 138–51.

Hoffmann-Riem, W. (1996), 'Möglichkeiten des Rechts bei der Bewirkung von Innovationen', in VDI Technologiezentrum (ed.), *Rechtliche Rahmenbedingungen für Forschung und Innovation*, Dokumentation eines Workshops veranstaltet vom Bundesministerium für Bildung, Wissenschaft, Forschung und Technologie 11–12 September 1995 in Köln, Cologne: VDI, pp. 10–25.

Holler, M.J. (1996), 'Die Rationalität strategischer Normung in Europa', in Schenk, K.-E., D. Schmidtchen and M.E. Streit (eds), *Vom Hoheitsstaat zum Konsensualstaat: Neue Formen der Kooperation zwischen Staat und Privaten*, Jahrbuch für Neue Politische Ökonomie, 15, Tübingen: Mohr-Siebeck, pp. 137–53.

Holler, M., G. Knieps and E. Niskanen (1997), 'Standardization in Transport Markets: a European Perspective', in *EURAS Yearbook of Standardization*, 1, München: Accedo-Verlags-Gesellschaft, pp. 371–90.

Hudson, J. and P. Jones (2001), 'Measuring the efficiency of stochastic signals of product quality', *Information Economics and Policy*, **13** (1), 35–49.

IFO Institut (1998), *Wissensverbreitung und Diffusionsdynamik im Spannungsfeld zwischen innovierenden und imitierenden Unternehmen – Neue Ansätze für die Innovationspolitik*, Studie im Auftrag des Bundesministeriums für Wirtschaft, München: IFO.

Institut der Deutschen Wirtschaft (1997*), Zahlen zur Wirtschaftlichen Entwicklung der Bundesrepublik Deutschland*, Köln: IW Köln.

ISO (1982), *Benefits of Standardization*, Genf: ISO.

ISO (2001), 'Servicing the service industry: taking the issue further down the road', *ISO Bulletin*, 5–11.

ISO and IEC (1990), *A Vision for the Future: Standards Needs for Emerging Technologies*, Genf: ISO.

Janz, N. and G. Licht (eds) (1999), *Innovationsaktivitäten in der deutschen Wirtschaft: Analyse der Mannheimer Innovationspanel im Verarbeitenden Gewerbe und im Dienstleistungssektor*, Baden-Baden: Nomos.

Janz, N., G. Licht and Th. Doherr (2001), 'Innovation Activities and European Patenting for Geman Firms', in proceedings (CD-ROM) of the Annual Conference of the European Association of Research of Industrial Economics (EARIE), Dublin: EARIE.

Japanese Industrial Standards Committee (1997), *JIS Yearbook*, Tokyo: JSA.

Japanese Industrial Standards Committee (1998), *JIS Catalogue*, Tokyo: JSA.

Jensen, R. and M. Thursby (1996), 'Patent races, product standards and international competition', *International Economic Review*, **37**, 21–49.

Johannsen, C.G. (1995), 'Application of the ISO 9000 standards of quality management in professional services: an information sector case', *Total Quality Management*, **6** (3), 231–42.

Jones, P. and J. Hudson (1996), 'Standardization and the costs of assessing quality', *European Journal of Political Economy*, **12**, 355–61.

Jungmittag, A. and P.J.J. Welfens (1996), *Telekommunikation, Innovation und die langfristige Produktionsfunktion: Theoretische Aspekte und eine Kointegrationsanalyse für die Bundesrepublik Deutschland*, Diskussionsbeitrag 20 des Europäischen Instituts für Internationale Wirtschaftsbeziehungen (EIIW),

Potsdam: EIIW.

Jungmittag, A., K. Blind and H. Grupp (1999), 'Innovation, standardization and the long-term production function: a co-integration approach for Germany 1960–96', *Zeitschrift für Wirtschafts- und Sozialwissenschaften*, **119**, 205–22.

Kamien, M.I. and I. Zang (2000), 'Meet me halfway: research joint ventures and absorptive capacity', *International Journal of Industrial Organisation*, **18**, 995–1012.

Karapetrovic, S. and W. Willborn (2001), 'ISO 9000 quality management standards and financial investment services', *The Service Industries Journal*, **21** (2), 117–37.

Karapetrovic, S., D. Rajaman and W. Willborn (1997), 'ISO 9000 for small business: do it yourself', *Industrial Management*, **39** (3), 24 ff.

Katz, M.L. and C. Shapiro (1985), 'Network externalities, competition and compatibility', *American Economic Review*, **75**, 424–40.

Katz, M.L. and C. Shapiro (1986), 'Technology adoption in the presence of network externalities', *Journal of Political Economy*, **94**, 822–41.

Katz, M.L. and C. Shapiro (1992), 'Product introduction with network externalities', *Journal of Industrial Economics*, **40** (1), 55–84.

Katz, M.L. and C. Shapiro (1994), 'Systems competition and network effects', *Journal of Economic Perspectives*, **8** (2), 93–115.

Kaufer, E. (1989), *The Economics of the Patent System*, Chur et al.: Harwood.

Kennedy, C. and A. Thirlwall (1972), 'Surveys in applied economics: technical progress', *Economic Journal*, **82**, 11–72.

Kindleberger, C.P. (1983), 'Standards as public, collective and private goods', *KYKLOS*, **36**, 377–96.

Kleinaltenkamp, M. (1995), 'Entwicklungsbegleitende Normung. Beschleunigung der Diffusion neuer Technologien oder schädlicher Eingriff in den Marktprozeß?', in Sadowski, D., H. Czap and H. Wächter (eds), *Regulierung und Unternehmenspolitik*, Wiesbaden: Gabler, pp. 81–101.

Kleinaltenkamp, M. and A. Marra (1994), 'Schaffen Normen Märkte? Die Entwicklungsbegleitende Normung im Laserbereich "auf dem Prüfstand"', *VDI-Zeitung*, **136**, 74–7.

Kleinemeyer, J. (1998), *Standardisierung zwischen Kooperation und Wettbewerb*, Frankfurt/Main et al.: Lang.

Klemperer, Paul, (1987), 'The competitiveness of markets with switching costs', *RAND Journal of Economics*, **18** (1), 138–50.

Köhler, H. (1985), 'Die haftungsrechtliche Bedeutung technischer Regeln', *Betriebs-Berater Beilage*, **4**, 10–15.

Korinek, K. (1996), 'Normung im Spannungsfeld von Effizienz und demokratischer Legitimation', in DIN (ed.), *Normung in Europa und das DIN – Ziele für das Jahr 2005*, Bericht über die außerordentliche Sitzung des DIN-Präsidiums am 23 April 1996 in Berlin, Berlin: DIN, pp. 57–62.

Kristiansen, E.G. (1998), 'R&D in the presence of network externalities: timing and compatibility', *RAND Journal of Economics*, **29**, 531–47.

Kristiansen, E.G. and M. Thum (1997), 'R&D incentives in compatible networks', *Journal of Economics*, **65**, 55–78.

Kruse, J. (1989), 'Ordnungstheoretische Grundlagen der Deregulierung', in Verein für Socialpolitik (ed.), *Deregulierung – eine Herausforderung an die Wirtschafts- und Sozialpolitik in der Marktwirtschaft*, Schriften des Vereins für Socialpolitik, 184, Berlin: Duncker & Humblot, pp. 9–36.

Kuhlmann, S., C. Bättig, K. Cuhls and V. Peters (1997), *Regulation und künftige Technikentwicklung*, Karlsruhe: Fraunhofer Institute for Systems and Innovation

Research ISI.

Langhammer, R.J. (1998), 'Die Weiterentwicklung der WTO', *Wirtschaft und Statistik*, **27**, 121–6.

Lecraw, D.J. (1984), 'Some economic effects of standards', *Applied Economics*, **16**, 507–22.

Lecraw, D.J. (1987), 'Japanese Standards: a Barrier to Trade?', in Gabel, H.L. (ed.), *Product Standardization and Competitive Strategy*, Amsterdam et al.: North-Holland, pp. 29–46.

Lefebvre, E. and L.-A. Lefebvre (2002), 'Innovative Capabilities as Determinants of Export Performance and Behaviour: a Longitudinal Study of Manufacturing SMEs', in Kleinknecht, A. and P. Mohnen (eds), *Innovation and Firm Performance: Econometric Explorations of Survey Data*, Houndmills: Palgrave Publishers.

Lehr W. (1992), 'Standardization: understanding the process', *Journal of the American Society for Information Science*, **43**, 550–55.

Leland H.E. (1979), 'Quacks, lemons, and licensing: a theory of minimum quality standards', *Journal of Political Economy*, **87**, 1328–46.

Levin, A. and C.-F. Lin (1992), *Unit Root Tests in Panel Data: Asymptotic and Finite Sample Properties*, Discussion Paper 92/23, San Diego: University of California.

Levin, R.C., A. Klevorick, R.R. Nelson and S.G. Winter (1987), 'Appropriating the returns from industrial research and development', *Brookings Papers on Economic Activity*, **3**, 783–820.

Licht, G., C. Hipp, M. Kukuk and G. Münt (1997), *Innovationen im Dienstleistungssektor*, Baden-Baden: Nomos.

Liebowitz S.J. and S.E. Margolis (1990), 'The fable of the keys', *Journal of Law and Economics*, **33**, 1–25.

Liebowitz, S.J. and S.E. Margolis (1994), 'Network externality: an uncommon tragedy', *Journal of Economic Perspectives*, **8**, 133–150.

Liebowitz, S.J. and S.E. Margolis (1999), *Winners, Losers & Microsoft. Competition and Antitrust in High Technology*, Oakland, USA: The Independent Institute.

Lim, A. (2002), *Standard Setting Processes in ICT: The Negotiations Approach*, Working Paper 2.19, Faculty of Technology Management, Eindhoven: Technische Universiteit Eindhoven.

Link, A.N. (1983), 'Market structure and voluntary product standards', *Applied Economics*, **15**, 393–401.

Liphard, K.G. (1998), 'Brennstoffnormen als Bausteine der Qualitätspolitik', *Glückauf*, **134**, 23–28.

Love, J.H. and S. Roper (1999), 'The determinants of innovation: R&D, technology transfer and networking effects', *Review of Industrial Organization*, **15**, 43–64.

Lukes, R. (1968), 'Überbetriebliche technische Normen im Recht der Wettbewerbsbeschränkungen', in Heuss, E. (ed.), *Wettbewerb als Aufgabe – Nach zehn Jahren Gesetz gegen Wettbewerbsbeschränkungen*, Bad Homburg et al.: Gehlen, pp. 165–75.

Malisius, G. and T. Weidner (1998), 'Die Normenabteilung – fit für die Zukunft!', *DIN-Mitteilungen*, **77**, 272–5.

Mansell, R. (1995), 'Standards, Industrial Policy and Innovation', in Hawkins, R., R. Mansell and J. Skea, (eds), *Standards, Innovation and Competitiveness*, Camberley, UK and Brookfield, US: Edward Elgar, pp. 213–27.

Maskus K.E. and M. Penubarti (1998), 'How trade-related are intellectual property rights?', *Journal of International Economics*, **39**, 227–48.

Maskus, K.E. and J.S. Wilson (2000), *Quantifying the Impact of Technical Barriers to Trade: a Review of Past Attempts and the New Policy Context*, World Bank Working Paper, Washington, DC: World Bank.

Matutes, C. and P. Regibeau (1987), 'Standardization in Multi-Component Industries', in Gabel, H.L. (ed.), *Product Standardization and Competitive Strategy*, Amsterdam: North-Holland, pp. 23–8.

Matutes, C. and Regibeau, P. (1996), 'A selective view of the economics of standardization: entry deterrence, technological progress and international competition', *European Journal of Political Economy*, **12**, 183–206.

Matutes, C., P. Regibeau and K. Rockett (1996), 'Optimal patent design and the diffusion of innovations', *RAND Journal of Economics*, **27**, 60–83.

Mazzoleni, R. and R.R. Nelson (1998), 'The benefits and costs of strong patent protection: a contribution to the current debate', *Research Policy*, **27**, 273–84.

Meeus, M.T.H., J. Faber and L.A.G. Oerlemans (2002), *Why Do Firms Participate in Standardization? An Empirical Exploration of the Relation Between Isomorphism and Institutional Dynamics in Standardization*, Working Paper Department of Innovation Studies, Utrecht: University of Utrecht.

Metcalfe, J.S. and I. Miles (1994), 'Standards, selection and variety: an evolutionary approach', *Information Economics and Policy*, **6**, 243–68.

Meyer, R. (1995), *Parameter der Wirksamkeit von typenreduzierenden Normungsvorhaben*, DIN-Normungskunde Band 34, Berlin et al.: Beuth-Verlag.

Michaelis, J. (1990a), 'Mutual recognition of national regulations in the EC', *Inter-Economics*, **25**, 215–19.

Michaelis, J. (1990b), 'Lebensmitteletikettierung? Die Sichtweise eines Ökonomen', *Zeitschrift für das gesamte Lebensmittelrecht*, **17**, 233–51.

Michaelis, J. and C. Borrmann (1990), *Lebensmittel im europäischen Binnenmarkt*, Hamburg: Verlag Weltarchiv.

Molsberger, J. (1977), 'Kriterien zur Beurteilung des volkswirtschaftlichen Nutzens der Normung', *DIN-Mitteilungen*, **56**, 399.

Muehlbauer, H. (2001), 'Standards for the service industry: Europe as a case study', *ISO Bulletin*, 12–15.

Muehlbauer, H. and G. Cornelissen (1998), 'Normung von Dienstleistungen', *DIN-Mitteilungen*, **77** (11), 809–17.

Münt, G. (1996), *Dynamik von Innovation und Außenhandel*, Heidelberg: Physika-Verlag.

National Research Council, International Standards, Conformity Assessment and U.S. Trade Policy Project Committee (1995) (ed.), *Standards, Conformity Assessment, and Trade: Into the 21st Century*, Washington, DC: National Academy Press.

Neisen, H. (1977), 'Die Bedeutung der überbetrieblichen technischen Normen für die Gesetzgebung und Rechtssprechung', *Siemens-Zeitschrift*, **51**, 439–45.

Nelson, P. (1970), 'Information and consumer behavior', *Journal of Political Economy*, **78**, 311–29.

Nelson, R.R. and S.G. Winter (1982), *An Evolutionary Theory of Economic Change*, Cambridge, MA et al.: Belknap Press.

Nicolas F. and J. Repussard (1988), *Common Standards for Enterprises*, Luxembourg: Office for Official Publications of the European Communities.

Nohr, H. (1997), 'Internationale Normenklassifikation (ICS)', *Nachrichten für Dokumentation*, **48**, 87–90.

Normann, R. (1991), *Service Management – Strategy and Leadership in Service Business*, 2nd edition, Chichester et al.: John Wiley & Sons.

Nunnenkamp, P. (1985), 'Technische Handelshemmnisse – Formen, Effekte und Harmonisierungsbestrebungen', *Aussenwirtschaft*, 373–97.

OECD (1996), *Technology and Industrial Performance*, Paris: OECD.

OECD (1997a), *DSTI (ANBERD Database)*, Paris: OECD.

OECD (1997b), *Annual National Accounts 1960–95*, Paris: OECD.

OECD (1997c), *Proposed Guidelines for Collecting and Interpreting Technological Innovation Data*, Paris: OECD.

OECD (1997d), *STAN Database*, Paris: OECD.

OECD (1997e), *The OECD Report Regulatory Reform. Volume II: Thematic Studies*, Paris: OECD.

OECD (1998a), *International Trade by Commodities Statistics SITC – Rev. 2 1988–97*, Paris: OECD.

OECD (1998b), *International Trade by Commodities Statistics Historical Statistics SITC – Rev. 2 1960–90*, Paris: OECD.

Ordover, J.A. (1991), 'A patent system for both diffusion and exclusion', *Journal of Economic Perspectives*, **5** (1), 43–60.

Page, Th. and R.A. Spreng (2002), 'Difference scores versus direct effects in service quality management', *Journal of Service Research*, **4** (3), 184–92.

Pecher, R. (1996), 'ATV-Regelwerk und Ökonomie in der Abwassertechnik', *Korrespondenz Abwasser*, **5**, 720–29.

Pepels, W. (1999), 'Marketingrelevante Besonderheiten von Dienstleistungen', *WISU*, **5**, 699–704.

Pfau, W. (1991), 'A vision for the future: Globale Wirkungen von Forschung und neuen Technologien', *DIN-Mitteilungen*, **70**, 67–9.

Pindyck, R.S. and D.L. Rubinfeld (1991), *Econometric Models and Economic Forecasts*, 3rd edition, New York et al.: McGraw-Hill.

Pokorny, F. (1974), 'Die Kosten der Normung', *DIN-Mitteilungen*, **53**, 227–30.

Prebezac, D. (1997), 'The quality of air transport services in function of improving the total quality of tourism offer', *Tourism and Hospitality Management*, **3** (2), 381–92.

Prins, C. and M. Schiessl (1993), 'The new European telecommunications standards institute policy: conflicts between standardisation and intellectual property rights', *European Intellectual Property Review*, **15**, 263–6.

Reihlen, H. (1998), 'Deutsche und Europäische Normung in einem weltweiten Horizont', *Glückauf*, **134**, 11–17.

Reimers, K. (1995*), Normungsprozesse: Eine transaktionskostentheoretische Analyse*, Wiesbaden: Gabler.

Resetarits, P. (1997), 'Implementing the ISO 9000 Standards in Connecticut's Small and Medium Sized Enterprises', in *Proceedings of the First Interdisciplinary Workshop on Standardization Research*, Hamburg, pp. 97–110.

Ritterbusch, G.H. (1999), 'Normung im Hinblick auf das WTO-Übereinkommen über technische Handelshemmnisse', *DIN-Mitteilungen*, **78**, 4–9.

Rogers, E.M. (1995), *Diffusion of Innovations*, 4th edition, New York: The Free Press.

Rohlfs, J. (1974), 'A theory of interdependent demand for a communications service', *Bell Journal of Economics*, **5** (1), 16–37.

Röller, L.-H., M.M. Tombak and R. Siebert (1998), *The Incentives to Form Research Joint Ventures: Theory and Evidence*, Discussion Paper FS IV 98–15, Berlin: Wissenschaftszentrum Berlin.

Romer, P.M. (1990), 'Endogenous technological change', *Journal of Political Economy*, **5**, 71–102.

Roßnagel, A. (1993), *Rechtswissenschaftliche Technikfolgenforschung*, Baden-Baden.

Sachverständigenrat 'Schlanker Staat' (ed.) (1997), *Abschlußbericht Band 1*, Bonn.

Saloner, G. and A. Shepard (1995), 'Adoption of technologies with network effects: an empirical examination of the adoption of automated teller machines', *RAND Journal of Economics*, **26**, 479–501.

Salop, S.C. and D.T. Scheffman (1983), 'Raising Rivals' Costs', *American Economic Review*, **73** (2), 267–71.

Salop, S.C. and D.T. Scheffman (1987), 'Cost-raising strategies', *Journal of Industrial Economics*, **36** (1), 19–34.

Saviotti, P.P. (1991), 'The Role of Variety in Economic and Technological Development', in Saviotti, P.P. and J.S. Metcalfe (eds), *Evolutionary Theories of Economic and Technological Change*, Chur and Philadelphia: Harwood, pp. 172–208.

Saviotti, P.P. (1996), *Technological Evolution, Variety and the Economy*, Cheltenham, UK and Brookfield, US: Edward Elgar.

Scharf, A. (1999), 'Standardisierung von Schnittstellen könnte Millionenbeträge einsparen', *Handelsblatt*, 10 March 1999, p. B32.

Scharfstein, D. and J. Stein (1990), 'Herd behavior and investment', *American Economic Review*, **80**, 465–79.

Scherer, F.M. (1965), 'Firm size, market structure, opportunity and the output of patented inventions', *American Economic Review*, **55**, 1097–125.

Scherer, F.M. (1967), 'Research and development resource allocation under rivalry', *Quarterly Journal of Economics*, **81**, 359–94.

Scherer, F.M. (1998), 'The size distribution of profits from innovation', *Annales d'Economie et de Statistique*, **49–50**, 495–516.

Scherer, F.M. and D. Ross (1990), *Industrial Market Structure and Economic Performance*, 3rd edition, Boston: Houghton Mifflin Company.

Schlag, C.-H. (1997), 'Die Kausalitätsbeziehung zwischen der öffentlichen Infrastrukturausstattung und dem Wirtschaftswachstum in der Bundesrepublik Deutschland', *Konjunkturpolitik*, **43**, 82–106.

Schmenner, R.W. (1992), 'How Can Service Businesses Survive and Prosper?', in Lovelock, C.H. (ed.), *Managing Services – Marketing, Operations and Human Resources*, New Jersey: Prentice-Hall, pp. 31–42.

Schröer, G. and Stahlecker, P. (1996), 'Ist die gesamtwirtschaftliche Produktionsfunktion eine Kointegrationsbeziehung? Empirische Analyse vor und nach der Wiedervereinigung', *Jahrbücher für Nationalökonomie und Statistik*, **215**, Stuttgart: Lucius & Lucius, pp. 513–25.

Schumpeter, J. (1934), *The Theory of Economic Development*, Cambridge, MA: Harvard University Press

Schwitalla, B. (1993), *Messung und Erklärung industrieller Innovationsaktivitäten – mit einer empirischen Analyse für die westdeutsche Industrie*, Heidelberg.

Senden, M.J. and I. Wöckel (1997), 'Gelebte Unternehmensphilosophie contra Normierungszwänge: ISO 9000 im DLR-Wissenschaftsmanagement', *Wissenschaftsmanagement*, **3**, 269–72.

Shapiro, C. (1983), 'Premiums for high quality products as returns to reputation', *Quarterly Journal of Economics*, 659–79.

Shapiro, C. (2001), 'Navigating the Patent Thicket: Cross Licenses, Patent Pools and Standard Setting', in Jaffe, A., J. Lerner and S. Stern (eds), *Innovation Policy and the Economy*, 1, Cambridge, MA: MIT Press, pp. 119–50.

Shapiro, C. and H.R. Varian (1999), 'The art of standards wars', *California Management Review*, **41** (2), 8–32.

Shavell, S. and T. van Ypersele (2001), 'Rewards versus intellectual property rights', *Journal of Law and Economics*, **44**, 525–47.

Shurmer M. (1993), 'An examination into sources of network externalities in the packaged PC spreadsheet market', *Information Economics and Policy*, **5** (3), 231–51.

Shurmer M. and G.M.P. Swann (1995), 'An analysis of the process generating de facto standards', *Journal of Evolutionary Economics*, **5** (2), 119–32,

Shy, O. (2001), *The Economics of Network Industries*, Cambridge, New York and Melbourne: Cambridge University Press.

Sinn, H.-W. (1996), *The Subsidiarity Principle and Market Failure in Systems Competition*, Discussion Paper No. 103, München: CES (Center for Economic Studies).

Sinn, H.-W. (1997), 'The selection principle and market failure in systems competition', *Journal of Public Economics*, **66**, 247–74.

Sirilli, G. and R. Evangelista (1998), 'Technological innovation in services and manufacturing results from Italian surveys', *Research Policy*, **27**, 881–99.

Skea J. (1995), 'Changing Procedures for Environmental Standards Setting in the European Community', in Hawkins, R.W., R. Mansell and J. Skea (eds), *Standards, Innovation and Competitiveness: the Politics and Economics of Standards in Natural and Technical Environments*, Camberley, UK and Brookfield, US: Edward Elgar.

Statistisches Bundesamt (1996), *Produzierendes Gewerbe*, Fachserie 4, Reihe 4.2.3: Konzentrationsstatistische Daten für den Bergbau und das Verarbeitende Gewerbe sowie das Baugewerbe 1993 und 1994, Wiesbaden.

Statistisches Bundesamt (1997), *Statistisches Jahrbuch 1997 für die Bundesrepublik Deutschland*, Wiesbaden.

Statistisches Bundesamt (verschiedene Jahrgänge), *Preise: Fachserie 17, Reihe 8: Preise und Preisindizes für die Ein- und Ausfuhr*, Wiesbaden.

Statistisches Bundesamt (verschiedene Jahrgänge), *Statistisches Jahrbuch für die Bundesrepublik Deutschland*, Wiesbaden.

Steffensen, B. (1997a), 'Die Verringerung von Innovationshindernissen durch freiwillige Vereinbarungen', in Akademie für Technikfolgenabschätzung in Baden-Württemberg (ed.), *Innovationen in Baden-Württemberg*, Stuttgart, pp. 152–67.

Steffensen, B. (1997b), *New Technologies and Standardization: a Useful but Problematic Combination*, paper for the interdisciplinary workshop on Standardization Research at the University of the Federal Armed Forces, Hamburg: Universität der Bundeswehr.

Steyer, R. (1997), 'Netzexternalitäten', *Wirtschaft und Statistik*, **26**, 206–10.

Stübler, H., G. Kohlrautz and S. Händel (1984), 'Kosten und Nutzen der Normung in der Bundesrepublik Deutschland', *DIN-Mitteilungen*, **63**, 593–7.

Stuurman, C. (1995), *Technische normen en het recht*, Dissertation an der Universität Amsterdam: Oostzaan.

Stuurman, C. (1997), 'Standardisation and Standards: a Legal Vacuum?', in Hesser, W. (ed.), *Interdisciplinary Workshop on Standardization Research Proceedings*, Hamburg: Universität der Bundeswehr, pp. 29–40.

Swann G.M.P. (1985), 'Product competition in microprocessors', *Journal of Industrial Economics*, **34** (1), 33–54.

Swann G.M.P. (1987), 'Industry Standard Microprocessors and the Strategy of Second Source Production', in Gabel, H.L. (ed.), *Product Standardisation and Competitive Strategy*, Amsterdam: North Holland.

Swann G.M.P. (1990), 'Standards and the Growth of a Software Network: a Case Study of PC Applications Software', in Berg, J. and H. Schumny (eds), *An Analysis of the IT Standardisation Process*, Amsterdam: Elsevier Science Publishers.

Swann G.M.P. (1998), *Are Network Externalities Linear?*, unpublished paper,

Manchester: Manchester Business School.

Swann, G.M.P. (1999), *The Economics of Measurement*, report for NMS Review, Manchester: University of Manchester.

Swann, G.M.P. (2000), *The Economics of Standardization*, final report for standards and technical regulations, Directorate Department of Trade and Industry, Manchester: University of Manchester.

Swann, G.M.P., P. Temple and M. Shurmer (1996), 'Standards and trade performance: the UK experience', *Economic Journal*, **106**, 1297–313.

Sykes, A.O. (1995), *Product Standards for Internationally Integrated Goods Markets*, Washington, DC: Brookings Institution.

Tamm Hallstrom, K. (1996), 'The production of management standards', *Revue d'Economie Industrielle*, **75**, 61–76.

Tassey G. (1982), 'The role of government in supporting measurement standards for high-technology industries', *Research Policy*, **11**, 311–20.

Tassey, G. (1995), 'The Roles of Standards as Technology Infrastructure', in Hawkins, R., R. Mansell and J. Skea (eds), *Standards, Innovation and Competitiveness*, Camberley, UK and Brookfield, US: Edward Elgar, pp. 161–71.

Tassey G. (2000), 'Standardization in technology-based markets', *Research Policy*, **29** (4/5), 587–602.

Temple, P. and G. Williams (2002), 'Infra-technology and economic performance: evidence from the United Kingdom measurement infrastructure', *Information Economics and Policy*, **14** (4), 435–52.

Tether, B.S., C. Hipp and I. Miles (2001), 'Standardisation and particularisation in services: evidence from Germany', *Research Policy*, **30**, 1115–38.

Thiard, A. and W.F. Pfau (1991), *Forschung & Entwicklung und Normung*, Luxembourg: Office for Official Publications of the European Communities.

Thierstein, A., K. Blind and Ch. Abbeg (2000), 'Normung: Wirkungen auf Aussenwirtschaft und Innovation', *Aussenwirtschaft*, **55** (4), 503–26.

Thum, M. (1994), 'Möglichkeiten und Grenzen staatlicher Standardsetzung', *Homo oeconomicus*, **11**, 465–99.

Thum, M. (1995), *Netzwerkeffekte, Standardisierung und staatlicher Regulierungsbedarf*, Tübingen: Mohr.

Tirole, J. (1989), *Theory of Industrial Organization*, Cambridge, MA: MIT Press.

Toth, B. (1997), 'Putting the US standardization system into perspective: new insights', *DIN-Mitteilungen*, **76**, 792–8.

Verspagen, B. and K. Wakelin (1997), 'Trade and technology from a Schumpeterian perspective', *International Review of Applied Economics*, **11**, 181–94.

Voelzkow, H. (1996), *Private Regierungen in der Techniksteuerung – Eine sozialwissenschaftliche Analyse der technischen Normung*, Frankfurt/Main et al.: Campus Verlag.

de Vries, H.J. (1997a), 'Standardization in Service Sectors – Exploration of Market Needs in the Netherlands', in *Proceeding of the First Interdisciplinary Workshop on Standardization Research*, Hamburg: Universität der Bundeswehr, pp. 311–33.

de Vries, H.J. (1997b), 'Standardization – what's in a name?', *Terminology*, **4** (1), 55–83.

de Vries, H.J. (1999), *Standardization. A Business Approach to the Role of National Standardization Organizations*, Boston, Dordrecht, London: Kluwer Academic Publishers.

Wakelin, K. (1997), *Trade and Innovation. Theory and Evidence*, Cheltenham, UK and Lyme, US: Edward Elgar.

Wakelin, K. (1998a), 'Innovation and export behaviour at the firm level', *Research*

Policy, **26**, 829–41.

Wakelin, K. (1998b), 'The role of innovation in bilateral OECD trade performance', *Applied Economics*, **30**, 1335–46.

Weise, P. (1994), 'Natur, Normen, Effizienz: Prozesse der Normbildung als Gegenstand der ökonomischen Theorie', in Biervert, B. and M. Held (eds), *Das Naturverständnis der Ökonomik. Beiträge zur Ethikdebatte in den Wirtschaftswissenschaften*, Frankfurt/Main: Campus Verlag, pp. 106–23.

Weiss, M.B.H. (1990), 'The standards development process: a view from political theory', *Standard View*, **1** (2), 35–41.

Weiss, M.B.H. and C. Cargill (1992), 'Consortia in the standards development process', *Journal of the American Society for Information Science*, **43**, 559–65.

Weiss, M.B.H. and M. Sirbu (1990), 'Technological choice in voluntary standards committees: an empirical analysis', *Economics of Innovation and New Technologies*, **1**, 111–33.

Weissinger, R. (1995), *ISO Standards and International Markets. A Preliminary Analysis of their Correlative Development*, Genf: ISO.

von Weizsäcker, C.C. (1982), 'Staatliche Regulierung – positive und normative Theorie', *Schweizerische Zeitschrift für Volkswirtschaft und Statistik*, **2**, 325–43.

Wey, Ch. (1999), *Marktorganisation durch Standardisierung: Ein Beitrag zur Neuen Institutionenökonomik des Marktes*, Berlin: edition sigma.

Willgerodt, H. and J. Molsberger (1978), *Der volkswirtschaftliche Nutzen der Normung*, Gutachten im Auftrage des Bundesministerium für Wirtschaft, Bonn: BMWi.

Wissenschaftsstatistik GMBH (1998), *Forschung und Entwicklung in der Wirtschaft 1995 bis 1997: Bericht über die FuE-Erhebung 1995 und 1996*, Essen: Stifterverband der deutschen Wirtschaft.

Withers, B.E., M. Ebrahimpour and N. Hikmet (1997), 'An exploration of the impact of TQM and JIT on ISO 9000 registered companies', *International Journal of Production Economics*, **53** (2), 209–16.

Woeckener, B. (1996), 'Standardisierung für die Informationsgesellschaft', *Jahrbücher für Nationalökonomie und Statistik*, **215**, pp. 257–73.

Woeckener, B. (1997), 'The European Standardization System: How Much in Need of Reform Is It?', *EURAS Yearbook of Standardization*, **1**, München: Accedo Verlags-Gesellschaft, pp. 391–410.

Wölker, Th. (1996), 'Technische Normen im Verlauf der Geschichte', *Homo oeconomicus*, **13**, 51–75.

Appendices

*Appendix I   Concordance between the International Classification of
Standards (ICS) and the International Patent Classification
(IPC)*

ICS Group	ICS Code	IPC Code
Health Care Technology[1]	11	A61 ohne A61K; H05G
Environmental/Health Potection; Safety	13	A62; B01D-053; B09; B65F; C02; E01F008; F01N; F16P; F23G; J; G08B; G10K011; G21F
Metrology and Measurement; Testing	17; 19	G01B; C; D; F; G; H; J; K; L; M; P; R; S; T; V; W; G04; G12
Mechanical Systems	21	F16B; C; D; F; G; H; M; N; S
Fluid Systems	23	F15; F16J; K; L; T; F17; F24F
Mechanical Engineering	25	B21; B23; B24; B25; B26; B30; C 23; F27
Energy; Heat Transfer Engineering	27	E04B; F01B; D; K; L; M; N; P; F02 ohne K; F03; F22; F23 ohne G und J; F24B; C; H; J; F25; G21 B; C; D; G; H; F28
Electrical Engineering	29	G05F; H01B; F; H; K; M; R; T; H02; H05B; C; F
Electronics	31	H01C; G; J; L; S; H05 H; K; G09F-009; G09G
Telecommunications	33	H01P; Q; H03; H04
Information Technology; Office Equipment	35	B41; B42; B43; G06; G10L; G11
Image Technology	37	B44D; G02; G03; G09F
Precision Mechanics; Jewelry	39	A44C; G04B; C; D
Road Vehicle Engineering	43	B60 ohne B60C; B62; F01L; M; N; P
Railway Engineering	45	B61

ICS Group	ICS Code	IPC Code
Shipbuilding; Marine Structures	47	B63 ohne G
Aircraft; Space Vehicle Engineering	49	B64; F02K
Materials Handling Equipment	53	B65G; B66
Packaging; Distribution of Goods	55	B65 ohne F; G; B67
Textile and Leather Technology	59	C14; D01; D02; D03; D04; D06B; C; G; H; J; L; M; N; P; Q
Clothing Industry	61	A41;A42;A43;A44B; D05
Agriculture	65	A01; A24; C05
Food Technology	67	A21; A22; A23; B02B; C; C12C; G; H; J; L; C13
Chemical Technology	71	B01; B04; C01; C06; C07; C08B; C09K; C11 ohne B; G01N;
Mining; Minerals	73	B03; B07B; E21;
Petroleum; Related Technology	75	B32B-011; C10; C11B
Metallurgy	77	B22; B32B-015; C21; C22; C25
Wood Technology	79	B27; B32B-021
Glass/Ceramics Industries	81	B32B-017;-018;-019; C03; C04B 033-037
Rubber/Plastics Industries	83	B29; B32B-023; -027; B60C; C08 ohne B; C09H; J;
Paper Technology	85	B31; B32B-028; D21
Paint/Color Industries	87	B05; C09B; C; D; F; G
Construction Materials; Building	91	B28; B32B-013; C04B 002-032 u. 038-041; E03; E04 ohne B; E05; E06; F21; F24B; C; D; F
Civil Engineering	93	E01 ohne E01F-008; E02
Military Engineering	95	B63G; B64D-007; F41; F42; G21J
Housekeeping; Entertainment; Sports	97	A45B; C; D; F; A46B; A47; A63; B44; B68; D06F; G10B; C; D; F; G; H; K

Appendix II   *Concordance between the International Classification of Standards (ICS) and the Industrial Sector Classification WZ1993*[2]

ICS Group	ICS Code	WZ 1993	Industrial Sector
Health Care Technology; Metrology and Measurement; Testing; Image Technology; Precision Mechanics; Jewelry	11 17 19 37 39	DL 33	Medical; measuring and control technology; optics
Mechanical Systems	21	DJ 28	Manufacturing of metal goods
Mechanical Engineering	25	DK 29.1-5	Mechanical engineering
Electrical Engineering	29	DL 31	Manufacturing of equipment to generate and distribute electricity
Electronics; Telecommunications	31 33	DL 32	Radio; television and communications engineering
Information Technology; Office Equipment	35	DL 30	Manufacture of office machines; data processing equipment and systems
Road Vehicle Engineering	43	DM 34	Manufacture of automobiles and automobile parts
Railway Engineering; Shipbuilding; Marine Structures; Aircraft; Space Vehicle Engineering	45 47 49	DM 35	Other vehicle engineering
Textile and Leather Technology; Clothing Industry	59 61	DB DC	Textile; clothing and leather industry
Food Technology	67	DA	Food industry; tobacco processing
Chemical Technology; Paint/Color Industries	71 87	DG ohne 24.4	Chemical industry (without pharmaceuticals)
Mining; Minerals	73	CB	Mining and extraction of stones and earth
Petroleum; Related Technology	75	CA	Coke; petroleum processing; production of breeding materials

ICS Group	ICS Code	WZ 1993	Industrial Sector
Metallurgy	77	DJ 27	Basic metal industries
Wood Technology	79	DD	Wood (without furniture manufacturing)
Glass/Ceramics Industries; Construction Materials; Building	81 91	DI	Glass; ceramics; handling of stones
Rubber/Plastics Industries.	83	DH	Manufacturing of rubber and plastic goods
Paper Technology	85	DE	Paper industry
Housekeeping; Entertainment; Sports	97	DN	Furniture; jewellery; musical instruments etc.

Appendix III   *Concordance between the International Classification of Standards (ICS) and International Industrial Sector Classification ISIC Rev. 2*

ICS Group	ICS Code	ISIC Code	ISIC Groups (2nd rev.)
Health Care Technology; Metrology and Measurement; Testing; Image Technology; Precision Mechanics; Jewelry	11 17 19 37 39	3850	Instruments
Mechanical Systems	21	3810	Fabricated metal products
Mechanical Engineering	25	382 – 3825	Non-electrical machinery
Electrical Engineering	29	383 – 3832	Electrical machinery
Electronics; Telecommunications	31 33	3832	Electronic equipment & components
Information Technology; Office Equipment	35	3825	Office machinery & computers
Road Vehicle Engineering; Railway Engineering;	43 45	3843 3842 + 3844 + 3849	Motor vehicles; Other transport equipment
Shipbuilding; Marine Structures;	47	3841	Shipbuilding
Aircraft; Space Vehicle Engineering	49	3845	Aerospace
Textile and Leather Technology; Clothing Industry	59 61	3200	Textiles; footwear & leather
Food Technology	67	3100	Food; drink & tobacco
Chemical Technology; Paint/Color Industries	71 87	351 + 352 – 3522	Industrial chemicals
Petroleum; Related Technology	75	353 + 354	Petroleum refining
Metallurgy	77	3700	Basic metal industries
Wood Technology	79	3300	Wood; cork & furniture
Glass/Ceramics Industries; Construction Materials; Building	81 91	3600	Stone; clay & glass
Rubber/Plastics Industries	83	355 + 356	Rubber & plastic products
Paper Technology	85	3400	Paper & printing
Housekeeping; Entertainment; Sports	97	3900	Other manufacturing

Appendix IV   Concordance between the International Classification
of Standards (ICS), the International Patent Classification
(IPC) and the Standard International Trade Classification
(SITC 2)

ICS Classified Group	ICS Code	IPC Code	SITC Code (Revision 2)
Health Care Technology	11	A61 without A61K; H05G	541.9; 774; 872; 899.6
Metrology and Measurement; Testing	17 19	G01B; C; D; F; G; H; J; K; L; M; P; R; S; T; V; W; G04; G12	873; 874
Mechanical Systems	21	F16B; C; D; F; G; H; M; N; S	694; 699.2; 699.3; 699.4; 749.1; 749.3; 749.9
Fluid Systems	23	F15; F16J; K; L; T; F17; F24F	742; 743; 749.2
Mechanical Engineering	25	B21; B23; B24; B25; B26; B30; C 23; F27	695; 696; 728.19; 728.48; 736; 737; 741.3; 745.1; 745.21; 745.27
Energy; Heat Transfer Engineering	27	E04B; F01B; D; K; L; M; N; P; F02 without K; F03; F22; F23 without G and J; F24B; C; H; J; F25; G21 B; C; D; G; H; F28	351; 711; 712; 713.8; 714.88; 714.99; 718; 741.1; 741.2; 714.4; 741.5; 741.6
Electrical Engineering	29	G05F; H01B; F; H; K; M; R; T; H02; H05B; C; F	693; 716; 771; 772; 773; 778; 893.5
Electronics	31	H01C; G; J; L; S; H05 H; K; G09F-009; G09G	776
Telecommunications	33	H01P; Q; H03; H04	761; 762; 763; 764
Information Technology; Office Equipment	35	B41; B42; B43; G06; G10L; G11	751; 752; 759; 893.94; 895.1; 895.2; 895.92; 895.93; 895.94; 895.95
Image Technology	37	B44D; G02; G03; G09F	871; 881; 882; 883; 884
Precision Mechanics; Jewelry	39	A44C; G04B; C; D	667; 885; 897; 899.1; 899.3
Road Vehicle Engineering	43	B60 without B60C; B62; F01L; M; N; P	663.82; 713.2; 713.9; 744.1; 781 782; 783; 784; 785; 786.11; 786.12; 786.8

ICS Classified Group	ICS Code	IPC Code	SITC Code (Revision 2)
Railway Engineering	45	B61;	791
Shipbuilding; Marine Structures	47	B63 without G	713.3; 793
Aircraft; Space Vehicle Engineering	49	B64; F02K	713.1; 714. 4; 714.81; 714.91; 718.88; 792
Materials Handling Equipment	53	B65G; B66	744.2; 744.9
Packaging; Distribution of Goods	55	B65 without F; G; B67	692; 745.22; 745.23; 745.24; 745.25; 745.26; 786.13; 893
Textile and Leather Technology	59	C14; D01; D02; D03; D04; D06B; C; G; H; J; L; M; N; P; Q	21; 26; 61; 65; 724
Clothing Industry	61	A41;A42;A43;A44B; D05	842; 843; 844; 845; 846; 847; 848; 851
Agriculture	65	A01; A24; C05	00; 03; 08; 12; 271; 56; 721; 722; 728.43; 941
Food Technology	67	A21; A22; A23; B02B; C; C12C; G; H; J; L; C13	01; 02; 04; 05; 06; 07; 09; 11; 22; 41; 42; 43; 592.1; 727
Chemical Technology	71	B01; B04; C01; C06; C07; C08B; C09K; C11 without B; G01N	51; 52; 55; 57;
Mining; Minerals; Civil Engineering	73 93	B03; B07B; E21; E01 without E01F-008; E02	32; 723
Petroleum; Related Technology	75	B32B-011; C10; C11B	33; 34
Metallurgy	77	B22; B32B-015; C21; C22; C25	274; 28; 67; 68; 699.1; 699.6; 699.7; 699.8; 699.9; 728.45;
Wood Technology	79	B27; B32B-021	24; 63; 728.12; 728.44
Glass/Ceramics Industries	81	B32B-017;-018;-019; C03; C04B 033-037	663.9; 664; 665; 666; 728.41
Rubber/Plastics Industries	83	B29; B32B-023; -027; B60C; C08 without B; C09H; J	23; 58; 592.2; 62; 728.42; 893.99
Paper Technology	85	B31; B32B-028; D21	25; 64; 725; 726;

ICS Classified Group	ICS Code	IPC Code	SITC Code (Revision 2)
Paint/Color Industries	87	B05; C09B; C; D; F; G;	53; 895.91
Construction Materials; Building	91	B28; B32B-013; C04B 002-032 and 038-041; E03; E04 without B; E05; E06; F21; F24B; C; D; F;	273; 278; 661; 662; 663.1; 663.2; 663.3; 663.5; 663.81; 691; 728.11; 728.3; 812
Military Engineering	95	B63G; B64D-007; F41; F42; G21J;	894.6; 951
Housekeeping; Entertainment; Sports	97	A45B; C; D; F; A46B; A47; A63; B44; B68; D06F; G10B; C; D; F; G; H; K;	697; 775; 821; 831; 893.2; 893.3; 893.91; 893.92; 893.93; 894.1; 894.2; 894.7; 898; 899.4; 899.7; 899.8; 899.9

NOTES

[1]   In this ICS group, the very few standards for pharmacy are integrated. However, we neglect the numerous patents for pharmaceuticals, since they would bias the whole concordance.

[2]   Cf. regarding the classification of German R&D expenditure Stifterverband für die Deutsche Wirtschaft (ed.) (1998).

Index